3rd International Conference on Mechanical Engineering and Applied Composite Materials (MEACM2019)

IOP Conference Series: Materials Science and Engineering
Volume 744

Singapore
22 - 23 November 2019

ISBN: 978-1-7138-2656-9
ISSN: 1757-8981

Printed from e-media with permission by:

Curran Associates, Inc.
57 Morehouse Lane
Red Hook, NY 12571

Some format issues inherent in the e-media version may also appear in this print version.

Institute of Physics
Dirac House, Temple Back
Bristol BS1 6BE UK

Phone: 44 1 17 929 7481
Fax: 44 1 17 920 0979

techtracking@iop.org

Additional copies of this publication are available from:

Curran Associates, Inc.
57 Morehouse Lane
Red Hook, NY 12571 USA
Phone: 845-758-0400
Fax: 845-758-2633
Email: curran@proceedings.com
Web: www.proceedings.com

TABLE OF CONTENTS

Preface

Dear Distinguished Authors and Guests,

This book consists of papers presented during the 3rd International Conference on Mechanical Engineering and Applied Composite Materials which was held on 22-23 November 2019 in Singapore. Like the first MEACM held in Hong Kong, China (2017), the 2nd in Harbin, China (2018) the conference focuses on the Engineer Materials, Mechanical Materials, polymers and Modern composites, Functional ceramics, Mechanical power engineering, Nanomaterials for the various applications, Materials Processing Engineering, Building materials. The conference and the publications arising from it will trigger further related research and technological improvements in these important fields. All papers submitted by the participants of the conference were subjected to a peer-review process by the conference program committee and international reviewers. On behalf of the organizing committee, the editor would like to acknowledge the support and help of those involved in reviewing the papers and thank all of the authors for their contributions to this proceeding book. MEACM2019 aims to present the latest research and results of scientists (professors, students, PhD Students, engineers, and post-doc scientist) related to Mechanical Engineering and Materials topics. This conference provides opportunities for the different areas delegates to exchange new ideas and application experiences face to face, to establish business or research relations and to find global partners for future collaboration.

We appreciated all the participants' contribution for submitting the excellent research works to the conference. At last, we would like to thank all the attendees for your participation. The papers presented in Journal of Materials Science and Engineering were the selected manuscripts associated Mechanical Engineering and Applied Composite Materials research areas from MEACM2019.

Conference Organizing Chair
Prof. Peng-Sheng Wei
National Sun Yat-Sen University, Kaohsiung, Taiwan

IOP Conf. Series: Materials Science and Engineering **744** (2020) 011002 doi:10.1088/1757-899X/744/1/011002

Committees

Conference Chair

Professor Peng-Sheng Wei, National Sun Yat-Sen University, Kaohsiung, Taiwan

Program Committee

Prof. Steven Y. Liang, Georgia Institute of Technology, USA

Prof. Dr. Farid Nasir bin Ani, Universiti Teknologi Malaysia, Malaysia

PROF. JACOB OPADEYI, The University of the West Indies, Trinidad and Tobago

PROF. DR. MOHAMMAD NAZRI BIN MOHD. JA'AFAR, Universiti Teknologi Malaysia, Malaysia

Dr. Norhayati Binti Ahmad, Universiti Teknologi Malaysia, Malaysia

PROF. DR. MD MUSTAFIZUR RAHMAN, Universiti Malaysia Pahang, Malaysia

ASSOC. PROF. JIYUN ZHAO, City University of Hong Kong, Hong Kong

ASSOC. PROF. YONGSHENG CHEN, The Chinese University of Hong Kong, China

ASST. PROF. YU (WILL) WANG, Washington State University, USA

ASST. PROF. ZHIQUAN SHU, Washington State University, USA

ASST. PROF. MOHAMMAD H. RAHMAN, University of Wisconsin-Milwaukee, USA

ASST. PROF. KOW KIEN WHO, University College Sedaya International, Malaysia

DR. MOHD FARID BIN MUHAMAD SAID, Universiti Teknologi Malaysia, Malaysia

DR. NORHAYATI BINTI AHMAD, Universiti Teknologi Malaysia, Malaysia

IOP Conf. Series: Materials Science and Engineering **744** (2020) 011002 doi:10.1088/1757-899X/744/1/011002

IOP Conf. Series: Materials Science and Engineering **744** (2020) 011002 doi:10.1088/1757-899X/744/1/011002

DR.M.K.MARICHELVAM ASSISTANT PROFESSOR (SR. GRADE), Mepco Schlenk Engineering College, India

DR. M. SREENIVASAN, National Technical University of Ukraine, Ukraine

DR. PADMAKUMAR MUTHUSWAMY, Hindustan University, Chennai, India

First Name	Surname
Nicolay A.	Yashtulov
Nina V.	Smirnova
Aleksander	Muc
Aleksander	Muc
WenFung	Pan
Marina V.	Lebedeva
Marina V.	Lebedeva
Malgorzata	Chwal
Augustyn	Marcin
Xi Mo	Qu
Pisutti	Dararutana
Zhi Hua	Zhu
Xue Jun	Wen
Semra	KURAMA
Dong Dong	Meng
Ye Xin	Zhou
Chew Beng	Soh
Wei Yang	Samuel Lim
Zhi Hua	Zhu
Thian Khoon	TAN
Thian Khoon	TAN
Ye Feng	Wu
Goloshchapov Dmit	Leonidovich
Sharapov Rashid	Rizaevich
Aleksei	Kalinin
Xin Ci	Deng
Guang Ming	Kong
Xu Tong	Yang
Shibo	Liu
Treetep	Saengow
Fei	Weng
Yoshitaka	Fujimoto
Gong Zhi	Zhu
Xiang Qian	Yang
Ying Nan	Dong
Ju Song	Zhang
Wen Long	Lv
Viktoriia	Chaika
Li Cheng	Huang
Jun Tong	Yun
Yu Ting	Liu
Yun Xue	Liu
Liang	Ma

IOP Conf. Series: Materials Science and Engineering **744** (2020) 011003 doi:10.1088/1757-899X/744/1/011003

Institution
MIREA - Russian Technological University
Platov South-Russian State Polytechnic University (NPI)
Cracow University of Technology
Cracow University of Technology
National Cheng Kung University
MIREA - Russian Technological University
MIREA - Russian Technological University
Cracow University of Technology
Cracow University of Technology
China University of Petroleum,Beijing
Royal Thai Army
Guangdong Mechanical & Electrical Polytechnic
Shanghai Municipal Engineering Design Institute (Group
Eskisehir Technical University
Qingdao Binhai University
Harbin Institute of Technology
Singapore Instittute of Technology
Singapore Institute of Technology
Guangdong Mechanical & Electrical Polytechnic
University of Nottingham Malaysia
University of Nottingham Malaysia
HuBei University
Voronezh State University
Moscow State University of Civil Engineering
The Central Aerohydrodynamic Institute named after N.E
Shenyang JianZhu University
Qingdao Branch of Naval Aeronautical Institute
Shenyang JianZhu University
Singapore Institute of Manufacturing Technology
King Mongkut's University of Technology Thonburi
Singapore Institute of Manufacturing Technology
Tokyo Institute of Technology, Japan
DALIAN MARITIME UNIVERSITY
CNOOC Research Institute Co., Ltd
Shenyang Institute of Engineering
Shenyang jianzhu University
Institute of Semiconductors, Chinese Academy of Scienc
IKBFU
Wuhan University of Science and Technology
Wuhan University of Science and Technology
Wuhan University of Science and Technology
Shenyang Jianzhu University
Tianhong Measurement Institute

Email address
yashtulovna@mail.ru
smirnova_nv@mail.ru
olekmuc@mech.pk.edu.pl
olekmuc@mech.pk.edu.pl
z7808034@email.ncku.edu.tw
lebedevamv@mitht.ru
lebedevamv@mitht.ru
malgorzata.chwal@pk.edu.pl
marcin.augustyn@mech.pk.edu.pl
779824706@qq.com
pisutti@hotmail.com
184823174@qq.com
huangming@tongji.edu.cn
skurama@eskisehir.edu.tr
mengdongmail@163.com
zhouyexin@hit.edu.cn
Chewbeng.Soh@SIngaporeTech.edu.sg
Samuel.lim@singaporetech.edu.sg
184823174@qq.com
TK.Tan@nottingham.edu.my
TK.Tan@nottingham.edu.my
948504227@qq.com
goloshchapov@phys.vsu.ru
ptdm_zavkaf@mail.ru
aleksey.kalinin@tsagi.ru
517499113@qq.com
kguangming@126.com
574904865@qq.com
liu_shibo@simtech.a-star.edu.sg
treetep.marksaengow@mail.kmutt.ac.th
weng_fei@simtech.a-star.edu.sg
fujimoto@stat.phys.titech.ac.jp
dlmuzgz@126.com
979997829@qq.com
dyn4870@qq.com
410194188@qq.com
lwl@semi.ac.cn
v_chayka8@rambler.ru
taoboq@wust.edu.cn
taoboq@wust.edu.cn
taoboq@wust.edu.cn
1413472465@qq.com
ljweisz@163.com

IOP Conf. Series: Materials Science and Engineering **744** (2020) 011004 doi:10.1088/1757-899X/744/1/011004

Peer review statement

All papers published in this volume of *IOP Conference Series: Materials Science and Engineering* have been peer reviewed through processes administered by the proceedings Editors. Reviews were conducted by expert referees to the professional and scientific standards expected of a proceedings journal published by IOP Publishing.

Formation technology of effective electrode materials based porous silicon with platinum metal nanoparticles for autonomous energy systems

A V Ragutkin[1], A P Antropov[2], M V Lebedeva[2], A S Sigov[3], N A Yashtulov[4]

1 Vice-rector for innovation development, MIREA – Russian Technological University, Moscow, Russia
2 Associate professor, MIREA – Russian Technological University, Moscow, Russia
3 President, MIREA – Russian Technological University, Moscow, Russia
4 Professor, MIREA – Russian Technological University, Moscow, Russia
E-mail: yashtulovna@mail.ru

Abstract. In this work the effective electrode materials for autonomous energy sources based on porous silicon modified by platinum nanoparticles have been obtained. The functional characteristics of the electrodes have been studied by transmission electron microscopy and voltammetry methods. The electrode materials based on porous silicon had demonstrated high catalytic activity in fuel cells with hydrogen oxidation.

1. Introduction

One of the promising ways to miniaturize the autonomous power sources such as fuel cells (FC) is to use standard microfabrication techniques that have been developed for decades for microelectronics applications. These techniques enable mass fabrication at low cost (very large number of devices on a very small area), which could lead to a reduction in the global cost of miniature fuel cells. As a consequence, a large part of the hydrogen micro-fuel-cell prototypes reviewed from the literature involves silicon [1-12].

In recent decades, porous silicon has been regarded as a means to further increase the functionality of autonomous and portable energy technology. It was found that silicon porosification is a simple and cheap way of nanostructuring and bestowing to silicon properties, which are markedly different from the bulk material properties. In this regard, increased interest in porous silicon appeared in various fields, including microelectronics, photonics, medicine, chemical sensing and bioengineering [1-12]. Especially it should also be stated that PS is usually used in microelectromechanical systems (MEMS) designing [1, 3, 9, 11, 12]. It depends on special PS parameters (porosity, thickness of the porous layer, resistance to chemical reagents, high active surface area). These properties affect specific characteristics of the current source created on the basis of PS electrodes.

We used in a number of our studies porous silicon (PS) as one of the most promising kinds of support matrices for autonomous micropower current sources compatible with microchips for electronics. We suggested a process for preparing and stabilizing catalytically active layers of platinum metals on PS, polymer and carbon supports in water-organic solutions in the presence of surfactants [2, 5, 10].

The aim of this work is formation of composite electrodes based on porous silicon with platinum electrocatalysts and functional specific characteristics evaluation in hydrogen oxidation reaction during model fuel cells operation.

2. Experimental

2.1. Platinum nanoparticles synthesis

Platinum nanoparticles were synthesized by chemical reduction in reverse microemulsions with sodium tetrahydroborate $NaBH_4$ (98%, Merck, Germany), using nonionic surfactant – Triton X-100. The method of solution preparation was as follows: a water-organic solution of platinum K_2PtCl_4 (Sigma Aldrich, USA) was prepared. To form a microemulsion medium, 0.15 M solution of Triton X-100 (Sigma Aldrich, USA) and propanol-2 was used. Then, a microemulsion of the similar composition, contained 0.1 M water solution of the reducing agent – $NaBH_4$, was added to this solution under ultrasonic action for 2-3 min. The molar water/surfactant ratio (ω) while making the experiments was changed from 1.5 to 8. In order to prevent the sunlight destruction of nanoparticles, the solutions of microemulsions were stored in darkness at room temperature.

2.2. Porous silicon formation

In this work, porous silicon of n- and p-type with different porosity degree (P) (40-80%) was prepared by anodic electrochemical etching of silicon wafers [2, 5]. Anodic etching of monocrystalline silicon had been carried out in water-alcoholic solutions of hydrofluoric acid in a cell, which is a fluoroplastic cylinder without a bottom, mounted on a silicon plate (figure 1). In this case, the silicon plate is clamped between the cylinder and the aluminum anode. A platinum electrode is used as the cathode and reference electrode. The cylinder is filled with an electrolyte solution containing hydrofluoric acid, deionized water and isopropyl alcohol (IPA) in the ratio $HF:H_2O:IPA$ = 1:3:1. Anodizing process was carried out in galvanostatic mode with varying of etching current density from 20 to 60 mA/cm^2 [2, 5]. After the anodizing process, the samples were washed in ethanol and dried in a muffle furnace at temperature of 150-200° C for 1 hour.

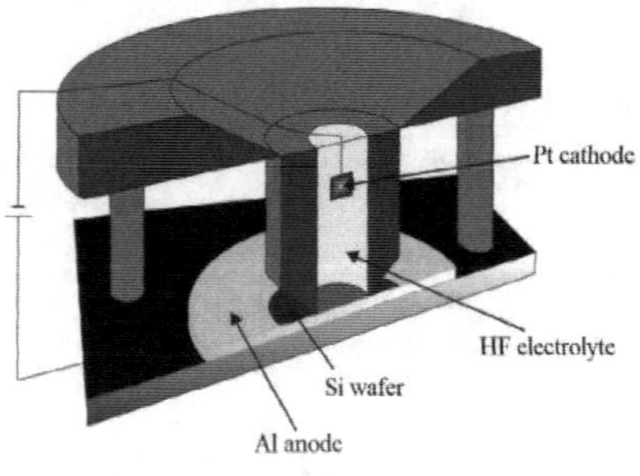

Figure 1. Installation scheme for anodic etching of monocrystalline silicon

2.3. Instrumentation

The morphology of the nanocomposite surface was studied by high-resolution transmission electron microscopy (HRTEM) on a Zeiss Libra 200FE device (Carl Zeiss, Germany). The catalytic activity and stability of the electrodes were estimated using cyclic voltammetry (CVA) on an IPC PRO M device (Tekhnopribor, Russia). The scanning rate was varied from 10 to 100 mV/s. The CVA data were fixed after the stabilization of the parameters of voltamperograms.

3. Results and discussions

3.1. Features of porous silicon production

The electrophysical properties of porous silicon are strongly affected by microimpurities present in it. With the introduction of small amounts of impurities of other elements the intrinsic semiconductor can be converted to an impurity, and depending on the type of impurity a semiconductor with a hole (p-type) or electronic (n-type) conductivity can be produced. During the

IOP Conf. Series: Materials Science and Engineering **744** (2020) 012001 doi:10.1088/1757-899X/744/1/012001

formation of a porous layer on single-crystal silicon plates, an important factor is the effect of illumination on the sample. This effect allows to generate an additional number of electron-hole pairs, increasing the concentration of holes on the surface of monocrystalline silicon. The illumination has a special effect on the n-type samples, allowing to increase the concentration of "holes" involved in the pore formation on the surface of the plate. However, the increased concentration of "holes" in the etching area leads to a sharp increase in the pore diameter of the sample, allowing to obtain layers of macroporous silicon. To obtain the layers of mesoporous n-type silicon in this work, we chose the values of the current density is higher than for samples p-type, at the same time the impact of current on the plate was reduced. The process of pore formation on the surface of a single-crystal silicon plate can be described by the following scheme (figure 2) [1, 3]:

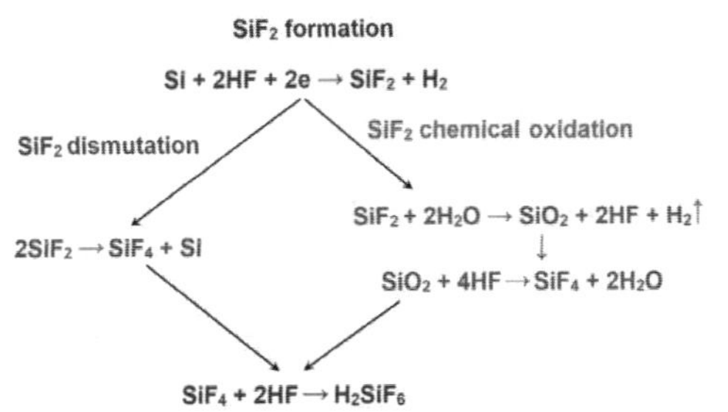

Figure 2. Installation scheme for anodic etching of monocrystalline silicon

As a result, the channels diameter of n-type PS was of 15-45 nm, p-type PS – 10-25 nm when samples porosity degree was 65±2% and 47±2%, respectively (table 1).

Table 1. Conditions of PS layers formation during anodic etching.

Conductivity type and porosity	Anodizing time, Sec	Current density, mA/cm^2	Specific internal surface area S_v, m^2/cm^3
n-type, 48	44	25	154,2
n-type, 66	32	53	171,4
p-type, 66	62	22	287,6

The specific area of the PS inner surface (S_V, table 1) has been determined with oxidative gravimetry method [3]. The method has a high accuracy and allows to characterize samples with a large specific surface area. The method is based on thermal oxidation of the inner surface of the pores and measurement of the mass and thickness of the formed oxide layer. PS samples were weighed before and after oxidation, and the thickness of the porous layer was determined by the SEM method on the sample cleavage. Next, the S_V value was calculated by the formula [3]:

$$S_V = \frac{4S}{\pi D^2 l} \quad (1)$$

$$S = 1.88\frac{m_{ox} - m_0}{\rho_{SiO_2} t} - 2\left(\frac{M}{\rho_{Si} L} - \frac{\pi D^2}{4}\right) \quad (2)$$

In the future, the PS nanopore sizes were controlled by scanning electron microscopy and transmission electron microscopy methods.

3.2. Composites investigation

To create nanocomposite materials, the sorption process of micellar solutions with platinum nanoparticles on the PS under ultrasonic treatment influence was carried out. In order to remove solvent and surfactant residues from the samples surface, a thorough washing with solutions of isooctane, ethanol and distilled water was carried out. The application of ultrasonic treatment while the formation of platinum–PS nanocomposites allows us, first, to initiate the decomposition of the associates of the reverse micelles and facilitate the destruction of micellar shells of nanoparticles. Second, localizing the PS samples in the zone of the formation of the platinum nanoparticles free from the micellar shells stimulates the localization of small nanoparticles on active PS centers and using an ultrasonic treatment prevents the agglomeration of nanoparticles both at the surface and in the pores of the matrix according to the HRTEM (figure 3) data.

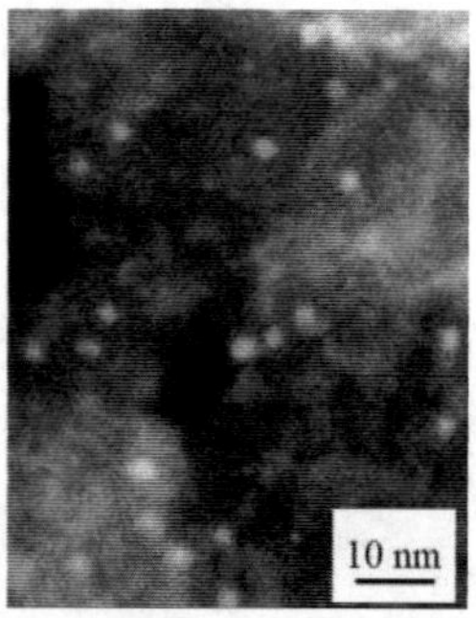

Figure 3. Microphotograph HRTEM of Pt nanoparticles in the depth of n-type PS pores.

According to the HRTEM data (figure 3), the synthesized platinum nanoparticles possess sizes in the range from 7 to 2 nm and smaller and are characterized by a homogeneous distribution at the surface and in the bulk of pores.

The catalytic activity of the synthesized Pt/PS composites was evaluated by cyclic voltammetry (CVA). The electrocatalytically active surface area (ECSA) was determined by the standard procedure of evaluation of CVA parameters, taking into account the catalyst content and the potential scanning rate. Figure 4 shows the potentiodynamic curves (CVA) for Pt nanocomposites on n- and p-type porous silicon, taken at the solubilization coefficient $\omega = 1.5$ and metal content $m_s(Pt) = 0.08$ mg cm^{-2}. Characteristic peaks corresponding to hydrogen adsorption/desorption (0-0.3 V) and oxygen reduction (>0.6 V) can be seen in the curves. At the Pt content lower than 0.03 mg cm^{-2}, these peaks are weakly pronounced. This is associated with relatively low platinum content of the samples.

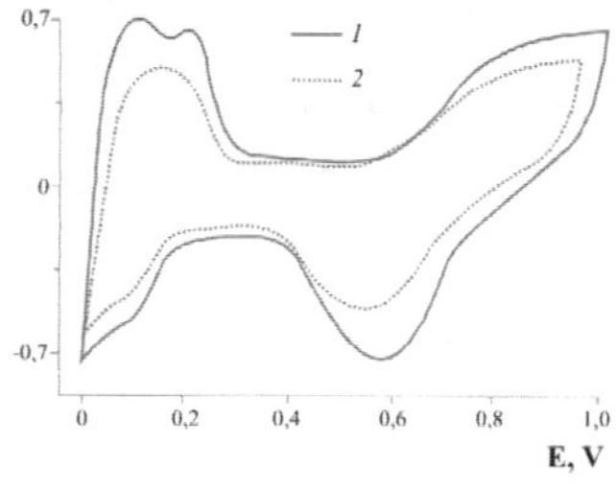

Figure 4. Potentiodynamic curves of Pt nanocomposites on (*1*) n- and (*2*) p-type of porous silicon at $\omega = 1.5$ and metal content $m_s(Pt) = 0.08$ mg cm^{-2}

Table 2 shows the parameters of the catalytic activity (current density, ECSA, stability) of platinum/PS nanocomposite samples containing 0.08 mg cm^{-2} Pt, prepared at solubilization coeffi cients ω of 1.5 and 5. These parameters were determined by CVA in reactions of oxygen electroreduction and hydrogen oxidation.Comparison of the results given in Table 2 shows that enhanced activity is exhibited by electrodes characterized by n-type of PS conductivity, 66% porosity, and minimal nanoparticle size and synthesized from water-organic solutions with $\omega = 1.5$. The nanoparticle size (d) range calculated from the CVA data [2, 5] is 1-4 nm. The CVA data agree with the HRTEM data (figure 3).

Table 2. Characteristics of Pt/PS nanocomposites with the metal content of 0.08 mg cm^{-2}, according to CVA data

№	PS type	Porosity degree, P, %	ω	ECSA, m^2/g	j, A/m^2	d, nm
1	n	66	1.5	214	121	1.3
2	n	66	5	202	117	1.4
3	p	66	1.5	188	110	1.5
4	p	66	5	165	105	1.7
5	n	48	1.5	119	82	2.4
6	n	48	5	104	76	2.7
7	p	48	1.5	91	70	3.1
8	p	48	5	84	67	3.3

4. Conclusion

Effective electrode materials based on porous silicon with platinum nanoparticles had been prepared. Received nanocomposites exhibit high catalytic activity in hydrogen oxidation and oxygen reduction. The enhanced activity showed the PS electrodes with n-type conductivity, porosity of 66% and with minimal nanoparticle size synthesized from water-organic solutions with ω = 1.5. These materials are of great interest in the autonomous energy sources constructing with high specific characteristics.

Acknowledgments
The research was carried out with financial support of the State task of the Russian Federation (project No. 2019-0801).

References

[1] Korotcenkov G 2016 *Porous silicon. From formation to application. Optoelectronics, microelectronics, and energy technology application* (CRC Press, Taylor & Francis Group, New York).
[2] Yashtulov N A, Ragutkin A V, Smirnov S E, Lebedeva M V 2017 *Tsvetnye Metally* **5** 58-63.
[3] Ischenko A A, Fetisov G V, Aslalnov L A 2014 *Nanosilicon: properties, synthesis, applications, methods of analysis and control* (CRC Press, Taylor & Francis Group, New York).
[4] Andersen S M, Larsen M J 2017 *Journal of Electroanalytical Chemistry* **791** 175-184.
[5] Yashtulov N A, Lebedeva M V, Ragutkin A V, Zaitsev N K 2018 *Russian Journal of Applied Chemistry* **91** 280-285.
[6] Yashtulov N A, Lebedeva M V 2017 *Russian technological journal* **5** 58-73.
[7] Boaks M, Schubert P J 2019 *Journal of Catalysis* **371** 81-87.
[8] Cui Y, Liu Y, Wu J, Zhang F, Liu X 2018 *Journal of Power Sources* **403** 118-126.
[9] Ervin M H, Piekiel N W, Isaacson B 2018 *Sensors and Actuators A: Physical* **280** 132-138.
[10] Lebedeva M V, Antropov A P, Ragutkin A V, Yashtulov N A 2018 *International Journal of Applied Engineering Research* **13** 16774-16777.
[11] Alarcón-Salazar J, Vázquez G.V, González-Fernández A A, Zaldívar-Huerta I E, Pedraza-Chávez J, Aceves-Mijares M 2018 *Advanced Materials Letter* **9** 116-122.
[12] Merazga S, Cheriet A, M'hammedi K, Mefoued A, Gabouze N 2019 *International Journal of Hydrogen Energy* **44** 9994-10002.

Automatic Lab scale equipment for continuous HMF production in biphasic systems

V A Klushin[1], A A Chizhikova[1], V V Gorchakov[2] and N V Smirnova[1]
1 Platov South-Russian State Polytechnic University, Novocherkassk, 346428, Russia
2 LLC Scientific and Production Enterprise "Intor", Novocherkassk, 346428, Russia
E-mail: smirnova_nv@mail.ru

Abstract. An automatic lab scale equipment for continuous HMF production by the dehydration of fructose with homogeneous catalysis in biphasic systems (water solution - MIBK) was designed, manufactured and tested. The dependencies of the HMF yield on the process parameters were studied. The feed rate of fructose varied from 50 to 100 g·h^{-1}, the extractant (MIBK) flow rate from 0.5 to 1.4 L·h^{-1} and the process temperature was maintained in the range from 50 to 87 °C. The maximum productivity of the automatic lab scale equipment 33.6 g·h^{-1} with a high HMF yield 48 % is achieved with the following process parameters - 87 °C, the fructose feed rate 100 g·h^{-1} and the extractant flow rate 1 L·h^{-1}. This automatic lab scale equipment may be prospectively used for the kilogram scale HMF synthesis.

1. Introduction

Interest in the use of bio-renewable raw materials for the production of fuels and valuable chemicals in the last decade has increased significantly [1]. Unlike fossil analogs (oil, gas, coal), biomass contains an excess of oxygen. This complicates its direct use as an energy source or source of raw materials in chemical technology. To reduce the amount of oxygen, the biomass must be dehydrated and hydrogenated [2-4]. One of the most important intermediates of biomass dehydration is 5-hydroxymethylfurfurol (HMF). The products of its further oxidation / reduction (2,5-diformylfuran, 2,5-furandicarboxylic acid, dimethylfuran, etc.) in the near future will be a substitute for the raw materials currently used for the production of polymers, drugs and fuels.

Currently, the most efficient and rational pathways for the synthesis of HMF are based on the use of fructose as a raw material [4-6]. To date, many options have been proposed for synthesis of HMF from a carbohydrate-containing raw material: homogeneous (acidic dehydration in an aqueous or anhydrous medium, in ionic liquids) and heterogeneous that used solid acids as catalysts. When-ionic liquids or aprotic solvents are used as a reaction medium, high yields of HMF (more than 90%) are achieved. However, the high cost of reagents and the difficulty of isolation of HMF from the reaction mixtures lead to a high product cost [3-7]. The introduction of a heterogeneous catalyst makes it possible to increase the yield of HMF in aqueous solutions, but the synthesis proceeds at high temperatures (up to 200 ° C) and pressure [3].

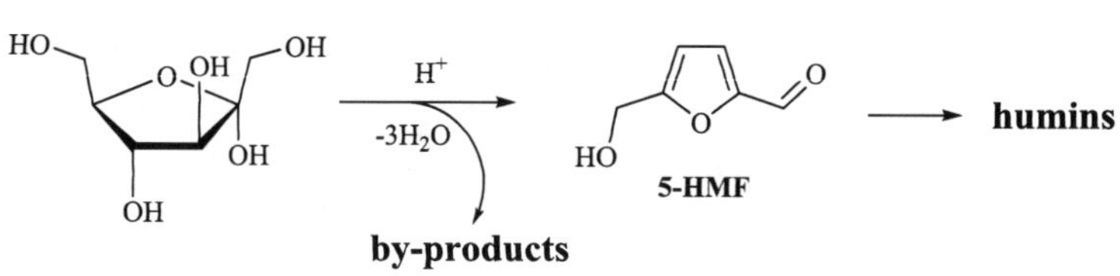

(1)

Regardless of the synthesis approach used, the isolation of HMF is carried out by its extraction with low-boiling water-immiscible organic solvents. Entering the extractant into the reaction zone during the synthesis in aqueous medium allows to increase the yield of HMF by times due to its removal from the reaction zone, thereby partially preventing the occurrence of side reactions. The most effective extractant is methyl isobutyl ketone (MIBK) [2, 9]. The production of HMF in two-phase aqueous-organic systems is commercially more justified compared to the synthesis in ionic liquids [6,7].

Most of the published studies on the production of HMF from monomeric sugars (fructose) or from other raw materials were carried out using laboratory equipment (autoclaves, test tubes) under periodic conditions. Even a few studies on the production of HMF in kilogram scales were also focused on the periodic action rectors [10]. Continuous synthesis methods are investigated only for systems with a heterogeneous catalyst [11].

Earlier, we have proved the prospects of $Na_2SO_4 \times 10H_2O$ and concentrated H_2SO_4 system as catalyst for the HMF synthesis in a two-phase an aqueous solution - MIBK system [6, 12]. $Na_2SO_4 \times 10H_2O$, reacting with H_2SO_4, with formation of $NaHSO_4$, that is a selective catalyst for the dehydration of fructose and fructose carbohydrates (sucrose, inulin) to HMF [2]. We optimized the process of fructose dehydration in this two-phase system in a batch reactor. We determined: (i) the duration of the process, (ii) the influence of the sulfuric acid content, (iii) temperature and (iv) the amount of extractant in terms of the maximum of HMF yield.

In this work, we are present a developed automatic lab scale equipment for continuous HMF production in biphasic systems and results of the optimization process of the dehydration of fructose during continuous synthesis to create a technologically, environmentally and economically efficient method for producing HMF.

2. Methods

The process of dehydration of fructose proceeds relatively slowly. For carrying out of the process under periodic conditions, the optimal synthesis parameters are: fructose concentration of 15%; the volume ratio of an aqueous solution of $NaHSO_4$ / MIBK equal to 3: 1; the temperature is 85 °C during 2 hours [12].

To implement the process in continuous conditions, automatic lab scale equipment was designed and manufactured, the general scheme and photo of which are presented in Figure 1. The reactor is made of chemically resistant stainless steel because of the reaction medium contains both sodium hydrosulfate and free sulfuric acid and are very corrosive. In addition formic acid and other by-products are also formed.

To separate the water-organic emulsion formed in the reactor 4, it is necessary to provide a side branch in the form of a collector 11. Due to the absence of mixing in the collector, conditions for the separation of two liquid phases are provided. Collector is equipped with a transparent part made of glass to fix the liquid level by the operator and prevent extractant overflow in emergency cases.

Heating of the reactor is carried out using electric heating elements. The temperature is controlled by an electronic TIR controller with accuracy maintain temperature 1 °C. To prevent heat from dispersing into the environment and to prevent the operator from contacting the heated surface, the reactor is wrapped in a heat insulating material and covered with a protective cover.

$Na_2SO_4 \times 10H_2O$ and concentrated H_2SO_4 are loaded into reactor 4 (Figure 1). After stirring the mixture for 15 minutes using a stirrer 3, MIBK is poured into the reactor. After heating of the reactor to a predetermined temperature from the hopper 5 using a dispenser for bulk materials 6, raw materials (fructose) are continuously fed into the reactor. MIBK also flows continuously from tank 1 using a peristaltic pump 2. Vapors formed in the reactor are condensed in refrigerator 7 and returned back. The finished product in the form of the HMF solution in MIBK continuously removed from the reactor using a peristaltic pump 8. To remove free acids the exit solution is passed through an absorber 9 mixture of calcium carbonate and sodium bicarbonate. The freed from acids HMF solution is collected in the tank 10, then it is sent for evaporation on a laboratory rotary vaporizer. The isolation of HMF requires removal of MIBK by distillation and re-extraction of the product with methyl tert-butyl ether to remove the main part of tars. As a result, the product has a purity of 95-96%.

IOP Conf. Series: Materials Science and Engineering **744** (2020) 012002 doi:10.1088/1757-899X/744/1/012002

The concentration of HMF in the reaction mixtures was determined by HPLC on an Agilent 1260. Infinity chromatograph equipped with a diode array detector and an Eclipse PAH reversed phase column (250×4.6 mm); eluent acetonitrile– water (70 : 30 by volume), flow rate 0.5 mL min^{-1}; column temperature 30°C; detection wavelength λ 284 nm; retention time of HMF 4.5 min.
The yield of HMF (wt. %) was determined using the equation given below.

$$Y(HMF) = \frac{C(HMF) \cdot V \cdot M(Fructose) \cdot 100}{W(Fructose) \cdot M(HMF)}$$

Here, C refers to concentration ($g \cdot L^{-1}$), M is the molecular weight ($g \cdot mol^{-1}$), V extractant consumption ($L \cdot h^{-1}$) and W fructose mass flow ($g \cdot h^{-1}$).

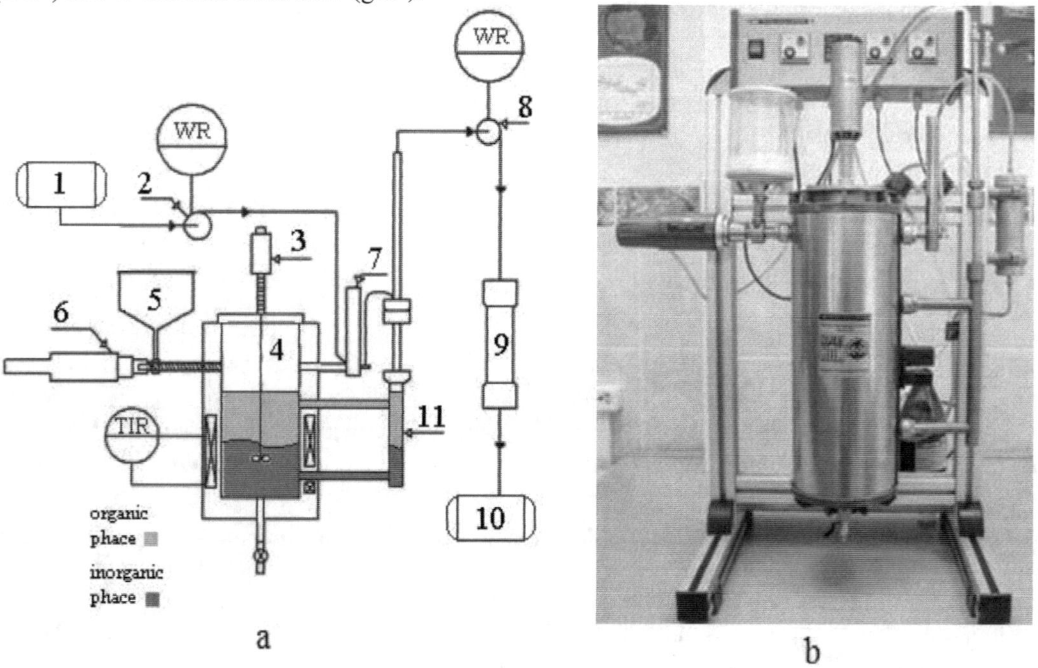

Figure 1 - Scheme (a) and photo (b) of the automatic lab scale equipment for continuous HMF production

1- tank with MIBK; 2,8 - pumps; 3 - mixing device; 4 - reactor; 5 - tank with raw materials; 6 - dispenser; 7 - refrigerator; 9 - acid absorber; 10 - tank for the HMF solution in MIBK; 11 - collector. WR - peristaltic pomp; TIR - temperature TIR-controller.

3. Optimization of synthesis conditions

When carrying out the process in laboratory conditions according to the method described in our previous article [11], in this was found that the presence of an extractant has a great effect. HMF accumulates in the reaction medium. Reaching the maximum concentration, it begins to react with fructose, which leads to the formation of by-products (humins) [13] (Scheme 1). To avoid such an outcome, it is necessary to organize the continuous removal of products from the reaction zone. This is achieved by carrying out the process in a two-phase system. HMF has a greater solubility in MIBK than in water; as a result it is extracted into the organic phase. Therefore, it can be expected that the continuous removal of the extractant, saturated with HMF, from the reaction system will contribute to an increase in the yield of the target product as compared with the process carried. It is very important to establish the optimal ratio of the feed rates of the reagent and flow rate of the extractant. These parameters were determined by calculation; however, they needed experimental verification and possibly refinement.

We studied the effect on the HMF yield and humins accumulation under following process parameters: the feed rate of fructose, the flow rate of the extractant and the temperature (Table 1).

With an increase in the fructose feed rate, the yield of HMF is decreases, and the HMF productivity of automatic lab scale equipment has extreme character. At low feed rates (50-70 g·h^{-1}) the concentration of fructose in the reaction medium does not reach the optimum, and at high feed rates, fructose does not have time to be converted and accumulates in the reactor. This leads to an increase in the rate of accumulation of insoluble humins in the aqueous phase (Table 1, lines 1-5).

When the flow rate of the extractant into the reactor is increase up to 1 L·h^{-1}, the yield of HMF increases too, and then almost does not change (Table 1, lines 6-10). Apparently, with an increase in the MIBK flow rate, the HMF concentration in the organic phase decreases that lead to increasing of the HMF extraction from water phase to organic phase rate. The optimal flow rate of the extractant 1.0 L·h^{-1} provides to get the highest yield of HMF 48% (Table 1, line 3).

With increasing of temperature from 50 to 87 °C, the yield of HMF increases from 3 to 48%. The largest yields of HMF was obtained at temperatures of 85-87 °C for 8 hours with MIBK flow rate 1 (L·h^{-1}), and fructose mass flow 100 g · h^{-1} (Table 1 line 3).

Table 1. Optimization of conditions of HMF synthesis in two phase system of aqueous solution of NaHSO4 / MIBK (1: 3 volumes ratio) using automatic lab scale equipment

№	Fructose feed rate (g·h^{-1})	MBIK flow rate (L·h^{-1})	Temperature (°C)	HMF yield (%)	HMF productivity (g·h^{-1})	Humins accumulation rate (g·h^{-1})	MIBK volume needs for extraction of 1 g HMF (mL)
1	50	1.0	87	63	22	0.5	-
2	70	1.0	87	58	28	2.1	-
3	100	1.0	87	48	34	5.0	29.3
4	120	1.0	87	37	31	13.2	-
5	140	1.0	87	28	27	25.2	-
6	100	0.5	87	27	-	-	32.5
7	100	0.7	87	33	-	-	30.3
8	100	1.0	87	48	-	-	29.8
9	100	1.2	87	50	-	-	34.3
10	100	1.4	87	53	-	-	37.7
11	100	1.0	50	3	-	-	-
12	100	1.0	60	7	-	-	-
13	100	1.0	70	14	-	-	-
14	100	1.0	80	45	-	-	-

4. Conclusion

The automatic lab scale equipment for continuous 5- HMF production in biphasic systems was developed, manufactured and tested for dehydrating fructose in a homogeneous catalytic system. It was established that the maximum productivity of the equipment is 33.6 g·h^{-1} with a high HMF yield of 48 % is achieved with the following process parameters - 87 °C, the fed rate of fructose 100 g·h^{-1}and the flow rate of the extractant 1 L·h^{-1}. This automatic lab scale equipment may be prospectively used for the kilogram scale HMF synthesis .

Acknowledgement

This work was financially supported by the Russian Science Foundation (Project No. 16-13-10444).

References

[1] Maity S K 2015 *Renewable Sustainable Energy Rev.* **43** 1427; Caes B R, Teixeira R E, Knapp K G and Raines R T 2015 *ACS Sustainable Chem. Eng.* **3** 2591; Sheldon R A 2016 *J. Mol. Catal. A: Chem* **3** 422.
[2] Putten R-J, van der Waal J C, Jong E, Rasrendra C B, Heeres H J and Vries J G 2013 *Chem. Rev.*, **113** 1499.
[3] Mukherjee A, Dumont M-J and Raghavan V 2015 *Biomass Bioenergy.* **72** 143.
[4] Rout P K, Nannaware A D, Prakash O, Kalra A and Rajasekharan R 2016 *Chem. Eng. Sci.* **142** 318; Gandini A, Lacerda T M, Carvalho A J F and Trovatti E 2016 *Chem. Rev.* **116** 1637; Esposito D and Antonietti M 2015 *Chem. Soc. Rev.,* **44** 5821; Kashparova V P, Khokhlova E A, Galkin K I, Chernyshev V M and Ananikov V P 2016 *Russ. Chem. Bull.* **64** 1069
[5] Krawielitzki S and Kläusli T M 2015 *Ind. Biotechnol.,* **11** 6
[6] Klushin V A, Galkin K I, Kashparova V P, Krivodaeva E A, Kravchenko O A, Smirnova N V, Chernyshev V M and Ananikov V P 2016 *Russian Journal of Organic Chemistry*, **767** 52(6)
[7] Chernyshev V M, Kravchenko O A, and Ananikov V P 2017 *Russian Chemical Reviews* **86(5)**, 357
[8] Khokhlova E A, Kachala V V and Ananikov V P 2014 *Russ. Chem. Bull.,* **62** 830; Kashin A S, Galkin K I, Khokhlova E A and Ananikov V P 2016 *Angew. Chem., Int. Ed. Engl.*, **55**, 2161; Yi Y-B, Lee J-W and Chung C-H 2015 *Environ. Chem. Lett.* **13** 173; Zhang J, Yu X, Zou F, Zhong Y, Du N and Huang X 2015 *ACS Sustainable Chem. Eng.* **3** 3338; Zhou J, Huang T, Zhao Y, Xia Z, Xu Z, Jia S, Wang J and Zhang Z C 2015 *Ind. Eng. Chem. Res.* **54** 7977
[9] Egorova K S and Ananikov V P 2014 *ChemSusChem*, **7** 336; Saha B and Abu-Omar M M 2014 *Green Chem.* **16** 24; Blumenthal L C, Jens C M, Ulbrich J, Schwering F, Langrehr V, Turek T, Kunz U, Leonhard K and Palkovits R 2016 *ACS Sustainable Chem. Eng.* **4** 228; Sindermann E C, Holbach A, de Haan A and Kockmann N 2016 *Chem. Eng. J.* **283** 251; Ma H, Wang F, Yu Y, Wang L and Li X 2015 *Ind. Eng. Chem. Res.* **54** 2657
[10] Rapp K M 1987 (Süddeutsche Zucker-Aktiengesellschaft) U.S. Patent 4740605; M'Bazoa C; Raymond F; Rigal L and Gaset A 1990 (Furchim SaRL) Fr. Patent FR 2669635
[11] Tuercke T; Panic S and Loebbecke S 2009 *Chem. Eng. Technol.* **32** 1815
[12] Klushin V A, Kashparova V P, Chizhikova A A and Smirnova N V 2016 *University News. North-Caucasian Region. Technical Sciences Series.* **4** 110
[13] Chernysheva D V, Chus Y A, Klushin V A, Lastovina T A, Pudova L S, Smirnova N V, Kravchenko O A, Chernyshev V M and Ananikov V P 2018 *ChemSusChem*, **11(20)** 3599

IOP Conf. Series: Materials Science and Engineering **744** (2020) 012003 doi:10.1088/1757-899X/744/1/012003

Description of delaminations in composite multilayered structures – comparison of numerical and experimental results for compressed plates

A. Muc

Institute of Machine Design, Faculty of Mechanical Engineering, Cracow University of Technology, Kraków, Poland.
E-mail: olekmuc@mech.pk.edu.pl

Abstract. The buckling loads and the post-buckling behavior of the perfect and imperfect delaminated plate are investigated in the paper. For this purpose the concise formulations enabling to derive pre- and post-buckling behaviour of plates with embedded delaminations is presented. The computations are made using FE package NISA II and Mathematica package. Obtained results for buckling loads and the post-buckling behaviour are compared with the experimental results. The Digital Image Correlation (DIC) technique is employed to verify the uniformity and homogeneity of deformations.

1. Introduction

The influence of delaminations on the deformations, failure modes and loads have been studied by the author for many years. The analysis started in 1995 when the buckling loads of doubly-curved shells with delaminations were considered [1]. Then, various problems connected with the behaviour of composite structures with delaminations were discussed by the author with coworkers in Refs [2-12]. The discussed problems dealt with the analysis of composite plates and panels with elliptic and circular delaminations and subjected to static and cyclic (fatigue) loads. The considerations were not only theoretical (numerical – FE studies) but also experimental with the use of the SHM and thermography. The cited above references give the review of the literature in this area. The supplemented information from the existing literature is also given in Ref [13].

The aim of the present paper is following:

- to present the results of experimental investigations by the analysis of the deformation process characterized by the force-displacement curves and with the use of the digital image correlation (DIC) technique – see Ref. [14];
- to compare the experimental results with the finite element (FE) computations.

2. Experimental Studies

The composite laminated plates were made of eight layers of glass/epoxy prepreg (TVR 380M12/26%R). The symmetric laminates were studied and each of the layers were oriented at $\pm 45^0$ referring to the axis of loading. Seven rectangular plates were tested:

- two plates without delaminations (L_x=200, L_y=300);
- two plates (L_x=200, L_y=300) with a central circular delamination (D_{del}=142.5) located between fourth and fifth layer (see Fig. 1a);

IOP Conf. Series: Materials Science and Engineering **744** (2020) 012003 doi:10.1088/1757-899X/744/1/012003

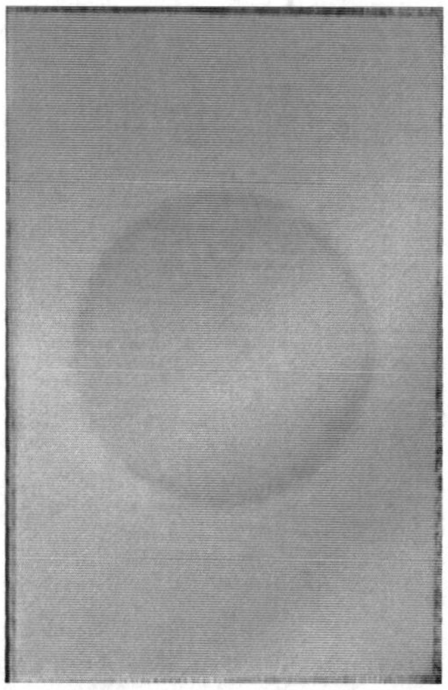

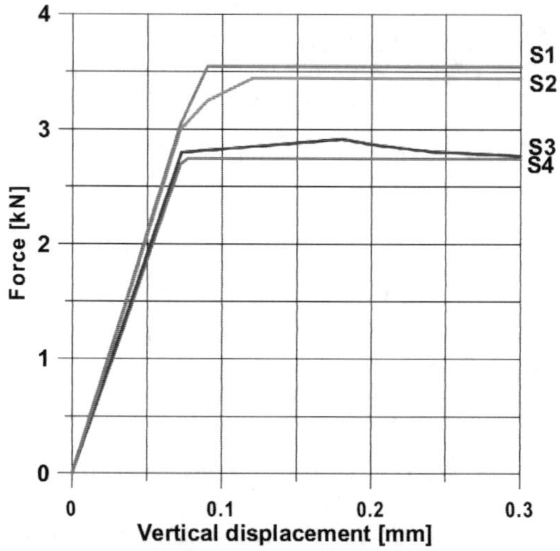

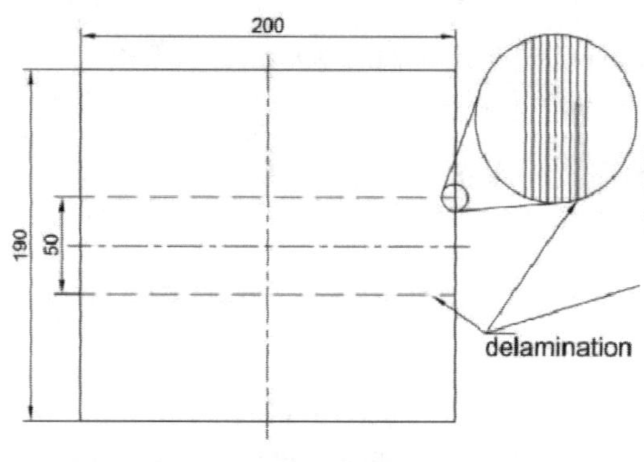

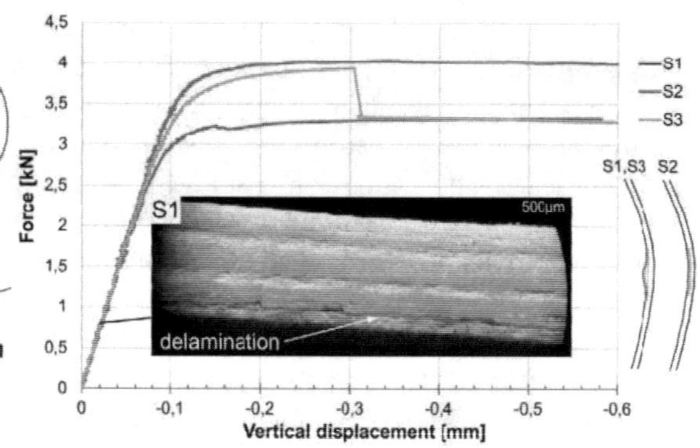

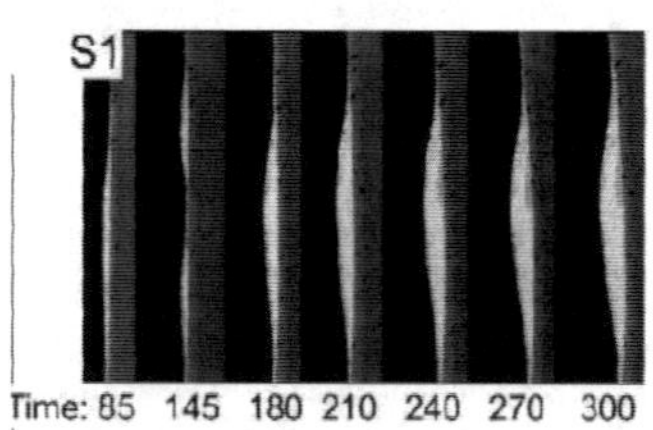

IOP Conf. Series: Materials Science and Engineering **744** (2020) 012003 doi:10.1088/1757-899X/744/1/012003

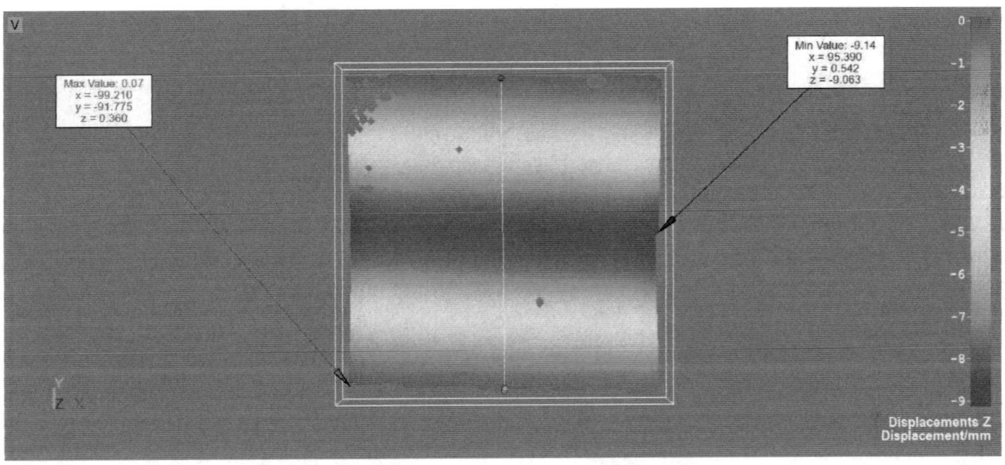

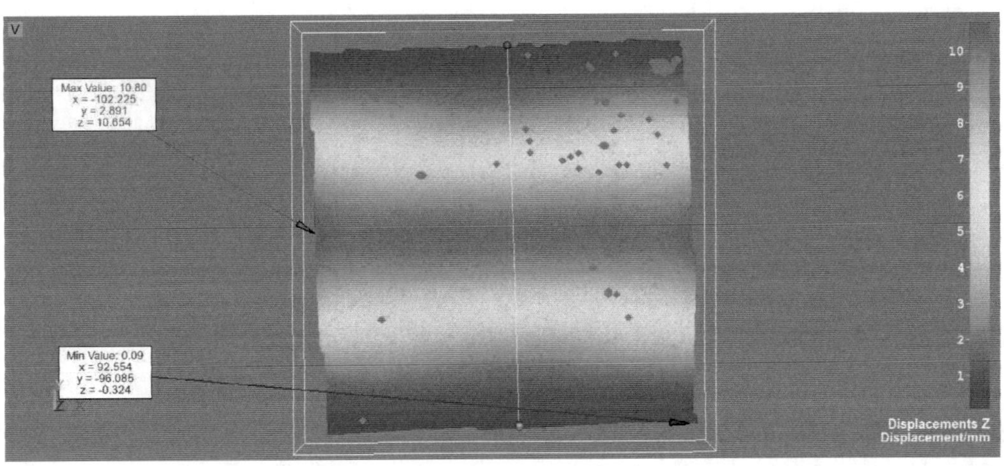

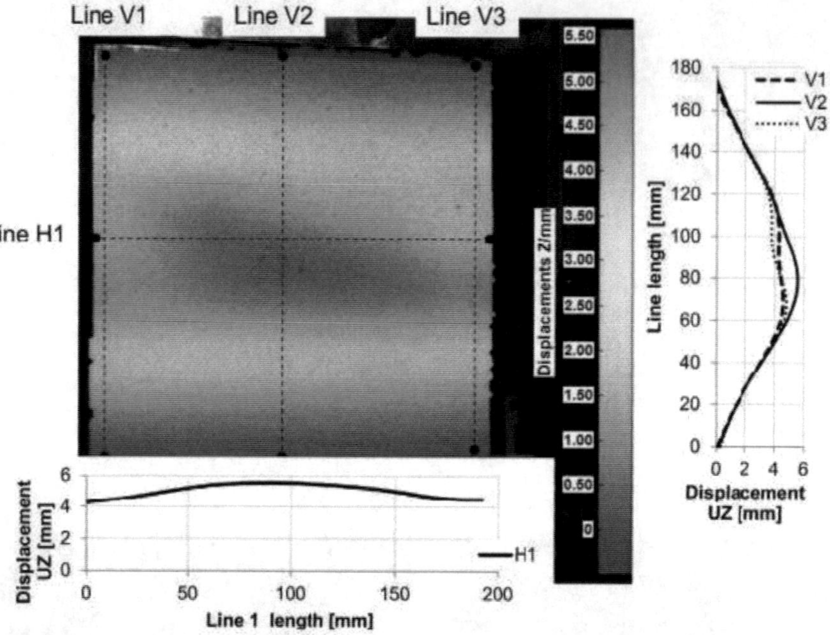

Figure 5. The normal w deflection distribution for plate with rectangular delamination (specimen S1).

4. Numerical (Finite Element) modelling

Then, with the aid of the numerical FE package NISA II the buckling loads and the post-buckling behaviour (the Riks method) were computed. The example of the mesh is shown in Figures 6 a, b. The analysis was carried out with the use of 3D FE layered orthotropic elements. In the thickness direction the plate was divided into four layers – the thickness of the first layer is equal to the thickness of the first sublaminate. The results were compared with the results presented by Ricci [15] for perfect delaminated plates. Reducing the delamination ratio ($\delta \to 0$) one can obtain the results for perfect plates (Figure 6 c).

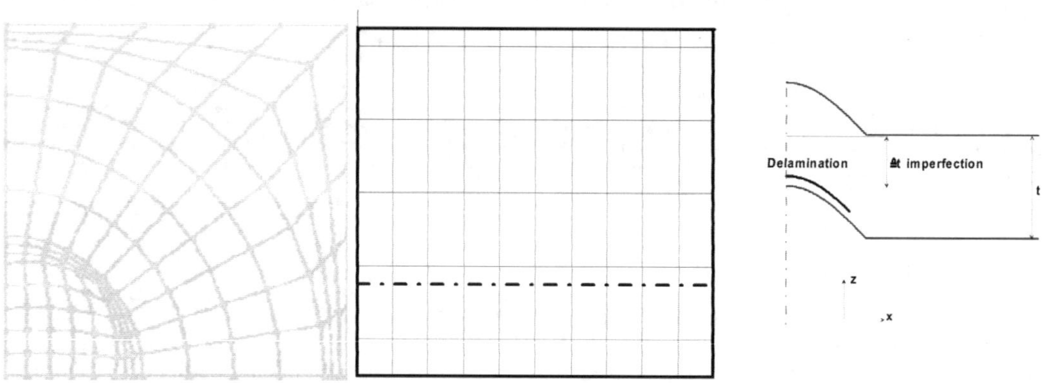

Figure 6. Example of the mesh used for the discretization of the delaminated plates (low density) a) circular delamination, b) rectangular delamination, c) form of the imperfection.

5. Discussion of Results and Concluding Remarks

The experimental (average for the identical type of specimens) and numerical results are compared in Table 1. Let us note that the difference of dimensions between the tested plates is observed (see Figures 1 and 2). As it may be observed in Figure 7 the prescribed imperfection reduces the values of buckling loads. However, the experimental values of buckling loads are lower than theoretical. The discrepancy is mainly due to random scatter of geometrical parameters characterizing the tested plates.

IOP Conf. Series: Materials Science and Engineering **744** (2020) 012003 doi:10.1088/1757-899X/744/1/012003

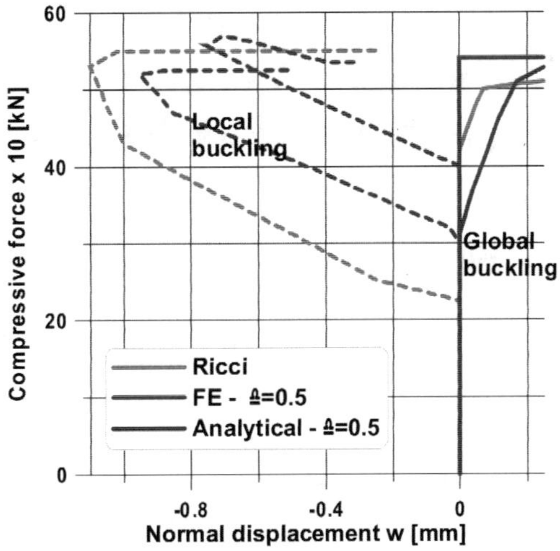

	Without delamination	Circular delaminatiom	Rectangular delamination
Experimental (average)	3.35 [kN]	2.75 [kN]	3.55 [kN]
Numerical (perfect plates)	5.5 [kN]	3.2 [kN]	4.25 [kN]

IOP Conf. Series: Materials Science and Engineering **744** (2020) 012004 doi:10.1088/1757-899X/744/1/012004

Free vibrations of cylindrical laminated shells – introduction to optimal design in divergence and flutter problems

A. Muc, M. Augustyn

Institute of Machine Design, Faculty of Mechanical Engineering, Cracow University of Technology, Kraków, Poland.
E-mail: olekmuc@mech.pk.edu.pl

Abstract. Aeroelastic properties, such as free vibrations of multilayered composite cylindrical shells, are investigated in the present paper. It is well known that the eigenmode (eigenvalue) analysis is a crucial point in the estimations/calculations of the divergence and flutter speeds characterizng dynamic instability. In this study the maximization problem of eigenfrequencies and eigenmodes is discussed, especially from the point of view of stacking sequences of composite panels. A special attention is focused on two types of the shell wall construction: a) angle-ply $\pm\theta$ and b) symmetrical discrete laminates built of layers oriented at 0^0. $\pm45^0$ and 90^0. The analytical formulae is derived with the use of the Rayleigh-Ritz method for simply supported cylinders. It allows us to solve the optimum problem easily for angle-ply fibre orientations. For symmetrical laminates with discrete fibre orientations the optimal solutions are searched for using finite element analysis combined with genetic or evolutionary algorithms. The influence of geometrical parameters describing the cylindrical panel is also demonstrated herein.

1. Introduction

A flutter phenomenon is defined as a dynamic aeroelastic instability. Aeroelasticity denotes the combination of aerodynamic, inertial and elastic forces in such a way that the structure and the flow around it interact with each other. As a result, an energy transfer between the fluid and the structure takes place and, according to the nature of this transfer, will lead to either stable or unstable motion. Actually flutter phenomenon strongly depends on the flexibility of the structure and appears when the mechanical work is lower than the aerodynamic work, i.e. when the mechanical damping is too small to overcome the aerodynamic excitations. In external flows, flutter appears when two vibrating modes interact together at distinct frequencies. In general the flutter phenomenon can lead and result in the appearance of structural failure such as fatigue damage or progressive fracture of structural components. A detailed knowledge in this area might have prevented dangerous accidents, particularly in the case of aircrafts.

The number of aeroelastic problems and the complexity of analysis depend on the type of structure considered (plates, shells – laminated composites, sandwiches, functionally graded materials - FGM), speed regime (the value of the Mach number – subsonic and/or supersonic speed) and the stages of design. The solution of the flutter problems actually starts with the finite element modeling and free vibration analysis.

For free vibration analysis, damping matrix is assumed to be zero and there no external force acting on the system. This is a generalized eigenvalue problem; the solution can be obtained using eigenvalue extraction routine since the solutions are always real and positive if the aerodynamic pressure forces are equal to zero. For increasing dynamic pressure forces the eigenfrequencies are not constant some

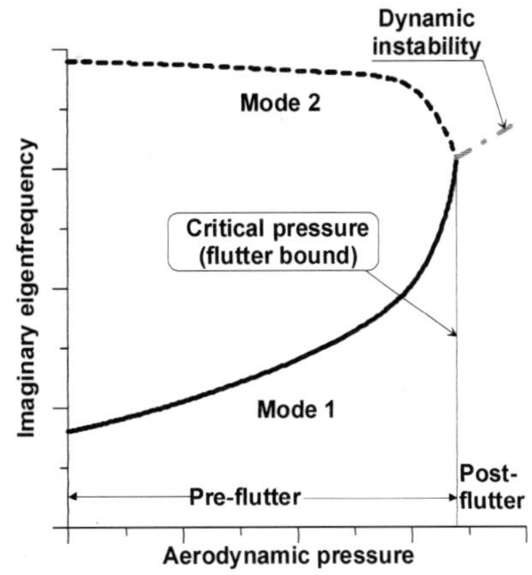

$$\underset{\mathbf{s}}{Max}\ \omega(\mathbf{s})$$

$$\delta H = \delta \int (K - \Pi) d\tau$$

$$\{\rho^{(0)}, \rho^{(1)}, \rho^{(2)}\} = \sum_{k=1}^{N} \int_{z_{k-1}}^{z_k} \rho_{(k)} \{1, z, z^2\} dz$$

$$\left[\mathbf{r}^{L-K}\right]^{Tr} = [u, v, w] \quad L_{1j}^{L-K} r_j^{L-K} = \rho^{(0)} \ddot{r}_1^{L-K} - \rho^{(1)} \frac{\partial \ddot{r}_3^{L-K}}{\partial x} \quad L_{2j}^{L-K} r_j^{L-K} = \rho^{(0)} \ddot{r}_2^{L-K} - \rho^{(1)} \frac{\partial \ddot{r}_3^{L-K}}{\partial y}$$

$$L_{3j}^{L-K} r_j^{L-K} = \rho^{(0)} \ddot{r}_3^{L-K} + \rho^{(1)} \left(\frac{\partial \ddot{r}_1^{L-K}}{\partial x} + \frac{\partial \ddot{r}_2^{L-K}}{\partial y} \right) - \rho^{(2)} \nabla^2 \ddot{r}_3^{L-K}$$

$$\rho t \omega^2 \alpha_m^2 = D_{11} \alpha_m^4 + 2(D_{12} + 2D_{66}) \alpha_m^2 \beta_n^2 + D_{22} \beta_n^4 + p(A_{rs}) \alpha_m = \frac{m\pi}{L} \beta_n^2 = \frac{n}{R}$$

$$p(A_{rs}) = \frac{\alpha_m^2}{R^2} \frac{\alpha_m^2 (A_{11} A_{22} - A_{12}^2) A_{66}}{\alpha_m^4 A_{11} A_{66} + \alpha_m^2 \beta_n^2 (A_{11} A_{22} - A_{12}^2 - 2 A_{12} A_{66}) + \beta_n^4 A_{22} A_{66}}$$

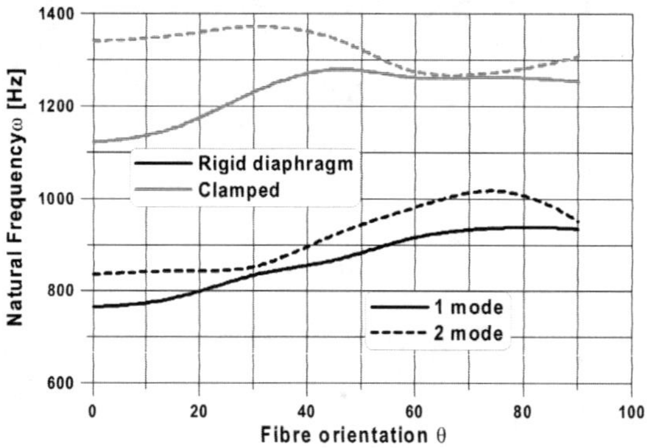

only one half of the cylinder should be considered in the circumferential direction with the symmetry (anti-symmetry) conditions at the edges. The form of the critical vibration depends on the orientation of the laminate, as shown in Figure 3d. The change in the form of the basic vibrations can also be seen in Figure 1. This corresponds to the approach of the curves characterizing the first two forms of the vibrations. Therefore, the solution to the optimization problem (1) always exists, although the application of the specific numerical values requires the use of optimization algorithms.

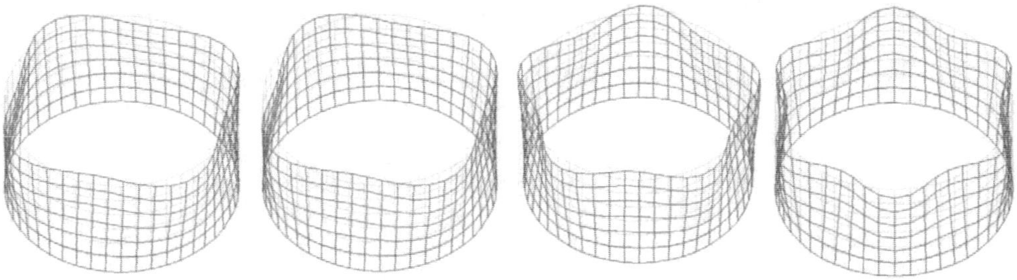

(a) (b) (c) (d)

Figure 3. The natural frequencies for clamped cylindrical shells (the analyzed shell in the longitudinal direction): (a) orientation θ =0° - 1 mode shape (frequency = 1.12254E+03 Hz); (b) orientation θ =0° - twin 1 mode shape (frequency = 1.12254E+03 Hz), (c) orientation θ =0° - 2 mode shape (frequency = 1.34058E+03 Hz), (d) orientation θ =90° - 1 mode shape (fre-quency = 1.25436E+03 Hz).

Searching for the optimal laminate configuration as a given discrete set of orientation of the individual layers (0°, ±45°, 90°) is considered the optimization problem cannot be solved using standard FEM software. It is necessary to introduce the external optimization procedures (e.g. genetic or evolutionary algorithms). The objective function is changed here, because we are currently looking for the stacking sequences of the laminate that corresponds to the minimum natural frequencies. In the analysis it is assumed that the laminate is composed of 64 individual layers, so the variable chain (chromosome) contains 16 elements (design variables).

The problem was solved by looking for the minimum frequency for specific (prescribed in advance) modes of vibrations *m*, *n*. The numerical computations of the objective function (FEM) was used. In the calculations, 1/8 cylinder was discretized using the four node quadrilateral shell elements (NKTP 32).

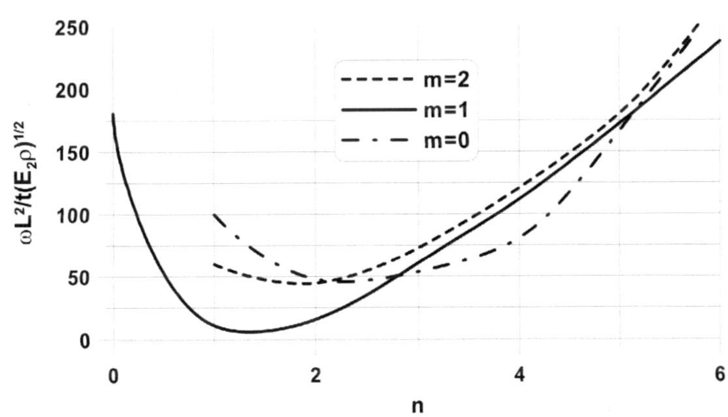

Figure 4. The optimal values of the natural frequencies (L/D=1, t/R=0,02, E_2/E_1 =1/6).

The distributions presented in Figure 4 demonstrate, that for the value of m=n=1 the minimum is achieved and it corresponds to the string in the form: $[3333222222111111]_s$ (1-0^0, 2-±45°, 3-90°). The form of the optimal chromosome clearly shows, that in this case the junction between the flexural/bending effects (terms of the [D] matrix) and the membrane effects (terms of the [A] matrix) occurs. It can be verified easily that in the case of only flexural effects, the problem of searching for an optimum is reduced to the analysis of the plate free vibrations. In the case of the domination of flexural effects (high value *n*) for cylindrical shells, the optimal configuration is to made of the fibers oriented at the angle 0° – the chain of ones.

4. Concluding Remarks

For angle-ply laminated cylindrical shells the optimal fibre configurations can be found easily plotting the relation (5) versus fibre orientations for the prescribed geometrical parameters and the number of modes in buckling.

Using the discrete fibre orientations (0°, ±45°, 90°) the numerical FE analysis is required connected with the application of genetic or evolutionary algorithms described in Refs [19-21]. It is necessary to emphasize that the use of them does not allow to obtain the global optimum but allows us to obtain better solution than starting ones. The set of the optimization problems allows comparing the effectiveness of a developed version of the optimization algorithms. As the assessment criterion, the number of population generations required to complete the optimization process was adopted (for simplification in the drawings marked as iterations). It should be emphasized, that in each of the numerical examples analyzed in the section 3, the convergence of the process was achieved. However, it was always conditioned by the adoption of the appropriate values of the indicators regulating the convergence process. It was noted, that:

- in the genetic algorithms (GA) the values of the probabilities must be low, close to zero and cross points may not be too large (the best results were obtained for values 0.4-0.6),
- in the simulated annealing (SA) algorithms it is advantageous to adopt a moderate cooling rate,
- in the genetic algorithms, increasing the number of the cross points (above one) does not improve the convergence of the process, but even makes it worse,
- both in the genetic algorithms and the simulated annealing algorithms, in-creased population size improves process convergence, but in the case of using FEM packets, this requires extending the calculation time (in the examples studied the maximum number of individuals was equal to150).

5. References

[1] Leissa A W 1973 *NASA SP-288*
[2] Bert C W, Baker V and Egle D 1969 *J Compos Mater* **3** 79
[3] Kalnins A 1964 *J Acoust Soc Am* **36** 1355
[4] Cohen G A 1965 *AIAA J* **3** 2305
[5] Tornabene F Fantuzzi N Bacciocchi M and Viola E 2015 *Compos Part B* **81** 196
[6] Grandhi R 1993 *AIAA J* **31** 2296
[7] Szyszkowski W and King J M 1993 *AIAA J* **31** 2163
[8] Bert C W 1977 *J Sound Vib* **50** 229
[9] Reiss R and Ramachandran S 1987 *Computer Struct* **4** 476
[10] Muc A and Zuchara P 1999 *Arch Civil Eng* **45** 289
[11] Muc A and Zuchara P 2000 *Compos Struct* **48** 145
[12] Rand R A and Shen C N 1973 *Computer Struct* **3** 247
[13] Teters G A, Rikards R B and Narusberg V L 1978 *Optimization of composite shells* (Zinatnie Riga in Russian).
[14] Kayran A and Anlaş G 1999 *J Vib Control* **5** 355
[15] Baniczuk N V Iwanowa S Ju and Szaranjuk AW 1989 *Dynamics of the structure, Analysis and optimization* (Sience Moscow in Russian).
[16] Muc A 2018 *Latin Am J Solids Struct* **15** 87
[17] Muc A and Bondyra A 2013 *Proc 9 ICCST (Naples)* ed Meo M (Destech Publ Lancaster) p.765
[18] Muc A 2018 *Aerospace* **5** 95
[19] Muc A and Gurba W 2001*Composite Structures* **54** 275
[20] Muc A 2007 *Thin-Walled Struct* **45** 816
[21] Muc A and Muc-Wierzgoń M 2012 *Compos Struct* **94** 1461

IOP Conf. Series: Materials Science and Engineering **744** (2020) 012005 doi:10.1088/1757-899X/744/1/012005

The Influence of Diameter-to-thickness Ratios on the Response and Failure for Locally-dented Circular Tubes Submitted to Cyclic Bending

K. L. Lee[1], M. L. Weng[2] and W. F. Pan[3]

1 Professor, Department of Innovative Design and Entrepreneurship Management, Far East University, Taiwan
2 Graduate Student, Department of Engineering Science, National Cheng Kung University, Taiwan
3 Professor, Department of Engineering Science, National Cheng Kung University, Taiwan
E-mail: z7808034@email.ncku.edu.tw

Abstract. This paper describes the cyclic bending response and failure of 6061-T6 aluminum alloy locally-dented tubes (abbreviation: 6061-T6 LDT) with different diameter-to-thickness ratios (abbreviation: D_o/t ratio) of 16.5, 31.0 and 60.0. A small lateral local dent was created on the tube using a simple technique. A tube having a dent depth from very shallow to about 2.4 times the thickness of the tube wall is considered. From the first bending cycle, the moment-curvature relationship depicts a closed and stable loop. Due to small and localized dents, dent depth shows very small influence on its relationship. But, as the number of bending cycle increases, the ovalization-curvature relationship shows an increase and a ratchet-like distribution. However, the dent depth shows a dramatic effect to this relationship. Furthermore, for a certain D_o/t ratio, five non-paralleled straight lines of five different dent depths were found for the controlled curvature-number of cycles required to ignite failure relationship on a log-log scale. Finally, an empirical formula was introduced to describe the aforementioned relationship. As a result, the experimental and analytical data were found to agree well.

1. Introduction

As we all know that the bending of a circular tube will lead to tube's ovalization (i.e., the amount of change in outer diameter divided by the amount of original diameter, $\Delta D_o/D_o$). In cyclic bending loads, $\Delta D_o/D_o$ slowly increases with the number of bending cycles. Finally, when a certain number of bending cycles is reached and $\Delta D_o/D_o$ reaches a certain critical value, the circular tube will undergo buckling or fracture failure. Thus, to completely understanding the behavior of circular tubes submitted to cyclic bending is a very important topic for the industry.

Research on the cyclic bending of smooth circular tubes has yielded significant results so far. In 1985, Professor Kyriakides and his research team designed a mechanical device to conduct the cyclic bending tests for various materials of circular tubes (6061-T6 aluminum alloy tube, 1018 steel tube, 304 stainless steel tube, and 1020 steel tube). Many experimental and theoretical studies have been carried out under monotonic or cyclic bending with or without internal or external pressure [1-6].

In 2010, Pan et al. began investigations on the response of the notched tube submitted to a cyclic bending load. Lee et al. [7] examined the changes in $\Delta D_o/D_o$ of SUS304 stainless steel sharp-notched tubes undertaken cyclic bending. In another study, Lee et al. [8] determined the collapse of SUS304 stainless steel sharp-notched tubes undertaken cyclic bending at different curvature rates; changes in both notch depth and curvature rate were also examined in this work. Chung et al. [9] investigated the stability of 6061-T6 aluminum alloy sharp-notched tubes submitted to cyclic bending. However, all of the above studies are circular circumferential notched tubes. Later, Lee et al. [10] studied the deterioration and failure of 6061-T6 aluminum alloy local sharp-notched tubes submitted to cyclic bending. In addition, the moment (M)-curvature (κ) and $\Delta D_o/D_o$-κ relationships were analysed by the finite element analysis software "ANSYS". Lee et al. [11] examined the behavior of SUS304 stainless steel local sharp-notched tubes with different notch depths at different notch directions undertaken cyclic bending. In their study, the M-κ and $\Delta D_o/D_o$-κ relationships were also analysed by ANSYS. In addition, the controlled curvature-number of cycles required to ignite buckling relationship was carried out related descriptions. Experimental study of 6061-T6 LDTs with different D_o/t ratios of 16.5, 31.0 and 60.0 submitted to cyclic bending was demonstrated. This paper used the tube-bending device and curvature-ovalization measurement apparatus to perform the curvature-controlled cyclic bending tests. The quantities of M, κ and $\Delta D_o/D_o$ were measured by sensors on testing facilities. Additionally, the number of bending cycles required to ignite failure (N_f) was also recorded.

2. Experiments

2.1. Experimental device

Fig. 1(a) schematically shows the experiment executed by a specially built tube-bending machine. This facility was set up to conduct monotonic, and cyclic bending tests. A detailed description of the experimental setup is included in some of the papers (e.g., Lee et al. [8], Chung et al. [9]). Pan et al. [12] designed a new apparatus to measure the κ and $\Delta D_o/D_o$ of the tube shown in Fig. 1(b). Two side-inclinometers in the apparatus were used to detect the changes in the angle of the tube during cyclic bending. The magnitude of κ can be obtained by an easy calculation according to angle changes. The version of the calculation can be discovered in the paper by Pan et al. [12]. The value of $\Delta D_o/D_o$ can also be measured in the central portion of the apparatus that included a magnetic detector and a magnetic block.

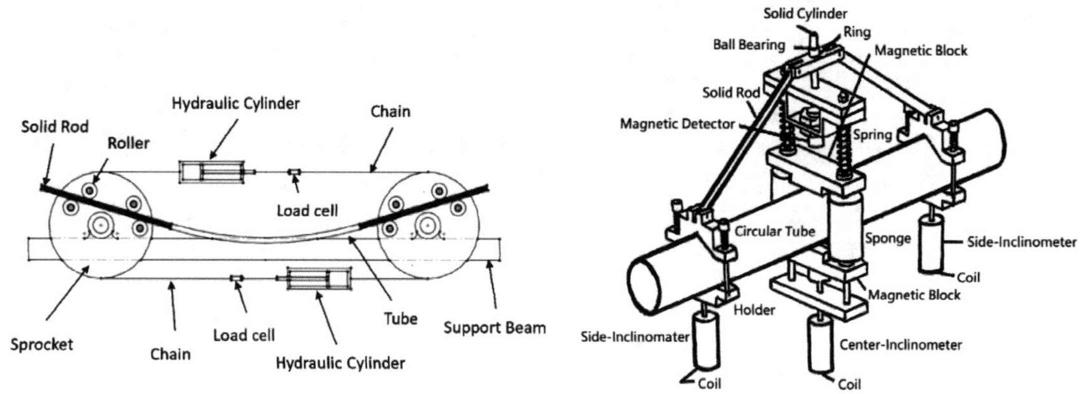

(a) (b)

Figure 1. Schematic drawings of (a) the tube-bending machine and (b) curvature-ovalization measurement apparatus.

2.2. Material and Specimens

A round tube made of 6061-T6 aluminum alloy is used in this study and its chemical composition is Mg 0.916, Si 0.733, Cu 0.293, Ti 0.268, Fe 0.256, Mn 0.132, Zn 0.0983, Cr 0.0682, Ni 0.0056, Pb 0.005, Sn < 0.001, and Al remainder. The mechanical properties of this material are: ultimate tensile strength of 310 MPa, 0.2% offset yield strength of 286 MPa, percent elongation of 23%.

The original smooth 6061-T6 aluminum alloy tube has a length L_o = 800 mm, an outer diameter D_o = 35.0 mm, and a wall thickness t = 3.0 mm. The original circular tubes were processed on the outside surface to give the desired D_o/t ratios of 16.5, 31.0 and 60.0, as shown in Fig. 2. However, the inner radiuses of all tested tubes were intact (29.0 mm). Next, the tube with a certain D_o/t ratio was again treated on the outer surface to achieve the desired dent shape and depth. Fig. 3(a) and Fig. 3(b) respectively depict a picture and a schematic drawing of dent production. The upper mould was in contact with the surface of the tube and the pressure was applied to create a dent. In present study, five dent depths (a) were took into account: 0.0, 0.3, 0.6, 0.9 and 1.2 mm. Note that a = 0.0 mm indicates a tube with a smooth surface.

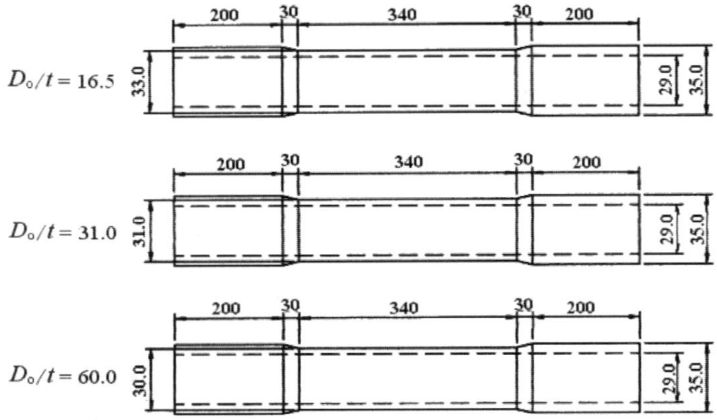

Figure 2. A schematic drawing of tube's dimensions for D_o/t ratios of 16.5, 31.0 and 60.0.

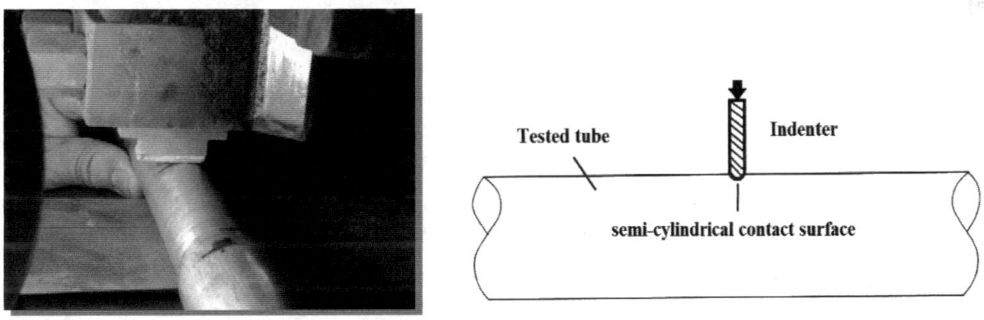

(a) (b)

Figure 3. Process a dent on the 6061-T6 aluminum alloy tube.

2.3. Test Procedures

The bending experiment was carried out under curvature control conditions with a loading curvature of of 0.03 m^{-1}/s. Two load cells (Fig. 1(a)) settled in the tube-bending machine were used to measure M. The amounts of κ and $\Delta D_o/D_o$ were measured by the curvature-ovalization measurement apparatus, as shown in Fig. 1(b). In addition, the sum of N_f was also recorded.

3. Experimental Result and Discussion

3.1. Mechanical Response

Figs. 4(a)-4(o) respectively show a typical experimentally determined M-κ curves for 6061-T6 LDTs, with D_o/t = 16.5, 31.0, 60.0 and a = 0.0, 0.3, 0.6, 0.9 and 1.2 mm, when submitted to cyclic bending. The tubes were cycled between $\kappa = \pm$ 0.4 m^{-1}. As can be seen from Figs. 4(a)-4(e) that the M-κ relationships were summarized as an almost closed and stable hysteresis loops from the first bending cycle. However, it was observed Figs. 4(f)-4(j) that from the first bending cycle, the M-κ relationships were considered to be a closed and stable hysteresis loop for a = 0.0, 0.3 and 0.6 mm. The M-κ relationships were closed hysteresis loops, but became a stable loop after several cycles of bending for other a. Additionally, from Figs. 4(k)-4(o), the M-κ relationships were closed and stable hysteresis loops from the first bending cycle for a = 0.0 and 0.3 mm. However, the M-κ relationships were closed hysteresis loops, but became stable after a few bending cycles for other a.

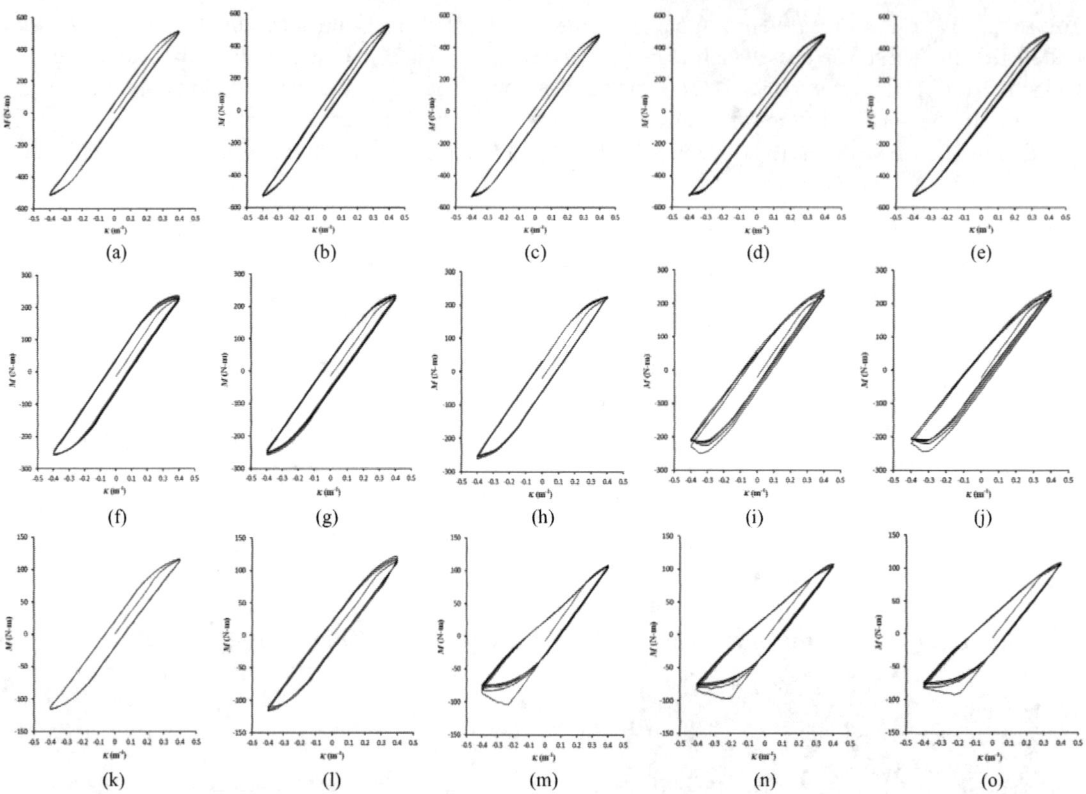

Figure 4. Experimental M-κ curves for 6061-T6 LDTs with D_o/t = 60.0 and a = (a) 0.0, (b) 0.3, (c) 0.6, (d) 0.9 and (e) 1.2 mm, D_o/t = 31.0 and a = (f) 0.0, (g) 0.3, (h) 0.6, (i) 0.9 and (j) 1.2 mm, D_o/t = 16.5 and a = (k) 0.0, (l) 0.3, (m) 0.6, (n) 0.9 and (o) 1.2 mm submitted to cyclic bending.

Figs. 5(a)-5(o) respectively show a typical experimentally determined $\Delta D_o/D_o$-κ curves for 6061-T6 LDTs, with D_o/t = 16.5, 31.0, 60.0 and a = 0.0, 0.3, 0.6, 0.9 and 1.2 mm, when submitted to cyclic bending. The tubes also cycled between $\kappa = \pm$ 0.4 m^{-1}. It was demonstrated that the $\Delta D_o/D_o$-κ relationships displayed a ratchet and an increasing trend when the number of bending cycles increased,. For a certain D_o/t ratio, larger a caused an asymmetrical appearance of the $\Delta D_o/D_o$-κ relationship. Moreover, a larger D_o/t ratio or a in the dented tubes caused larger $\Delta D_o/D_o$.

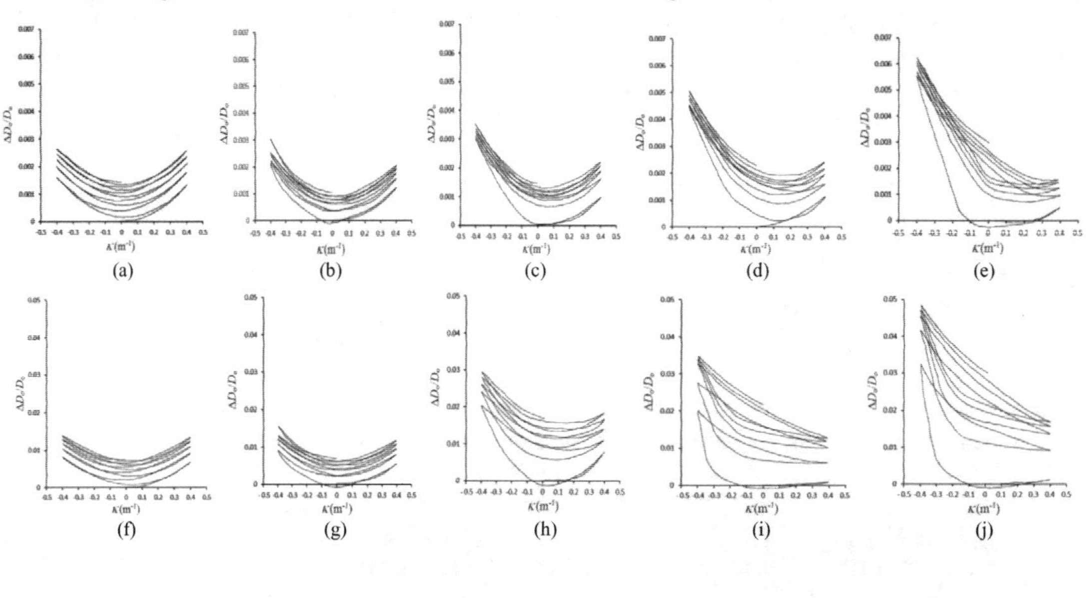

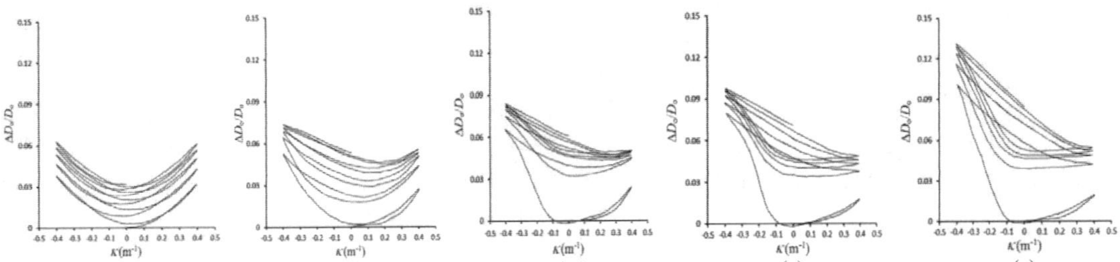

(k) (l) (m) (n) (o)

Figure 5. Experimental $\Delta D_o/D_o$-κ curves for 6061-T6 LDTs with D_o/t = 60.0 and a = (a) 0.0, (b) 0.3, (c) 0.6, (d) 0.9 and (e) 1.2 mm, D_o/t = 31.0 and a = (f) 0.0, (g) 0.3, (h) 0.6, (i) 0.9 and (j) 1.2 mm, D_o/t = 16.5 and a = (k) 0.0, (l) 0.3, (m) 0.6, (n) 0.9 and (o) 1.2 mm undertaken cyclic bending.

3.2. Failure

Figs. 6(a)-6(c) respectively show the experimental κ_c/κ_o-N_f curves for 6061-T6 LDTs with D_o/t = 16.5, 31.0 and 60.0 under cyclic bending. For a certain κ_c/κ_o and a, tubes with a larger D_o/t ratios led to a lower N_f. In addition, for a certain D_o/t ratio and κ_c/κ_o, tubes with a larger a led to a lower N_f. Subsequently, the experimental data in Figs. 6(a)-6(c) were plotted on a log-log scale, and five straight dotted lines were observed in Figs. 7(a)-7(c). Note that the dotted lines were obtained by the least squares method. It was found that for each D_o/t ratio, five non-paralleled dotted lines corresponded to five different a. In addition, the values of κ_c were normalized using $\kappa_o = t/D_o^2$ (Kyriakides and Shaw [1]).

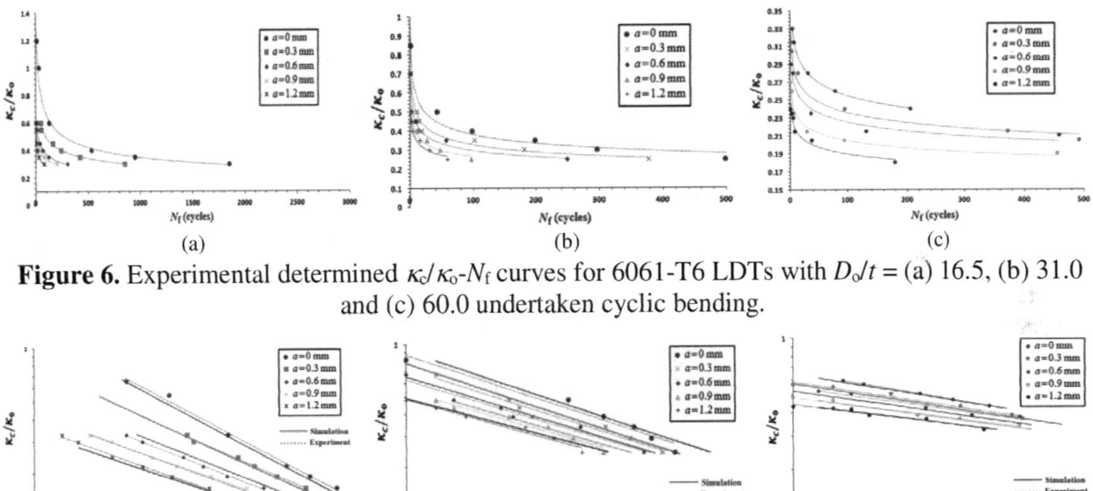

(a) (b) (c)

Figure 6. Experimental determined κ_c/κ_o-N_f curves for 6061-T6 LDTs with D_o/t = (a) 16.5, (b) 31.0 and (c) 60.0 undertaken cyclic bending.

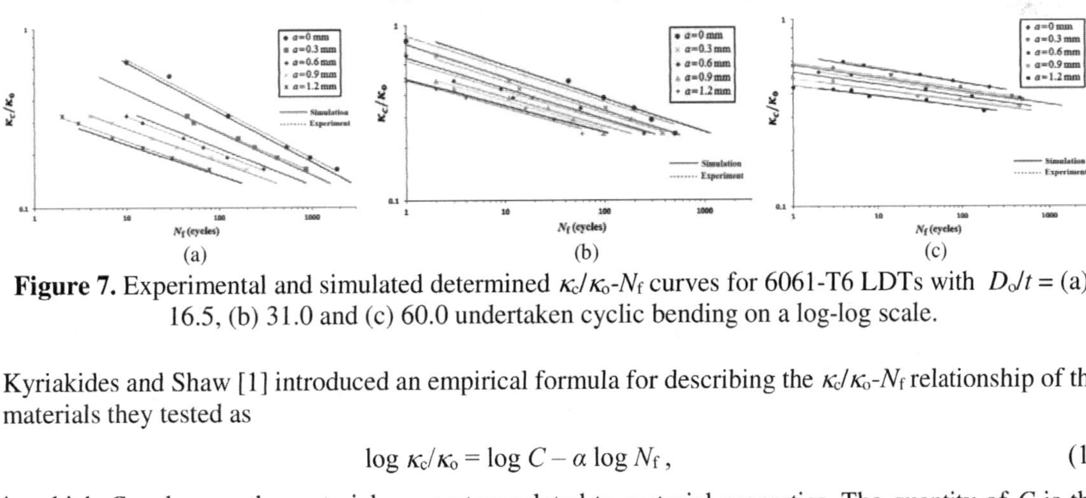

(a) (b) (c)

Figure 7. Experimental and simulated determined κ_c/κ_o-N_f curves for 6061-T6 LDTs with D_o/t = (a) 16.5, (b) 31.0 and (c) 60.0 undertaken cyclic bending on a log-log scale.

Kyriakides and Shaw [1] introduced an empirical formula for describing the κ_c/κ_o-N_f relationship of the materials they tested as

$$\log \kappa_c/\kappa_o = \log C - \alpha \log N_f , \tag{1}$$

in which C and α are the material parameters related to material properties. The quantity of C is the amount of κ_c/κ_o when N_f = 1, and α is the slope of the line in the log-log plot. In this study, the liner relationships between ln$C$ and a/t and lnα and a/t were respectively found in Figs. 8(a)-8(c) and Figs. 9(a)-9(c). Therefore, we write

$$\ln C = C_o - \beta\,(a/t) \tag{2}$$

and

$$\ln\alpha = \alpha_o - \gamma(a/t), \tag{3}$$

in which C_o, β, α_o and γ are material parameters. The quantities of C_o and β were the intercepts and slopes in Figs. 8(a)-8(c) for D_o/t = 16.5, 31.0 and 60.0, respectively, and quantities of α_o and γ were the intercepts and slopes in Figs. 9(a)-9(c) for D_o/t = 16.5, 31.0 and 60.0, respectively.

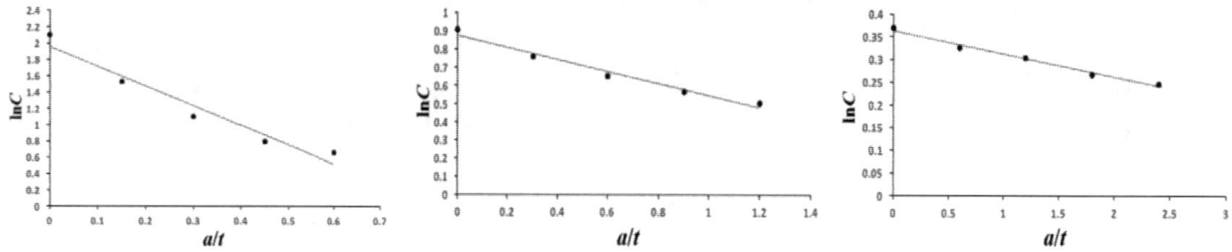

(a) (b) (c)

Figure 8. ln$C$-a/t relationship for D_o/t = (a) 16.5, (b) 31.0 and (c) 60.0.

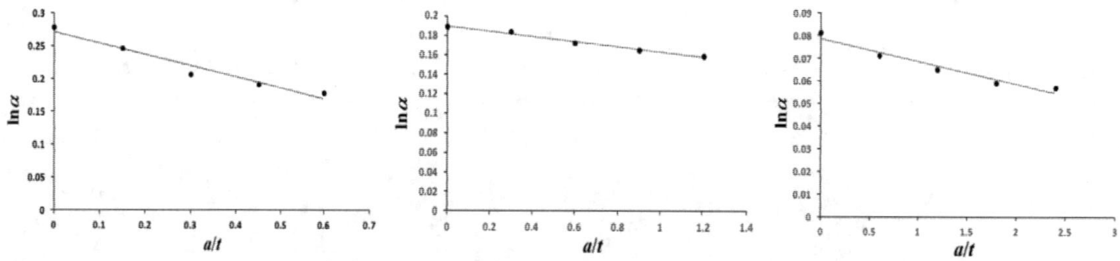

(a) (b) (c)

Figure 9. lnα-a/t relationship for D_o/t = (a) 16.5, (b) 31.0 and (c) 60.0.

Next, the material parameters C_o, β, α_o and γ were assumed to be related to the D_o/t ratios. In this study, we tried to find the linear relationships between the above material parameters and D_o/t ratios. After a lot of attempts, the linear relationships were obtained in Figs. 10(a)-10(d). Thus, the linear relationships can be written to be

$$\ln C_o = a_1 \ln D_o/t + a_2, \tag{4}$$

$$\ln\beta = b_1 \ln D_o/t + b_2 \tag{5}$$

$$1/\ln\alpha_o = c_1 \ln D_o/t + c_2 \tag{6}$$

and

$$1/\gamma = d_1(D_o/t) + d_2, \tag{7}$$

where a_1, a_2, b_1, b_2, c_1, c_2, d_1 and d_2 are material constants. The values of a_1, b_1, c_1 and d_1 were the slopes in Figs 10(a)-10(d), respectively. The amounts of a_2, b_2, c_2 and d_2 were the intercepts in Figs. 10(a)-10(d), respectively. These material constants, a_1, a_2, b_1, b_2, c_1, c_2, d_1 and d_2, were determined to be -1.356, 4.551, -3.098, 9.684, 0.287, -1.574, 2.161 and -29.657, respectively.

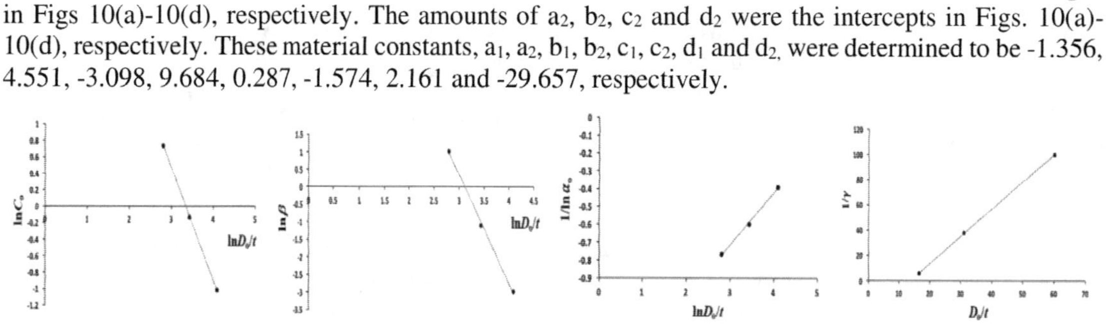

(a) (b) (c) (d)

Figure 10. (a) lnC_0-lnD_o/t relationship, (b) lnβ-lnD_o/t relationship, (c) 1/lnα_o-lnD_o/t relationship, and (d) $1/\gamma$-D_o/t relationship.

Finally, Eqs. (1)-(7) were used to simulate the experimental finding in Figs. 7(a)-7(c). The simulated results of κ_c/κ_o-N_f relationships for 6061-T6 LDTs with D_o/t = 16.5, 31.0 and 60.0 under cyclic bending on a log-log scale are respectively depicted in Figs. 7(a)-7(c) in solid lines. Good agreement has been achieved between the experimental and simulation results.

4. Conclusions

The response and failure of 6061-T6 LDTs with different D_o/t ratios and a subjected to cyclic bending were examined. According to the experimental and simulation results, this study draws the following conclusions:

(1) The experimental M-κ relationship for 6061-T6 LDTs with any D_o/t ratio or a displayed a closed and stable hysteresis loop from the first bending cycle or after a few bending cycles. When a is less than t, the magnitudes of M are almost equal at the maximum and minimum curvatures. When a is close to or larger than t, due to the two sides of the dent coming into contact during reverse bending, the magnitude of M at the minimum curvature may smaller than that at the maximum curvature.

(2) The experimental $\Delta D_o/D_o$-κ relationship for 6061-T6 LDTs with any D_o/t ratio or a revealed an increase and a ratchet-like distribution in the number of bending cycles. The $\Delta D_o/D_o$-κ relationships were symmetrical for a = 0.0 mm, but asymmetrical for $a \neq$ 0.0 mm. Moreover, the tubes with a larger D_o/t or a led to more asymmetrical trend and a larger $\Delta D_o/D_o$.

(3) The empirical formula of Eq. (1) introduced by Kyriakides and Shaw [1] was modified to describe the κ_c/κ_o-N_f relationship for 6061-T6 LDTs with different D_o/t ratios undertaken cyclic bending. According to the experimental data, the forms of the material parameters, C and α, were respectively introduced in Eqs. (2) and (3). In addition, The forms of the material parameters, C_o, β, α_o and γ, were respectively introduced in Eqs. (4)-(7). The analytical data described by Eqs. (1)-(7) were very consistent with the experimental findings (Figs. 7(a)-7(c)).

5. References

[1] Kyriakides S. and Shaw P. K. 1987 *J. Pres. Ves. Tech.* **109** 169
[2] Corona E. and Kyriakides S. 1991 *Thin-Walled Struct.* **12** 229
[3] Corona E. and Kyriakides S. 2000 *Int. J. Solids Struct.* **24** 505
[4] Limam A., Lee L. H., Corana E. and Kyriakides S. 2010 *Int. J. Mech. Sci.* **52** 37
[5] Bechle N. J. and Kyriakides S. 2014 *Int. J. Solids Struct.* **51** 967
[6] Jiang D., Kyriakides S., Bechle N. J. and Landis C. M. 2017 *Int. J. Solids Struct.* **124** 192
[7] Lee K. L., Hung C. Y. and Pan W. F. 2010 *J. Mech.* **26** 403
[8] Lee K. L., Hsu C. M. and Pan W. F. 2013 *Acta Mech. Solida Sinica* **26** 629
[9] Chung C. C., Lee K. L. and Pan W. F. 2016 *Int. J. Struct. Stab. Dyn.* **26** 1550035
[10] Lee K. L., Chang K. H. and Pan W. F. 2016 *Struct. Eng. Mech.* **60** 387
[11] Lee K. L., Chang K. H. and Pan W. F. 2018 *Int. J. Struct. Stab. Dyn.* **18** 1850099
[12] Pan W. F., Wang T. R. and Hsu C. M. 1998 *Exp. Mech.* **38** 99

Acknowledgments

The study was conducted under the authority of the project MOST 105-2221-E-006-073, supported by the Ministry of Science and Technology. Here, we are very grateful for the support provided by the Ministry.

Nafion-carbon containing nanocomposite materials with high catalytic activity for portable energy sources

M V Lebedeva[1], A V Ragutkin[2], A P Antropov[1], N A Yashtulov[3]

1 Associate professor, MIREA – Russian Technological University, Moscow, Russia
2 Vice-rector for innovation development, MIREA – Russian Technological University, Moscow, Russia
3 Professor, MIREA – Russian Technological University, Moscow, Russia
E-mail: lebedevamv@mitht.ru

Abstract. In this work the electrode materials on combined polymer-carbon matrix-supports modified by bimetal platinum-ruthenium nanoparticles have been formed. The functional characteristics of the electrodes have been studied by atomic force microscopy, scanning electron microscopy and voltammetry methods. The catalytic activity of the synthesized composites in the methanol oxidation reaction had been studied.

1. Introduction

The direct methanol fuel cells (DMFC) are advantageous devices due to easy transportation and storage of the fuel, reduced system weight, and high energy efficiency potential energy sources for portable devices, electronic and transportation purposes [1-7]. Methanol is a suitable fuel for fuel cells from the perspective of its high energy density, relatively easy and safe handling. Low-temperature DMFCs with proton exchange membrane (PEM), for example, Nafion-type membrane [7-10, 13] are usually operated at 50-80 °C with the liquid methanol fuel.

Pt is well known catalyst for methanol electro-oxidation reaction. However, Pt is poisoned by carbon monoxide CO and other reaction intermediates under the low-temperature reaction conditions of DMFCs. To solve the catalyst poisoning problem on methanol electro-oxidation reaction, the use of bimetallic anode catalysts is an effective way for construction of fuel cells with increased activity and stability. Therefore, the addition of Ru improves the methanol electro-oxidation activity of Pt catalyst due to bi-functional mechanism. Catalysts based platinum and ruthenium nanoparticles are of great interest as anode materials with superior performance in methanol oxidation reaction [1-7, 11-14].

For the formation of electrode materials for fuel cells it is necessary to select the optimal matrix-support [4, 7-10, 13]. One of such matrix-carriers are solid polymer membranes, to which the following requirements are presented:

- high proton conductivity – (impedance);
- low electronic component – (voltammetry);
- low membranes permeability on fuel and O_2;
- resistance to oxidizing media;
- good mechanical properties (flexibility, etc.).

Polymer membrane like Nafion, in general, meets these requirements. However, in order to achieve the best catalytic characteristics, many authors modify the polymer membrane with various inorganic additives, including nanoparticles of various metals [4, 7-10, 13].

It should be mentioned that carbon carriers (carbon black, graphene, carbon nanotubes) are also effective matrix carriers with a high surface area. The combined electrode materials which are composed of a polymer and a carbon component are of great interest because they allows to achieve high parameters in the studied fuel cells.

The aim of this work is formation of composite electrodes based on Nafion-carbon nanotubes support with platinum-ruthenium nanoparticles and evaluation of nanocomposite catalytic in methanol oxidation reaction.

2. Experimental

2.1. Platinum-ruthenium nanoparticles synthesis

Platinum-ruthenium nanoparticles were synthesized by chemical reduction in reverse microemulsions with sodium tetrahydroborate $NaBH_4$ (98%, Merck, Germany), using anionic surfactant – AOT in isooctane. The method of solution preparation was as follows: a water-organic solution of K_2PtCl_4 and $RuCl_3$ salts (Sigma Aldrich, USA) had been prepared. To form a microemulsion medium a 0.2 M solution of AOT (Sigma Aldrich, USA) and isooctane was used. Then, a microemulsion of the similar composition, contained 0.1 M water solution of the reducing agent – $NaBH_4$, was added to this solution under ultrasonic action for 2-3 min. The molar water/surfactant ratio (ω) while making the experiments was changed from 1.5 to 8. In order to prevent the sunlight destruction of nanoparticles, the solutions of microemulsions were stored in darkness at room temperature. Pt:Ru molar rations were 1:1, 1:3 and 3:1.

2.2. Nafion-carbon nanotube composites formation

Metal-polymer films of Nafion (Nf) membranes with Pt-Ru nanoparticles were synthesized as described in [9, 10]. The solubilization of solutions was performed on an Ultrasonis Cleaner UD150SH-6L ultrasonic disperser (Eumax, Germany). For obtaining the metal-polymer films, the samples of the Nafion membranes were placed in cells with a reverse microemulsion solution with Pt-Ru nanoparticles and solution was sonicated during 3-5 minutes. After that, the modified metal-polymer membrane was washed to remove the excess surfactant and the organic solvent.

To obtain combined metal-polymer films with CNT, the Nafion membrane modified with Pt-Ru nanoparticles was immersed in a water solution of propanol-2 and glycerol (Merck, Germany) with CNT. In work the single-wall (SW) and multi-walled (MW) carbon nanotubes (Sigma Aldrich, USA) were applied as main components of the mixed Nafion-CNT membrane. The diameter of single-walled nanotubes was 1.3-2.3 nm, multi-walled – 110-170 nm. The resulting suspension was sonicated for 10-15 min. The homogeneous suspension containing a required amount of platinum-ruthenium catalyst, the polymer, and CNT was dried on a glass substrate at 70°C in an argon atmosphere. The obtained Pt-Ru/Nf-CNT composites were then hot pressed at 120°C to assure good adhesion between CNT and the Nafion membrane. In order to clean the electrodes from traces of surfactant and solvent, Nafion-CNT samples with Pt-Ru nanoparticles were washed in isooctane, ethanol and distilled water.

2.3. Instrumentation

The sizes, shape, and distribution of platinum-ruthenium nanoparticles were studied by atomic force microscopy (AFM) on NTegra Prima scanning microscope (NT MDT, Russia). The morphology of the nanocomposite was studied by scanning electron microscopy (SEM) on a JSM-7401F microscope (Jeol, Japan). The catalytic activity of the electrodes was estimated by cyclic voltammetry (CVA) method on IPC PRO M device (Tekhnopribor, Russia). The scanning rate was varied from 10 to 100 mV/s. The CVA data were fixed after the stabilization of the parameters of voltamperograms.

3. Results and discussions

3.1. Studies of nanoparticles size and shape

On figure 1 the AFM image of Pt-Ru nanoparticles, obtained by the method of chemical reduction using anionic surfactant with solubilization coefficient ω = 1.5 and when the metals ratio of 3:1 is presented. As it can be seen from the microphotograph, Pt-Ru nanoparticles are primarily characterized by the spherical shape. For water organic solutions of Pt-Ru with a minimum value of the solubilization coefficient, the average particle size was 3-4 nm. The increase of the solubilization degree to 5 was contributed to the increase in size of nanoparticles at all the metals ratio.

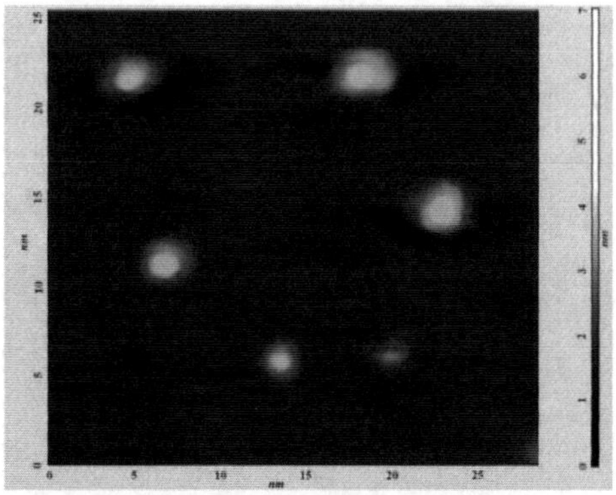

Figure 1. AFM image of Pt-Ru nanoparticles (metal ratio of 3:1) at ω = 1.5.

Feature of solid polymer membranes of the type Nafion is nanometer pore size with the membrane thickness less than 0.2 mm. Transport of hydrated protons is carried out through the membrane channels with a width of 1-2 nm, which are linked sulfacetamide hollow nanoscale clusters with an average diameter of 6 nm. The arrangement in the polymer matrix protects the nanoparticles from both desorption and agglomeration. On the surface of the membrane fragments of the internal cavities of the clusters of the polymer chain stimulate the adsorption of the low-dimensional nanoparticles, limiting their agglomeration, facilitate uniform distribution of nanoparticles in the film (Table 1). The size of nanoparticles in the membrane volume should probably not exceed 6 nm. Nanoparticles larger than 6 nm are fixed to the surface of the membrane. In addition, the presence of carbon nanotubes in the matrix-carrier contributes to additional stabilization and reduction of the size of nanoparticles in the composite membrane. The smallest particle size in the matrix is observed in nanocomposites formed at metals ratio of 3:1 and a minimum solubilization degree equal to 1,5.

Table 1. Size distribution of Pt-Ru nanoparticles support at different solubilization coefficient (ω) and metal molar ratio on Nafion-MWCNT according to SEM data

Pt:Ru	d, nm		
	ω = 1.5	ω = 3	ω = 5
3:1	2.3-3.2	2.8-3.6	3.4-4.1
1:1	2.9-3.7	3.3-4.5	3.8-4.9
1:3	3.1-4.0	3.7-5.0	4.4-5.7

3.2. Catalytic activity investigation

The CH_3OH oxidation reaction within a bi-functional mechanism with Pt-Ru catalyst can be described by the following reactions [11]:

$$Pt + CH_3OH = Pt(CH_3OH)_{ads} \quad (1)$$

$$Pt(CH_3OH)_{ads} - 4\bar{e} = Pt(CO)_{ads} + 4H^+ \quad (2)$$

$$Ru + H_2O - \bar{e} \rightarrow Ru\text{–}OH + H^+ \quad (3)$$

$$Pt(CO)_{ads} + Ru\text{–}OH - \bar{e} \rightarrow Pt + Ru + CO_2 + H^+ \quad (4)$$

As a result of the adsorption/dehydrogenation reaction of CH_3OH that takes place on Pt sites, an intermediate «CO»-type species is formed. This species is believed to be oxidized to CO_2 with the assistance of –OH species, which are formed by the partial oxidation of H_2O on Ru surface sites. Based on this reaction scheme, it is clear that the ratio and distribution of Pt and Ru catalyst surface sites influences the CH_3OH oxidation kinetics.

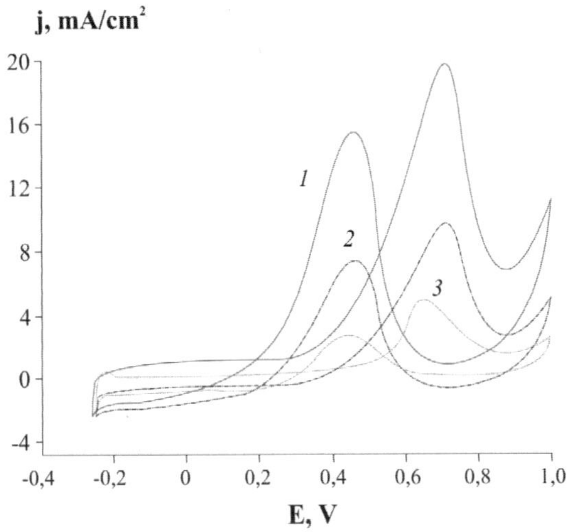

Figure 2. Cyclic voltammograms of Pt-Ru (3:1)/Nf-MWCNT (*1*), Pt-Ru (1:1)/Nf-MWCNT (*2*) and Pt/Nf-MWCNT (*3*) composites in 0,5 M H_2SO_4 + 1 M CH_3OH solution (the scan rate of 20 mV/s, Pt loading was 0.2 mg/cm^2)

The electrocatalytic activity of Pt-Ru electrodes in methanol oxidation reaction was investigated. Fig. 2 shows cyclic voltammograms (CVs) in 0.5 M H_2SO_4 + 1.0 M CH_3OH solution. The figure shows that the highest value of the peak of direct oxidation of methanol at 0.7 V exhibits Pt-Ru (3:1)/Nf-MWCNT nanocomposite. The peak of the reverse sweep at 0.44 V derives from the intermediates oxidation. According to Table 2, composites on multi-walled carbon nanotubes exhibit the highest current density than that on single-walled nanotubes, which is explained by the larger surface area of the MWCNT, as well as a larger number of π-electrons compared to the SWCNT.

Table 2. Catalytic characteristics of Pt-Ru(3:1) nanocomposites on various carriers type, according to CVA data (current density, j, is represented by direct oxidation peak at 0.7 V)

Support type	ω	j, A/m^2
Nf-MWCNT	1.5	19,8
	5	17,4
Nf-SWCNT	1.5	16,6
	5	12,7

4. Conclusion

Composite materials on combined polymer-carbon matrix carriers with bimetallic platinum-ruthenium nanoparticles had been obtained. The size and shape of nanoparticles had been investigated by electron microscopy methods. By cyclic voltammetry it was discovered that the higher catalytic activity in methanol oxidation reaction demonstrate the composites formed at a solubilization degree of 1.5 and Pt:Ru ratio of 3:1.

Acknowledgments
The research was carried out with financial support of the Russian Science Foundation (№ 18-79-00017).

References
[1] Klein J, Argast F, Engstfeld A K, Brimaud S et al. 2019 *Electrochimica Acta* **311** 244-254.
[2] Tedsree K, Thanatsiri A 2018 *Materials Today: Proceedings* **5** 10954-10963.
[3] Kuznetsov V V, Podlovchenko B I, Batalov R S, Filatova E A 2019 *Electrochimica Acta* **300** 274-283.
[4] Dicks A, Rand D A J 2018 *Fuel cell systems explained* Wiley 479 p.
[5] Tiwari J N, Tiwari R N, Singh G, Kim K S 2013 *Nano Energy* **2** 553-578.
[6] Battirola L C, Schneider J F, Torriani I C L, Tremiliosi-Filho G 2013 *International Journal of Hydrogen Energy* **38** 12060-12068.
[7] Sieben J M, Duarte M M E, Mayer C E 2010 *International Journal of Hydrogen Energy* **35** 2018-2024.
[8] Yashtulov N A, Lebedeva M V 2017 *Russian technological journal* **5** 58-73.
[9] Lebedeva M V, Antropov A P, Ragutkin A V, Yashtulov N A 2018 *International Journal of Applied Engineering Research* **13** 16774-16777.
[10] Lebedeva M V, Yashtulov N A, Flid V R 2019 *Kinetics and Catalysis* **60** 118-122.
[11] Watanabe M, Motoo S 1975 *Journal of Electroanalytic Chemistry* **60** 267-273.
[12] Borghei M, Scotti G, Kanninen P, Weckman T et al. 2014 *Energy* **65** 612-620.
[13] Lin C W, Lu Y S 2013 *Journal of Power Sources* **237** 187-194.
[14] Li X, Zhou Y, Du Y, Xu J et al. 2019 *International Journal of Hydrogen Energy* **44** 18050-18057.

IOP Conf. Series: Materials Science and Engineering **744** (2020) 012007 doi:10.1088/1757-899X/744/1/012007

The features of bimetallic Pt-Ru nanoparticles formation on the Nafion membrane for electrodes of chemical power sources

M V Lebedeva[1], A V Ragutkin[2], A P Antropov[1], N A Yashtulov[3]

1 Associate professor, MIREA – Russian Technological University, Moscow, Russia
2 Vice-rector for innovation development, MIREA – Russian Technological University, Moscow, Russia
3 Professor, MIREA – Russian Technological University, Moscow, Russia
E-mail: lebedevamv@mitht.ru

Abstract. In this work the electrode materials on Nafion modified by bimetal platinum-ruthenium nanoparticles have been formed. The functional characteristics of the electrodes have been studied by atomic force microscopy, scanning electron microscopy and voltammetry methods. The catalytic activity of the synthesized composites in the methanol oxidation reaction at different temperatures had been investigated.

1. Introduction

The performance and durability of low-temperature fuel cells (FC) heavily depend on the choice of the catalyst substrate matrix [1-13]. It provides electronic conductivity, uniform distribution of Pt-containing catalyst particles on the substrate surface and facilitates the transfer of gas to the formed porous catalyst bed. Catalysts deposited on carbon materials with a highly developed surface are widely used in low-temperature fuel cells. However, corrosion of catalysts based on carbonaceous materials, such as soot, was one of the main reasons for the reduction of fuel cell performance [1-5].

Over the last ten years, the research group started to look for corrosion-resistant ("non-carbon") of the matrix substrate for Pt electrocatalysts for FC. Various conductive and semiconductor materials were studied as functional matrix substrates for electrocatalysts in acidic media, as well as carbides, nitrides, etc [2-5].

Currently, among the solid polymer membranes used by researchers as electrolytes in chemical power sources, it is necessary to distinguish a perfluorinated polymer membrane developed by DuPont – Nafion (Nafion), which has high chemical resistance and proton conductivity [1-6].

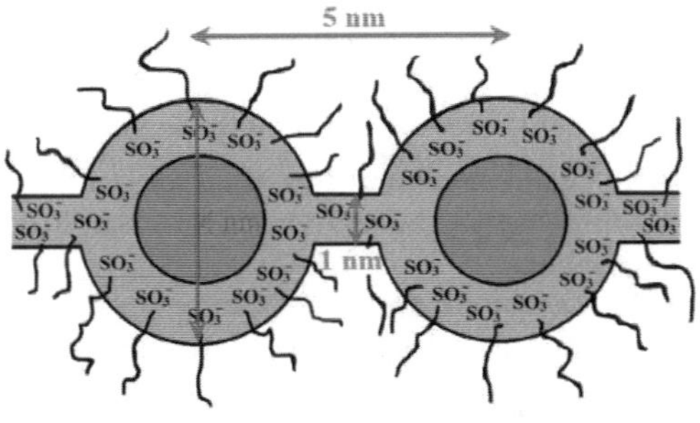

Figure 1. Nafion cluster model [6].

The cluster structure of Nafion membrane is shown in Figure 1 [6]. The polymer has a non-polar (hydrophobic) base, which provides mechanical strength and heat-resistant properties of the Nafion membrane. Side sulfo group (SO_3H) of the membrane is polar (hydrophilic) and can transfer protons.

The possibility of changing the physical and chemical properties of electrode materials by modifying the polymer matrix by nanoparticles [7, 9, 11, 13] of various metals stimulate the development of proton-conductive composite membranes suitable for use in FC, which can operate at temperatures above 100 ° C and relatively low humidity. Thus, studies are aimed at improving the performance of the membrane, suitable as an electrolyte for FC with high temperature and low humidity [6, 10].

The aim of this work is physico-chemical investigations of platinum-ruthenium/Nafion composites as effective electrode materials for chemical power sources.

2. Experimental

2.1. Platinum-ruthenium nanoparticles synthesis

Platinum-ruthenium nanoparticles were synthesized by chemical reduction in reverse microemulsions with sodium tetrahydroborate $NaBH_4$ (98%, Merck, Germany), using anionic surfactant – AOT in isooctane. The method of solution preparation was as follows: a water-organic solution of K_2PtCl_4 and $RuCl_3$ salts (Sigma Aldrich, USA) had been prepared. To form a microemulsion medium a 0.2 M solution of AOT (Sigma Aldrich, USA) and isooctane was used. Then, a microemulsion of the similar composition, contained 0.1 M water solution of the reducing agent – $NaBH_4$, was added to this solution under ultrasonic action for 2-3 min. The molar water/surfactant ratio (ω) while making the experiments was changed from 1.5 to 8. In order to prevent the sunlight destruction of nanoparticles, the solutions of microemulsions were stored in darkness at room temperature. Pt:Ru molar ration was 3:1.

2.2. Nafion composites formation

Metal-polymer films of Nafion (Nf) membranes with Pt-Ru nanoparticles were synthesized as described in [9, 10]. The solubilization of solutions was performed on an Ultrasonis Cleaner UD150SH-6L ultrasonic disperser (Eumax, Germany). For obtaining the metal-polymer films, the samples of the Nafion membranes were placed in cells with a reverse microemulsion solution with Pt-Ru nanoparticles and solution was sonicated during 3-5 minutes. After that, the modified metal-polymer membrane was washed to remove the excess surfactant and the organic solvent.

2.3. Instrumentation

The morphology of the nanocomposite was studied by scanning electron microscopy (SEM) on a JSM-7401F microscope (Jeol, Japan). The X-ray photoelectron spectrometer PHI 5500 ESCA (USA) was used to estimate the electrodes composition. The catalytic activity of the electrodes was estimated by cyclic voltammetry (CVA) method on IPC PRO M device (Tekhnopribor, Russia). The scanning rate was varied from 10 to 100 mV/s. The CVA data were fixed after the stabilization of the parameters of voltamperograms.

3. Results and discussions

3.1. Estimation of the electrodes composition

Figure 2 shows SEM-micrographs of bimetallic Pt-Ru with nanoparticles at the metals ratio of 3:1 and the solubilization coefficient ω = 1.5 on the surface of the Nafion membrane. The uniform nanoparticles distribution on the polymer film surface can be achieved by ultrasonic treatment at the formation stage of the metal polymer. As can be seen from the image, the main contribution to the formation of Pt-Ru nanocomposites is made by spherical nanoparticles with sizes from 7 to 9 nm. It can be noted that the large nanoparticle aggregates on the membrane surface are absence.

IOP Conf. Series: Materials Science and Engineering **744** (2020) 012007 doi:10.1088/1757-899X/744/1/012007

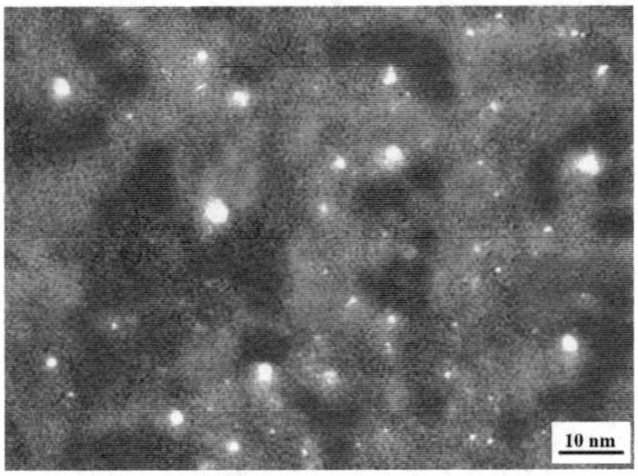

Figure 2. SEM image of Pt-Ru nanoparticles (metal ratio of 3:1) at ω = 1.5 on Nafion membrane surface.

To confirm the possible formation of oxide forms of palladium and ruthenium in the nanocomposites, the method of X-ray photoelectron spectroscopy (XPS) was used. Figure 3 shows fragments of high-resolution spectra for Pt(4f) in the XPS spectrum of bimetallic Pt-Ru/Nf nanoparticles (metal ratio 3:1) at ω = 1.5. The spectrum consists of two pairs of overlapping Lorentz curves. $Pt4f_{7/2}$ and $Pt4f_{5/2}$ lines with a binding energy of 71.30 eV (peak 1) and 74.57 eV (peak 3) can be attributed to metal platinum Pt^0. The second pair of curves was evident at 72.49 eV (peak 2) and 75.88 eV (peak 4), which can be attributed to Pt^{II} in the platinum oxide PtO and the platinum(II) hydroxide $Pt(OH)_2$. Comparison of peak heights indicates that metallic Pt^0 prevails in the Pt-Ru/Nf nanocomposite.

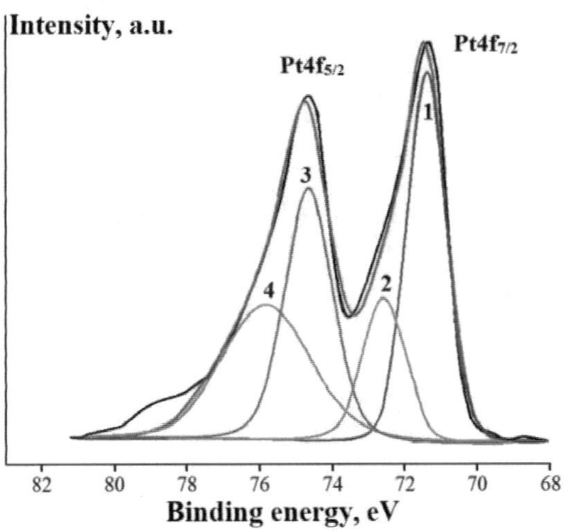

Figure 3. XPS-spectra of Pt 4f ($4f_{5/2}$ and $4f_{7/2}$) for Pt-Ru/Nf nanocomposite (metal ratio 3:1) at ω = 1.5.

The Ru(3p) XPS spectra in bimetallic Pt-Ru/Nf nanocomposite was also analyzed (figure 4). $Ru3p_{3/2}$ has three components with energies 461.32 eV (peak 1), 463.41 (peak 2) and 465.72 eV (peak 3), which can be attributed to metallic ruthenium Ru^0, Ru^{IV} (e.g., RuO_2) and ruthenium in a higher oxidation – Ru^{VI} in RuO_3. XPS results indicate the presence of oxides and metallic Ru in the nanocomposite. It is likely that the surface of the nanocatalyst consists mainly of Ru oxides and metallic Pt^0. Therefore, it seems that Ru in the composition of Pt-Ru/Nf nanocomposites undergoes noticeable oxidation under the influence of air oxygen, forming ruthenium oxide. However, the presence of ruthenium oxides in the nanocomposite can improve the catalytic properties in the CO oxidation reaction for methanol-air fuel cells.

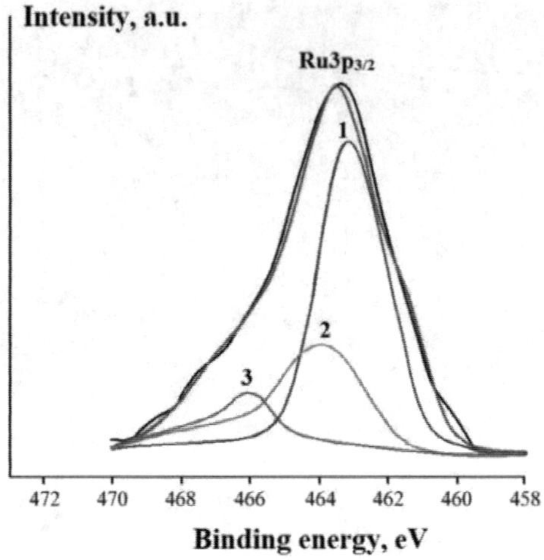

Figure 4. XPS-spectra of $Ru3p_{3/2}$ for Pt-Ru/Nf nanocomposite (metal ratio 3:1) at $\omega = 1.5$.

3.2. CH_3OH oxidation investigation at different temperature

The electrocatalytic activity of Pt-Ru electrodes in methanol oxidation reaction was investigated. Fig. 5 shows cyclic voltammograms (CVs) in 0.5 M H_2SO_4 + 1.0 M CH_3OH solution. The figure shows that the highest value of the peak of direct oxidation of methanol at 0.7 V exhibits Pt-Ru (3:1)/Nf nanocomposite. The peak of the reverse sweep at 0.44 V derives from the intermediates oxidation. To assess the effect of temperature on the current density parameters, a study was conducted at different temperatures: 25 °C, 60 °C and 80 °C.

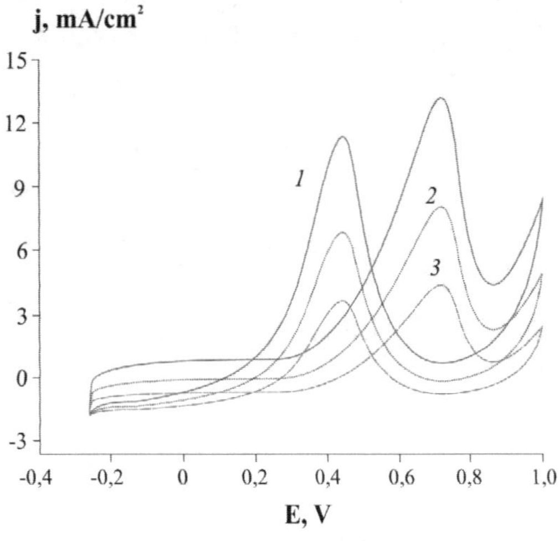

Figure 5. Cyclic voltammograms of Pt-Ru (3:1)/Nf composites in 0,5 M H_2SO_4 + 1 M CH_3OH solution (the scan rate of 20 mV/s, Pt loading was 0.2 $mg/cm)^2$ at different temperature: 25 °C (*2*), 60 °C (*1*), 80 °C (*3*).

4. Conclusion

In this paper anode materials with bimetallic Pt-Ru nanoparticles on the polymer membrane Nafion had been proposed. A study of the sizes of the nanoparticles in the polymer matrix by electron microscopy was carried out. The composites structure investigation made by X-ray photoelectron spectroscopy indicated a small number of ruthenium and platinum oxide forms in the nanocomposite. As a result of the study of catalytic activity, it was found that the optimal temperature for such electrode materials in the methanol oxidation reaction is 60 ° C. As the temperature rises, the current density decreases.

Acknowledgments
The research was carried out with financial support of the grant "University" of centralized Fund № NICH-45.

References

[1] Linse N, Gubler L, Scherer G G, Wokaun A 2011 *Electrochimica Acta* **56** 7541-7549.
[2] Noto V D, Negro E, Gliubizzi R, Gross S et al. 2007 *Journal of Electrochemical Society* **154** B745-B756.
[3] Cui G, Shen P K, Meng H, Zhao J et al. 2011 *Journal of Power Sources* **196** 6125-6130.
[4] Wickman B, Wesselmark M, Lagergren C, Lindbergh G 2011 *Electrochimica Acta* **56** 9496-9503.
[5] Liu Y, Ishihara A, Mitsushima S, Ota K 2010 *Electrochimica Acta* **55** 1239-1244.
[6] Spry D B, Goun A, Glusac K, Moilanen D E 2007 *Journal of American Chemical Society* **129** 8122-8130.
[7] Battirola L C, Schneider J F, Torriani I C L, Tremiliosi-Filho G 2013 *International Journal of Hydrogen Energy* **38** 12060-12068.
[8] Yashtulov N A, Lebedeva M V 2017 *Russian technological journal* **5** 58-73.
[9] Lebedeva M V, Antropov A P, Ragutkin A V, Yashtulov N A 2018 *International Journal of Applied Engineering Research* **13** 16770-16773.
[10] Goodarzi G A, Hayes J G 2018 *Electric powertrain : energy systems, power electronics & drives for hybrid, electric & fuel cell vehicles* John Wiley & Sons 557 p.
[11] Lebedeva M V, Yashtulov N A, Flid V R 2019 *Kinetics and Catalysis* **60** 118-122.
[12] Li X, Zhou Y, Du Y, Xu J et al. 2019 *International Journal of Hydrogen Energy* **44** 18050-18057.
[13] Kim T H, Yoo J H, Maiyalagan T, Yi S C 2019 *Applied Surface Science* **481** 777-784.

IOP Conf. Series: Materials Science and Engineering **744** (2020) 012008 doi:10.1088/1757-899X/744/1/012008

Modeling of atomic interactions in carbon nanotubes

M Chwał[1] and A Muc[1]

[1] Institute of Machine Design, Cracow University of Technology, Warszawska 24, 31-155 Cracow, Poland
E-mail: malgorzata.chwal@pk.edu.pl

Abstract: This work concerns atomic potentials applied in the description of carbon nanotubes mechanical behavior. The explicit relations of interatomic forces for different interatomic potentials have been derived. The fluctuations in the values of interatomic potential, interatomic force, elastic constant and Young's modulus have been put forward to elucidate the similarities and differences between potentials. The present analysis may be helpful in the equivalent continuum modelling of carbon nanotubes and nanocomposites.

1. Introduction

The specific properties of nanoscale carbon structures, mainly carbon nanotubes, draw the attention of scientific and commercial world because of wide range of potential applications [1]. In the description of carbon nanotubes properties various instruments are applied in both the experimental and theoretical field [2-5]. The creation of interatomic interaction models is the key to understand the phenomena at the atomic level. In quantum mechanics the electron cloud notation is applied (atoms and bonds fill the space through the electronic clouds) in which location and momentum of particle can be predicted only to the Heisenberg's constant – the uncertainty principle. In classical physics, the atomic orbitals definitions (the energetic levels) defined as the three-dimensional areas filled with electrons are applied. The interatomic bonds are being shown as lines and atoms as points to simplify the atomic structure models. The specific properties of carbon nanotubes are connected with the spatial arrangement of carbon atom bonds. Carbon has four valence electrons, which allows creating four bonds – two coming from *s* orbitals (spherical ones) and two coming from *p* orbitals (dumb-bells ones). Hybridization sp^3 creates a diamond structure. The graphite structure is created from partial hybridization sp^2. Hybridization sp^2 is known as the bond type σ placed angularly on plane (every 120°). The bonds type σ are strong covalent bonds that connect carbon atoms in plane. As a result, considering carbon nanotube as rolled-up graphite layers, a structure with high stiffness and strength is obtained. The remaining energetic level type *p* is perpendicular to the plane and creates bonds between neighboring graphite layers called bonds type π. π-bonds, also called van der Waals interactions, are much weaker than bonds type σ. The description of nanostructure's behavior starts from the consideration of interatomic interaction potential, i.e. special functions and physical constants based on overall rules of solid-state physics. Starting from the simplest description of interactions between two atoms, the interatomic potential may be really extended when very high accuracy in the atomic interactions is demanded. This paper is focused on the analysis of fluctuations in the values of interatomic potential, interatomic force, elastic constant and Young's modulus in the function of interatomic distance in carbon nanotubes. The value of elastic constant applied in the current studies is also compared with literature data.

2. Modeling of atomic interactions

The nature of interactions depends on complicated quantum effects that occur on the level of elementary parts of atom and control chemical properties such as valence and binding energy. The quantum effects also influence the interatomic bond topology, its formation and breakage. The potentials should explain the quantum effects even in the averaging sense to obtain reliable results in the numerical simulations basing on the solid-state physics. In general, an interatomic potential composes of the outside force field energy, potential energy between atom pairs, potential energy between three atoms and so on. Often to make the numerical simulations shorter, the atomic interactions between other than neighboring ones are omitted.

The description of the behavior of a nanostructure starts with the consideration of interatomic interactions potential $V(r_{ij})$ and physical constants. Based on the potential theory, an interaction force vector is described as follows:

$$\vec{F}(r_{ij}) = -\frac{\partial V}{\partial r_{ij}}\vec{n}_{ij} \tag{1}$$

where r_{ij} is the interatomic distance and n_{ij} is the unit vector between two atoms i and j. In the present work an equilibrium static states of the nanostructure is defined as the equilibrium state without any external forces and moments which can be described for example through the minimum of potential energy for equilibrium distance R_{ij} between two carbon atoms. The atomic force is expressed in the equivalent form [6]:

$$\vec{F}(r_{ij}) = k(r_{ij})*(R_{ij} - r_{ij})*\vec{n}_{ij} \tag{2}$$

where $k(r_{ij})$ is the elastic constant depending on an actual interatomic distance r_{ij}. The variations of the elastic constant $k(r_{ij})$ and Young's modulus $E(r_{ij})$ are computed as:

$$k(r_{ij}) = \frac{\partial^2 V(r_{ij})}{\partial r_{ij}^2}, \quad E(r_{ij}) = \frac{k(r_{ij})R_{ij}}{S_b}, \quad S_b = \frac{\pi d_b^2}{4} \tag{3}$$

where d_b and S_b is the diameter and the cross-section area of the atomic bond, respectively. To analyze the behavior of atomic bonds in a carbon nanotube, the relations (2) and (3) have been applied.

3. Interatomic potentials

The interatomic interactions are described as attractive or repulsive behavior. A state in which the attractive force is balanced with the repulsive force (the sum of the reaction forces are equal zero) is called the atomistic equilibrium state. In this state the reaction forces between atoms have the same values, but opposite directions and it results in the minimum of the potential energy. A decrease in the atomic distance causes an increase in the repulsive force whereas an increase in the atomic distance causes a decrease in the attractive force. In both situations changes in the potential energy are observed. The current analysis of carbon nanotube behavior involves Morse [7-8], Tersoff-Brenner [9-10] and Lennard-Jones [11-12] potentials.

Morse potential is written as:

$$V^M(r_{ij}) = D_e^M\left[e^{-2\beta^M(r_{ij}-R_{ij})} - 2e^{-\beta^M(r_{ij}-R_{ij})}\right], \qquad V^M(R_{ij}) = -D_e^M \tag{4}$$

where D_e^M is the minimum of potential and β^M is the constant correlated with energetic levels of atoms. Based on Eqs (1) and (4) the interatomic force for Morse potential is presented as:

$$F^M = 2\beta^M D_e^M\left[e^{-2\beta^M(r_{ij}-R_{ij})} - e^{-\beta^M(r_{ij}-R_{ij})}\right], \quad D_e^M = \frac{k_0}{2\beta^{M^2}} \tag{5}$$

For carbon atoms Tersoff-Brenner potential is as follow:

$$V^{TB}(r_{ij}) = V_R^{TB}(r_{ij}) - \overline{B}_{ij}V_A^{TB}(r_{ij}) \tag{6}$$

$$V_R^{TB}(r_{ij}) = f^{TB}(r_{ij})\frac{D_e^{TB}}{S^{TB}-1}e^{-\sqrt{2S^{TB}}\beta^{TB}(r_{ij}-R_{ij})}, \qquad V_A^{TB}(r_{ij}) = f^{TB}(r_{ij})\frac{D_e^{TB}S^{TB}}{S^{TB}-1}e^{-\sqrt{\frac{2}{S^{TB}}}\beta^{TB}(r_{ij}-R_{ij})}$$

where V_R^{TB} and V_A^{TB} describe the repulsive and the attractive parts, respectively, $f^{TB}(r_{ij})$ is the cut-off function, $\bar{B}_{ij}$ defines the "order" of atomic bonds, S^{TB} and β^{TB} are constants - see more details in [13]. The interatomic force in Tersoff-Brenner potential is written as:

$$F^{TB} = \frac{\beta^{TB} D_e^{TB}}{S^{TB}-1}\left[\sqrt{2S^{TB}}\, e^{-\sqrt{2S^{TB}}\beta^{TB}(r_{ij}-R_{ij})} - \bar{B}S^{TB}\sqrt{\frac{2}{S^{TB}}}\, e^{-\sqrt{\frac{2}{S^{TB}}}\beta^{TB}(r_{ij}-R_{ij})}\right] \tag{7}$$

In the current computations the following parameters for carbon have been used: R_{ij}=1.39 (Å), D_e^{TB}=9.613×10^{-19} (Nm), β^{TB}=2.1 (Å^{-1}), S^{TB}=1.22 according to paper by Brenner [10].

Lennard-Jones potential applied to describe van der Waals forces in carbon nanotubes is written as:

$$V^{LJ}(r_{ij}) = \frac{B^{LJ}}{r_{ij}^{12}} - \frac{A^{LJ}}{r_{ij}^{6}} \tag{8}$$

where A^{LJ} and B^{LJ} are positive constants. The left-hand side component of "6-12" potential describes the repulsive forces whereas the right-hand one the attractive forces. The attractive behavior is observed for long interatomic distances while for short, a strong repulsive core appears. Starting from Eq. (8) and based on Eq. (1) van der Waals force between atoms can be computed in the following form:

$$F^{LJ} = \frac{12B^{LJ}}{r_{ij}^{7}}\left(\frac{1}{r_{ij}^{6}} - \frac{1}{R_{ij}^{6}}\right), \quad A^{LJ} = \frac{2B^{LJ}}{R_{ij}^{6}}, \quad B^{LJ} = \frac{k_0}{72}R_{ij}^{14}, \quad D_e^{LJ} = \frac{k_0}{72}R_{ij}^{2} \tag{9}$$

The correlations between constants A^{LJ} and B^{LJ} were computed for equilibrium distance i.e. $r_{ij} = R_{ij}$, and $F(r_{ij}) = 0$.

4. Results and discussion

Based on the interatomic potentials and information about the analyzed structure the elastic constant can be computed. This information allows for evaluating Young's modulus values depending on the atom positions in the lattice. Considering three different interatomic potentials the carbon nanotube behavior was analyzed. In the equilibrium position, distance between carbon atoms R_{ij} = 1.42 (Å), atomic bond diameter d_w = 3.4 (Å), and bond stiffness E_0 = 1 (TPa) were adopted from [14]. The elastic constant in the equilibrium position was obtained from Eq. (3) and was k_0 = 639 (N/m), which is close to the elastic constant values reported by Odegard et al. [15] – see Table1. For Morse and Tersoff-Brenner potential the constants were applied according to [16] and [10] respectively. Applied values are presented in Table 2. In the analysis of Tersoff-Brenner potential it was assumed that $f^{TB}(r_{ij})$ =1 and $\bar{B}_{ij}$ =1 - see Eq. (6).

Table1. Comparison of elastic constant values.

Reference	Elastic constant k_0 (N/m)	Equilibrium distance R_{ij} (Å)
Belytschko et al. [16]	831	1.450
Brenner [10]	500	1.550
Odegard et al. [15]	652	1.421
Present value	639	1.420

IOP Conf. Series: Materials Science and Engineering **744** (2020) 012008 doi:10.1088/1757-899X/744/1/012008

Table 2. Values used in the computations of the interatomic potentials.

Lennard-Jones potential		Morse potential		Tersoff-Brenner potential	
D_e^{LJ}	1.789×10^{-19} (Nm)	D_e^{M}	4.637×10^{-19} (Nm)	D_e^{TB}	7.245×10^{-19} (Nm)
B^{LJ}	1203×10^{-140} (Nm^{13})	β^{M}	2.625 ($Å^{-1}$)	β^{TB}	2.1 ($Å^{-1}$)
A^{LJ}	293×10^{-80} (Nm^{7})			S^{TB}	1.22

In literature, different values of the elastic constant in the equilibrium position may be found because of the modification in the assumed initial interatomic distance R_{ij}. It is discussed in [14]. The elastic constant was applied to find the minimum value of potentials and also some constants for Lennard–Jones potential. In the present analysis it was assumed that only the interactions with the closest neighbors are taken into account. Using the elastic constant in the equilibrium position and the values from Table 2 the changes of the potential, the interatomic force, the elastic constant and Young's modulus were analyzed. The results are presented in Figs 1-4.

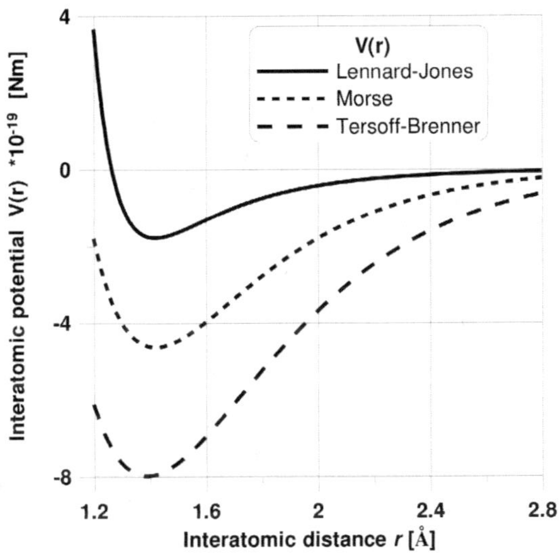

Figure 1. Interatomic potential V(r) vs interatomic distance.

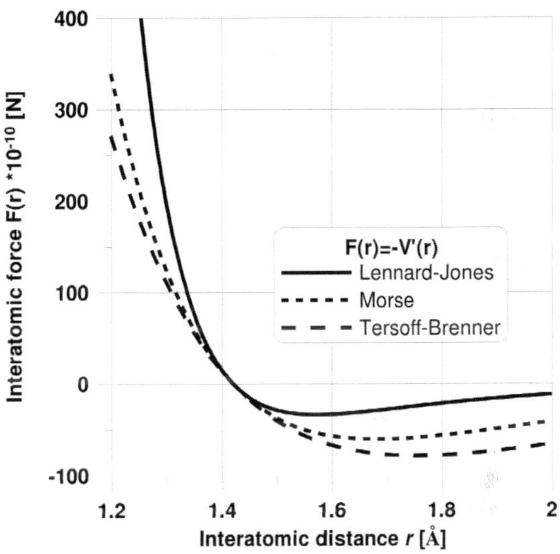

Figure 2. Interatomic force F(r) vs interatomic distance.

For the equilibrium position of atoms, the potential has the minimum value. The highest value is observed for Tersoff-Brenner potential. An increase in the interatomic distance causes an increase in potential values. The potential value is close to zero when the interatomic distance is a few times greater than the equilibrium position. A decrease in the atomic distance causes an increase in the repulsive force. In the equilibrium position of atoms the interatomic force is equal zero. In this position the transition between the repulsive and attractive forces is observed, and an increase in atomic distance causes an increase in the attractive force. This tendency is mainly visible for Lennard-Jones potential. The changes in the elastic constant and Young's modulus in the function of atomic distance are very similar. With an increase of the atomic distance a decrease of the elastic constant and Young's modulus can be seen. These two values remain in the range of positive value. It is visible that presented curves for different potential types have a very similar nature but different values.

IOP Conf. Series: Materials Science and Engineering **744** (2020) 012008 doi:10.1088/1757-899X/744/1/012008

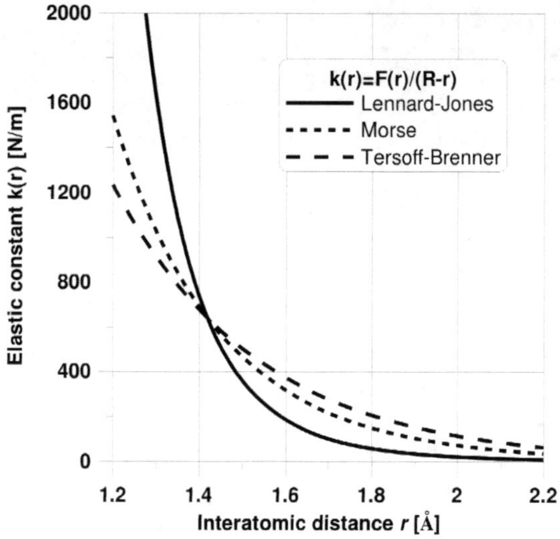

Figure 3. Elastic constant k(r) vs interatomic distance.

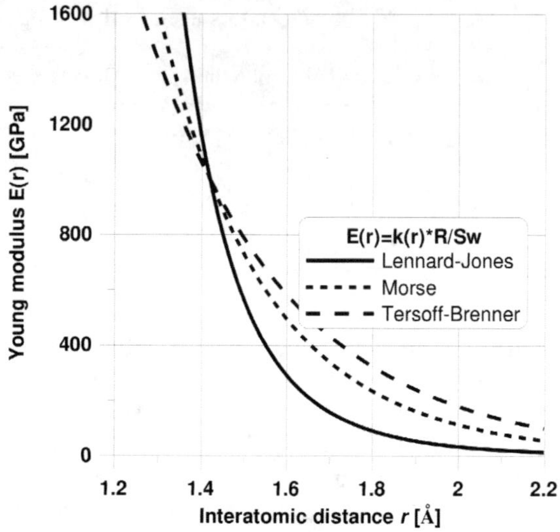

Figure 4. Young's modulus E(r) vs interatomic distance.

5. Conclusions

In the present work, various interatomic potentials used in the nanostructure characterization have been described. To analyze the behavior of carbon nanotubes Morse, Tersoff-Brenner and Lennard-Jones potentials have been chosen and the results for fluctuations of the interatomic potential, interatomic force, elastic constant, and Young's modulus in the function of interatomic distance have been presented in diagrams. The diagrams have revealed similarities in the shapes of distribution and differences in the values of the analyzed properties. The present analysis may be helpful in the equivalent continuum modeling of carbon nanotubes [17-18] and nanocomposites [19-20].

References

[1] Khan A, Jawaid M and Asiri AM 2019 *Nanocarbon and its Composites* Elsevier
[2] Attanasio A, Ceretti E, Fassi I and Pagano C 2017 *Adv Manufacturing Technol* **92** 1721
[3] Chwał M and Muc A 2016 *Compos B* **88** 295
[4] Tserpes KI and Papanikos P 2005 *Compos B* **36** 468
[5] Muc A and Jamróz M 2004 *Mech Compos Mater* **40** 101
[6] Chwał M 2011 *Procedia Eng* **10** 1579
[7] Morse PM and Stueckelberg ECG 1929 *Phys Rev* **33** 932
[8] Morse PM 1929 *Phys Rev* **34** 57
[9] Tersoff J 1986 *Phys Rev Lett* **56** 632
[10] Brenner DW 1990 *Phys Rev B* **42** 9458
[11] Jones JE 1924 *Proc Royal Society* **106** 441
[12] Jones JE 1924 *Proc Royal Society* **106** 463
[13] Chwał M 2017 *Acta Phys Polonica A* **131** 440
[14] Chwał M 2018 *J Mater Eng Perform* **27** 6087
[15] Odegard GM, Gates T, Nicholson LM and Wise KE 2002 *Compos Sci Technol* **62** 1869
[16] Belytschko T, Xiao SP, Schatz GC and Ruoff R 2002 *Phys Rev B* **65** 235430
[17] Muc A, Banaś A and Chwał M 2013 *Mechanics and Mechanical Eng* **17** 157
[18] Banaś A, Muc A and Chwał M 2013 *Key Engineering Mater* **542** 7
[19] Chwał M 2014 *Adv Mater Research* **849** 88
[20] Chwał M, Muc A 2019 *Mater* **12** 1474

IOP Conf. Series: Materials Science and Engineering **744** (2020) 012009 doi:10.1088/1757-899X/744/1/012009

Propelling torque of the self-adjusting vertical-axis rotor

M. Augustyn

Institute of Machine Design, Faculty of Mechanical Engineering, Cracow University of Technology, Kraków, Poland.
E-mail: marcin.augustyn@mech.pk.edu.pl

Abstract. The paper presents an innovative solution of the self-adjusting system of the wind turbine with horizontal axis, consisting of two-blade, carousel rotor with vertical axis and two-position self-adjusting mechanism. The procedure of computing power and propeller torque of rotor with self-adjusting system as a function of the angle of wind attack, the design and principles of operation of the adjusting turbine were presented. And also the proposal to use it as a self-adjusting system on the wind direction and at the same time as a protection of the main rotor HAWT (Horizontal Axis Wind Turbine) against excessive speed in strong winds. The proposed system can replace the most common solution like the directional rudder or the auxiliary wind motor, which generally requires the generator rotor brake.

1. Introduction

Analyzing the work of the self-adjusting system on wind direction of the wind turbine with vertical axis VAWT (Vertical Axis Wind Turbine) described in the patent of J. Ryś [1], it was noted that this system can also be used as a adjusting rotor in the popular today's horizontal axis wind turbine. Modification of this type of system by adding the turbine blades adjusting system, allows the free work of the turbine's main rotor, regardless of the changes of direction and speed of the wind stream. The goal of paper is the use of theoretical calculations and an analyze of the design and operation [2] for the proposed solution in order to eventually use of this turbine in exchange for a traditional adjusting system of turbine HAWT.

a) b)

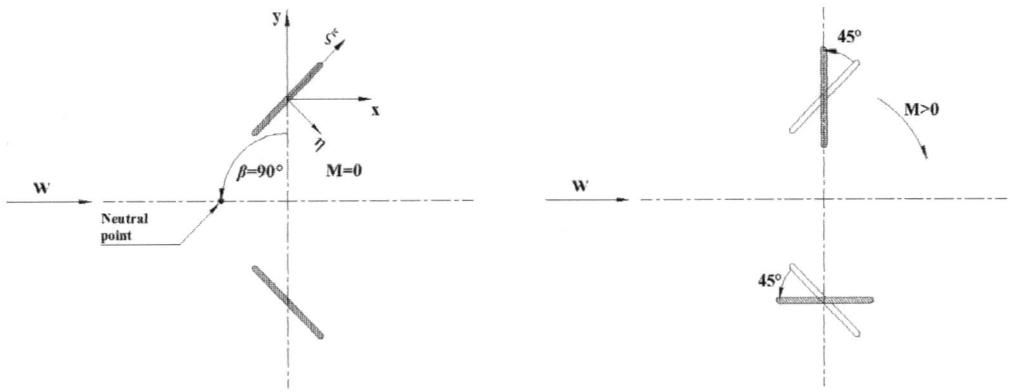

Figure 1. Neutral (a) and active (b) blades position of the self-adjusting turbine.

The theoretical calculations concerned definition and determine the value of the propeller torque M per 1m^2 of blade as a function of the angle of wind attack, the flat two-blades of rotor adjusting system of turbine, arranged relative to each other so that the axes of the cross-section blades are perpendicular to each other. It should be noted, that the blades are moving of planetary motion and the adjusting system changes the position of the blades in terms of yourself with the neutral position, which follow the wind direction (Figure 1a), onto the active position, (Figure 1b), which deflects the turbine when the main rotor speed exceed the limit value [3-10]. In this case the blades are changing position relative to each other via rotation around their own axis by an angle of 45°, and there are remaining in this position until fall of the speed of the main rotor onto the speed limit. Theoretical calculations for the active position are unnecessary.

Based on presented at the research of the work [11], the procedure of the propeller torque and the computing power and also the propeller torque of rotor with self-adjusting system as a function of the angle of wind attack were prepared. The resultant torque for the self-adjusting turbine:

$$M(\alpha_s) = \left(\left(M_{1\eta\xi}(\alpha_s) + M_{1\xi}(\alpha_s)\right) - M_{1\eta}(\alpha_s)\right) + \left(\left(M_{2\eta\xi}(\alpha_s) + M_{2\xi}(\alpha_s)\right) - M_{2\eta}(\alpha_s)\right), \quad (1)$$

is the sum of the propeller torque generated by the first and second rotor blade, where, the torques from the first blade in the moving coordinates ξ and η are described by equations:

$$M_{1\eta\xi}(\alpha_s) = 0{,}5\left(\frac{V_\eta(\alpha_s)}{W}\right)^2 \cdot k_{\eta\xi}, \quad (2)$$

$$M_{1\xi}(\alpha_s) = \left(\frac{V_\xi(\alpha_s)}{W}\right)^2 \cdot k_\xi \cdot \sin\left(\frac{\alpha(\alpha_s)}{2}\right) \cdot sign(V_\xi(\alpha_s)), \quad (3)$$

$$M_{1\eta}(\alpha_s) = s_{\eta\xi}\left(\frac{V_\eta(\alpha_s)}{W}\right)^2 \cdot k_\eta \cdot \cos\left(\frac{\alpha(\alpha_s)}{2}\right) \cdot sign\left(\cos\left(\frac{\alpha(\alpha_s)}{2}\right)\right), \quad (4)$$

and the torques generated from the second rotor blade:

$$M_{2\eta\xi}(\alpha_s) = 0{,}5\left(\frac{V_\eta(\alpha_s+180)}{W}\right)^2 \cdot k_{\eta\xi}, \quad (5)$$

$$M_{2\xi}(\alpha_s) = \left(\frac{V_\xi(\alpha_s+180)}{W}\right)^2 \cdot k_\xi \cdot \sin\left(\frac{\alpha(\alpha_s+180)}{2}\right) \cdot sign(V_\xi(\alpha_s+180)), \quad (6)$$

$$M_{2\eta}(\alpha_s) = s_{\eta\xi}\left(\frac{V_\eta(\alpha_s+180)}{W}\right)^2 \cdot k_\eta \cdot \cos\left(\frac{\alpha(\alpha_s+180)}{2}\right) \cdot sign\left(\cos\left(\frac{\alpha(\alpha_s+180)}{2}\right)\right). \quad (7)$$

The remaining values in the equations it the relative velocity components V_η and V_ξ [m/s] for the moving coordinates system associated with a turbine blade ξ and η, the wind speed W [m/s], the rotor speed V_o [m/s], the angle of the rotor rotation α_s and the angle of the wind attack β (wind direction). The coefficients associated with the components of the wind speed relative to the blade k_η, k_ξ, $k_{\eta\xi}$ and also the ratio of the frontal surface to the lateral surface of the rotor blade $s_{\eta\xi}$ (650x250x6 size of turbine's blade), as the average values, which have been set on the basis of experimental tests carried out in a wind tunnel the Wind Engineering Laboratory Cracow University of Science and Technology [12, 13] were taken: k_η=0,05; k_ξ=1; $k_{\eta\xi}$=0,05; $s_{\eta\xi}$=0,1.

The formula of the propelling torque at the end was described:

$$M = I_2 \cdot \rho \cdot R \cdot W^2, \quad (8)$$

where R is the radius of the turbine, ρ - density of the air [kg/m^3] and I_2 is the mean torque of the self-adjusting turbine with two blades [Nm]:

$$I_2 = \frac{\left(\int_0^{360} M(\alpha_s) d\alpha_s\right)}{360} \quad (9)$$

Power N [W] per one square meter of the self-adjusting turbine blade (ω – angular velocity of the rotor):

$$N = M \cdot \frac{V_o}{R} = M \cdot \omega . \quad (10)$$

In the calculations two cases were considered. The first for the speed rotor self-adjusting turbine equal to zero $V_o=0$ with including three different angles of wind attack β and the second for the speed of the turbine rotor which shall be different from zero $V_o=1$ and also the four different wind directions (the angle β). In both cases, the wind speed W=5.
The examples values of torque and power for the computing procedure below in tabular form were shown.

Table 1. The values of torque M and power N the self-adjusting turbine onto the direction wind, depending on the wind direction β for rotor turbine speed $V_o=0$ (ω=0) and $V_o=1$ (ω>0). The negative value of the power means braking power rotor.

β	0°	78,5°	90°	101,5°	180°
M [Nm] (V_o=0)	6,11155	1,218444	0	-1,218444	-6,11155
N [W] (V_o=0)	0	0	0	0	0
M [Nm] (V_o=1)	3,911392	-0,003785	-1,246515	-2,582774	-8,800632
N [W] (V_o=1)	13,037973	-0,012618	-4,155051	-8,609248	-29,335439

2. Interpolation of the propelling torque of the self-adjusting turbine

For as adopted form of the calculations, the propeller torque of the self-adjusting turbine, for the constant value of the radius of rotor equal R=0,3m as a function of the angle of wind attack β can be formulated as follows M(β):

$$M(\beta) = 0{,}2445 \cdot W^2 \cdot \left(1 - \frac{5}{4} \cdot \frac{V_o}{W}\right) \cdot \cos \cdot \left(\beta + arctg \frac{V_o}{W}\right), \quad (11)$$

The propeller torque of the self-adjusting turbine M(β) for any value of the rotor radius R at the function of the angle of the wind attack on the rotor β:

$$M(\beta) = 0{,}2445 \frac{R}{0{,}3} W^2 \cdot \left(1 - \frac{5}{4} \cdot \frac{V_o}{W}\right) \cdot \cos \cdot \left(\beta \pm arctg \frac{V_o}{W}\right), \quad (12)$$

The characters in the formula (10) are indicating on the rotor speed: (+) refers to ω>0, (–) refers to ω<0. In the case, when 90° > β > 0° the torque on the directional rotor is positive (right rotation of the rotor), when 180° > β > 90° is negative (left rotation of the rotor).
With this arrangement, any change of the wind direction causes tilt of the resultant force and the emergence of the torque on the axis of turbine. The torque of the self-adjusting gear, transferred to the bearing structure, gives the adjusting displacement in the direction of reducing the angle between the axis of the main rotor jib of the turbine nacelle, to which is attached the rotor of self-adjusting turbine, and the wind direction.
Adjust the blade position at an angle of 45° from the neutral position Figure 1a, an the active position Figure 1b, result in the self-adjusting system positive or negative value of the propeller torque, thanks to that the system can adjust the main rotor on new neutral position by changing the position of the main rotor.
The obtained results allow to illustrate the principle operation of the self-adjusting rotor, as were shown in Figure 2a and Figure 2b.
a) b)

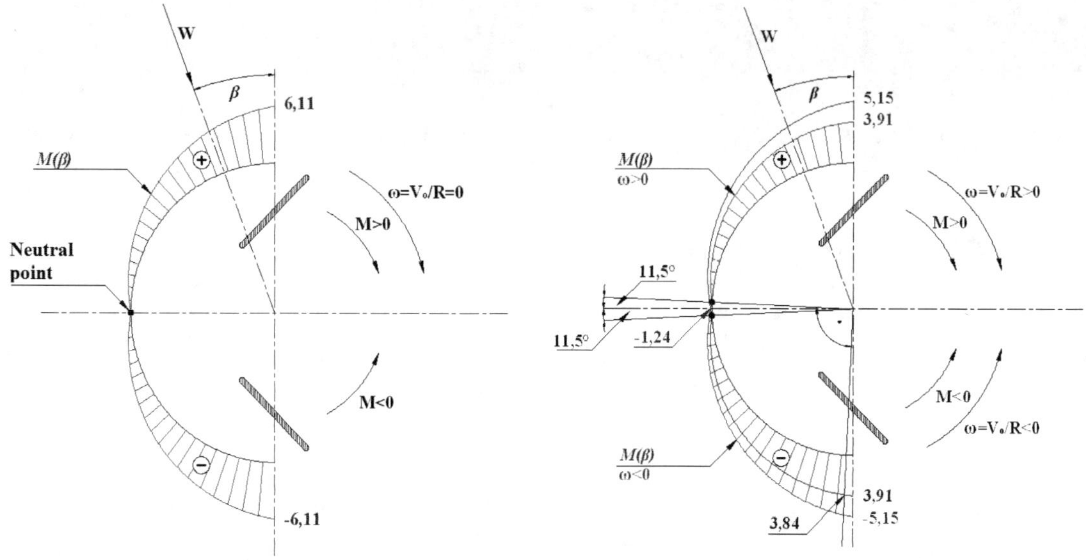

Figure 2. Graph of the propeller torque for two flat symmetrical blades of the self-adjusting system at a function of wind direction M(β): W=5 [m/s]; R=0,3 [m]; V_o=0 [m/s]; V_o/W=0 (a) and V_o=1 [m/s]; V_o/W=0,2 (b).

3. Example of use of the self-adjusting turbine

The properties described above allows to protect the main rotor HAWT against excessive speed in strong winds. The design solution of the self-adjusting system in detail in the patent application were discussed [2].

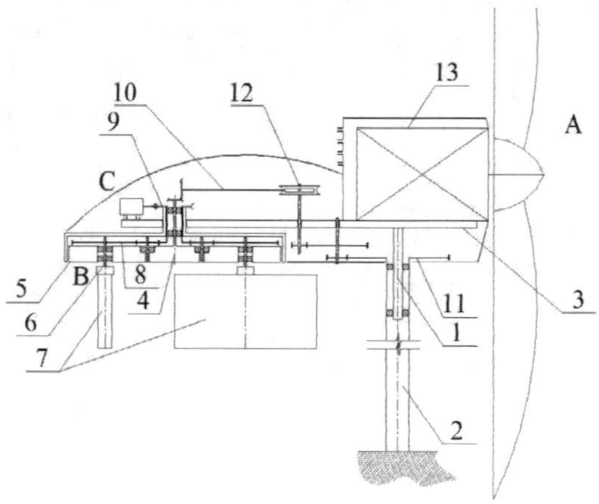

Figure 3. Kinematic scheme of the turbine with the self-adjusting system to protection the main rotor turbine against exceeding the speed limit, where:

A - main rotor of the horizontal axis wind turbine, *B* - rotor of self-adjusting turbine with the planetary motion of the two-blades to adjusting on wind direction the horizontal axis wind turbine, *C* – electromagnetic actuator or stepper motor of the self-adjusting system, which is forcing change the position of blades, *1* – shaft of the nacelle mounted with bearings on the supporting structure, *2* - supporting structure of the horizontal axis wind turbine (flagpole), *3* – jib of the self-adjusting system, *4* – axis of the adjusting rotor, *5* – supporting beam of the adjusting rotor, 6 – axis of the lobe of the adjusting rotor, *7* – lobe (blade) of the adjusting rotor, *8* – transmission gears with a ratio 1:2, *9* – self-locking worm gear, *10* – bevel gear, *11* – gear wheel mounted on the supporting structure, *12* – self-locking worm gear, *13* – generator of the main rotor.

In Figure 3 the scheme of the self-adjusting turbine on the wind direction were shown, equipped with an even number of blades with have a symmetrical cross-section, combined with the worm gear and the protection system of the main rotor HAWT against excessive speed in strong winds. These connections provide the perpendicular arrangement to each other the opposite aerofoils and change their position from neutral to active. In this case, the leaning gear is propels from the axis of the rotor, which on, with the end part is finds the bevel gear, which is connects to self-locking worm gear, and coupling here with the support structure wheel of the mast of the main rotor turbine, allows the main rotor axis deviation by a certain angle relative to the wind direction.

4. Conclusions

The obtained results allowed to analyze work of the rotor, used as the adjusting turbine of the main rotor of horizontal axis wind turbine. The self-adjusting vertical-axis turbine described in the paper, provides continuous setting of the main rotor turbine HAWT to the wind direction at the same time securing it against a strong wind, which could lead to damage the turbine. This solution is not required the brake of the generator rotor, as it is generally used in conventional constructions.

Changing the setting of the main rotor turbine to the wind direction is in the manner set without negative impact to the turbine, having a low gyroscopic torque. The self-adjusting system with the system of turbine protection against exceeding the speed limit does not require additional electric power from outside. At the time when the terminal voltage of the generator equal 230V, at the time when the terminal voltage of the generator equal 230V, the electromagnetic actuator started work via the contactor powered with phase of the generator.

Under conditions of strong wind (about 10m/s) self-adjusting system provides a practical constant voltage on the generator, in the case of an asynchronous generator - voltage 3x400V.

We anticipate build the prototype of the mobile turbine 3.5kW, based on prepared the documentation, for heating water in agricultural. The apply solution of the self-adjusting system according to the patent application [2] will significantly reduce the cost of the turbine due to the simplicity of design and ease of assembly and disassembly.

5. References

[1] Ryś J 1993 Arrangement for self-positioning a wing motor in wind's eye position. *Patent PAT.170515* [In Polish]
[2] Ryś J and Augustyn M 2015 Homing a wind turbine arrangement with a horizontal axis to the rotation wind direction. *Patent PAT.219834* [In Polish]
[3] HAU E 2013 Wind turbines: Fundamentals, Technologies, Application, Economics *Springer Verlag*
[4] Burton T, Jenkins N, Sharpe D and Bossanyi E 2001 Wind Energy Handbook. *John Wiley & Sons, Ltd.*
[5] Flaga A 2012 Wind turbines. *Published by Cracow University of Technology* [In Polish]
[6] Rossander M, Goude A and Eriksson S 2017 *MDPI Energies* 10(11), **1-21**
[7] Novak P, Ekelund T, Jovik I and Schmidtbauer B 1995 *IEEE Control Syst.*, 15, **28–38**
[8] Bossanyi EA 2000 *Wind Energy* 3, **149–163**
[9] Yang J, Song D, Dong M, Chen S, Zou L and Guerrero JM 2016 *Energy* 109, **294–309**
[10] Goude, A.; Bülow, F. Robust VAWT control system evaluation by coupled aerodynamic and electrical simulations. Renew. Energy 2013, 59, 193–201.
[11] Ryś J and Augustyn M 2012 *Technical journal MECHANIKA, Published by Cracow University of Technology,* 6-M/2012 Issue 11, **93-115** [In Polish]
[12] Augustyn M 2012 The study of aerodynamic forces on the turbine blades with use aerodynamic balance. *Published by Cracow University of Technology*, [In Polish].
[13] Flaga A 2012 Modeling of aerodynamical and mechanical characteristic of different type of wind rotors *Polish Associations for Wind Engineering* [In Polish].

IOP Conf. Series: Materials Science and Engineering **744** (2020) 012010 doi:10.1088/1757-899X/744/1/012010

Numerical Simulation of Hydraulic Fracturing Effect Based on Abaqus

Xinyong Li[1],Ximo Qu[2],Ting Sun[2],Jie Wang[2],Haoqing Fang[1],Zhiyuan Liu[1], Pandeng Luo[1],Bing Zhao[1] and Wenzhe Zhang[3]

1 Sinopec North-West,China
2 China University of Petroleum,Beijing,China
3 Petrochina Oil&Gas Pipeline Control Center,China
Email:779824706@qq.com

Abstract. This study simulates the hydraulic fracturing process based on ABAQUS, and obtains the results of hydraulic fracturing fracture propagation under the influence of multi-parameters.Our research shows that:The effect of reservoir modification increases with the increase of injection rate of fracturing fluid. Under the condition of $8m^3$/min or above, the fracture can communicate with the fault zone 100m away from the bottom of the well. Under the condition of 10cP, the effect of fracture propagation in formation is good. Under the condition of 50cP, the effect of fracture propagation in fracture zone is good. Under the condition of 200cP, the effect of crack propagation is poor. When the injection volume of fracturing fluid is between 0 and $1440m^3$, the effect of fracture propagation increases gradually. When the injection volume of fracturing fluid is larger than $1440m^3$, the contribution of continuous injection become weak. The maximum vertical crustal stress results in the fracture extending to the depth. The maximum horizontal crustal stress results in the horizontal propagation of fractured fractures.This study analyzed the influence of formation parameters and hydraulic fracturing parameters on fracture propagation. It also plays an active role in predicting the effect of hydraulic fracturing.

1. Introduction

This paper establishes a fracture type carbonate reservoir formation model, simulates the hydraulic fracturing process, and obtains the formation state after the hydraulic fracturing process based on Abaqus.

2. Background information

There are large reserves of carbonate rocks in a region of Northwest China. Experience shows that the effect of fracturing on stimulation is obvious. The following is a brief introduction of carbonate reservoir and hydraulic fracturing.

2.1. Carbonate Reservoir

Carbonate reservoir is one of many kinds of reservoirs, which is widely distributed in Northwest China. Carbonate reservoirs are often characterized by large reserves and high production. Its reservoir space can generally be divided into three categories (primary pore, fracture and karst cave), so carbonate reservoirs are often divided into dissolution type, pore type and fracture type. Compared with sandstone reservoirs, carbonate reservoirs have many types and regional differences[1].

2.2. Hydraulic fracturing
Hydraulic fracturing is a reservoir modification method that uses surface high-pressure pump to squeeze high viscosity fracturing fluid into reservoir through wellbore[2]. When the injection rate of fracturing fluid exceeds the absorptive capacity of the reservoir, a high pressure will be formed on the bottom of the well. When the pressure exceeds the fracture pressure of the reservoir rock near the bottom of the well, the reservoir will be fractured and fracture will occur. If the fracturing fluid is continuously squeezed into the reservoir, the fracture will continue to expand into the reservoir. In order to keep the fractures open, the sand carrying fluid with proppant (usually quartz sand) is then squeezed into the reservoir. After the sand carrying fluid enters the fracture, on the one hand, the fracture can continue to extend forward, on the other hand, the fractured fracture can be supported so as not to close. Then the replacement fluid is injected, and all the sand-carrying fluid in the wellbore is replaced into the fracture, and the fracture is supported by quartz sand. Finally, the injected high-viscosity fracturing fluid will degrade and drain out of the wellbore automatically, leaving one or more long, wide and high fractures in the reservoir, thus creating a new fluid passage between the reservoir and the wellbore.

3. Question
We know little about the effect of hydraulic fracturing on formation. In order to strengthen the prediction of hydraulic fracturing effect, we use numerical simulation method to solve this problem. We established a formation model considering fracture zone and natural fracture. By modifying various parameters, we obtained the influence of several parameters on hydraulic fracturing effect.

4. Simulation results
The effect of hydraulic fracturing is related to formation and hydraulic fracturing. We change the viscosity of fracturing fluid, the volume of fracturing fluid and the velocity of injecting fracturing fluid in hydraulic fracturing, and get the formation condition after hydraulic fracturing under these different conditions[3-5].

Table 1. Formatting sections, subsections and subsubsections.

Parameter	Parametric values
Viscosity of fracturing fluid	10cP,50Cp,100cP
Volume of fracturing fluid	$480m^3$,$960m^3$,$1440m^3$,$1920m^3$,$2400m^3$,$2880m^3$
Velocity of fracturing fluid injection	$6m^3$/min, $8m^3$/min, $12m^3$/min
Size of natural cracks	0m,10m,20m,50m,80m
Crustal stress	Vertical crustal stress is maximum Horizontal crustal stress is maximum

4.1. Viscosity of fracturing fluid

The pore pressure between well bottom and the fault zone under each viscosity are shown in the figure below.

IOP Conf. Series: Materials Science and Engineering **744** (2020) 012010 doi:10.1088/1757-899X/744/1/012010

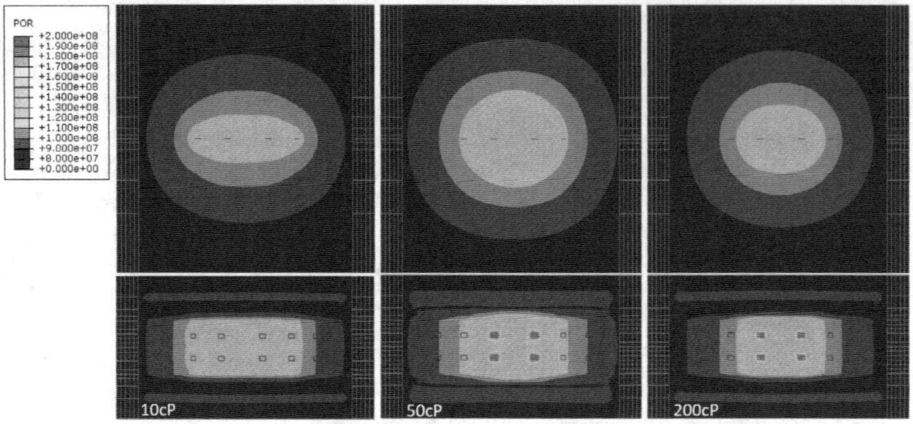

Figure 1. Pore pressure with viscosity of 10cP(Left); Pore pressure with viscosity of 50cP (Middle); Pore pressure with viscosity of 200cP (Right)

High pore pressure region of 50cP and 200cP is concentrated in the first fracture zone, and the high pore pressure zone is distributed in the first and second fracture zone at 10cP; When the viscosity of is reduced from 50cP to 10cP, the affected area of the formation increases, but the maximum pore pressure decreases.

4.2. Volume of fracturing fluid

The pore pressure between well bottom and the fault zone under each volume are shown in the figure below.

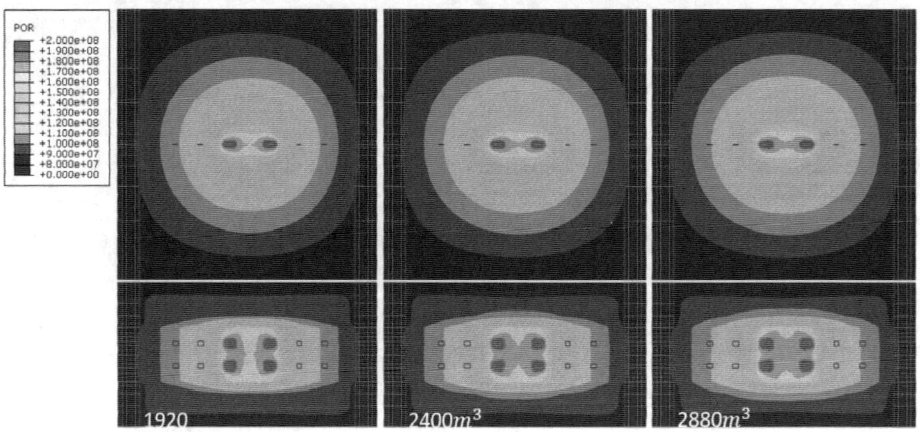

Figure 2. Pore pressure with volume of 480m^3(Left); Pore pressure with volume of 960m^3 (Middle); Pore pressure with volume of 1440m^3 (Right)

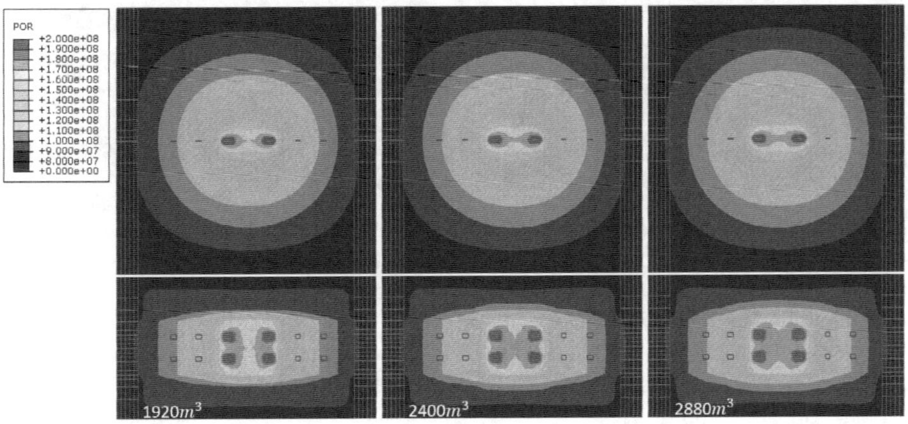

Figure 3. Pore pressure with volume of 480m^3(Left); Pore pressure with volume of 960m^3 (Middle); Pore pressure with volume of 1440m^3 (Right)

With the increase of injection volume, the affected area is gradually increased. When the injection volume reaches 1440m^3, the effect of pore pressure increase becomes worse. The injection volume is from 1440m^3 to 2880m^3, and the injection mainly strengthens the communication of the well bottom to the first crack zone。

4.3. Velocity of fracturing fluid injection

The pore pressure between well bottom and the fault zone under each velocity are shown in the figure below.

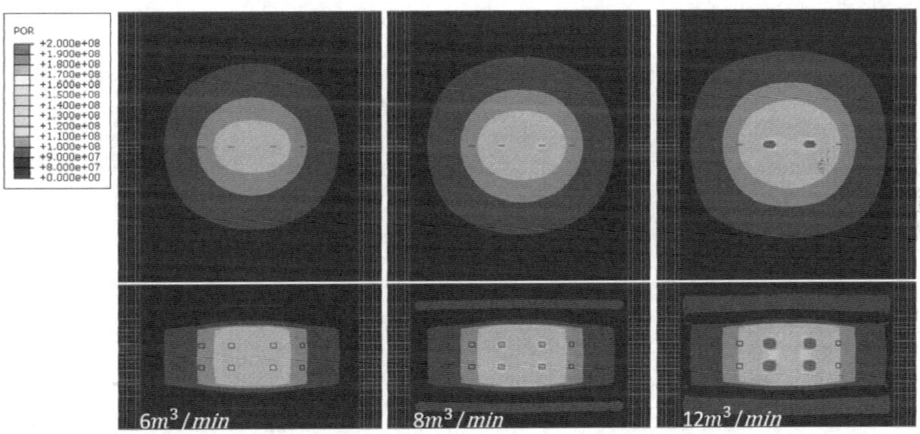

Figure 4. Pore pressure with velocity of 6m^3/min(Left); Pore pressure with velocity of 8m^3/min(Middle); Pore pressure with velocity of 12m^3/min(Right)

The higher velocity, the larger the volume of the pore pressure field.The high pore pressure regions are mainly concentrated in the first fracture zone, and as the velocity increases, the volume of high pore pressure velocity increases.

4.4. Crustal stress

The pore pressure, displacement, crack propagation between well bottom and the fault zone under each velocity are shown in the figure below.

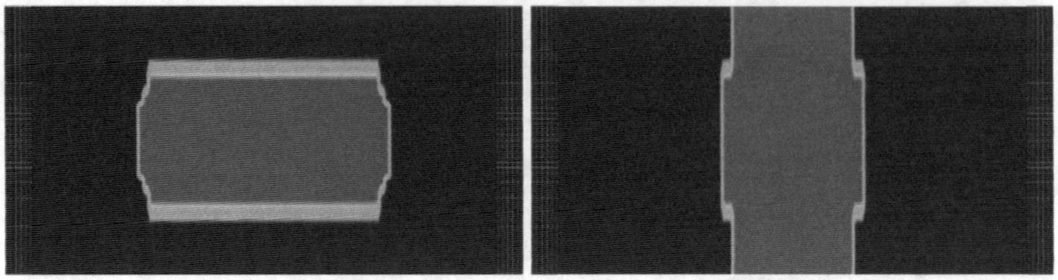

Figure 5. Crack when the horizontal crustal stress is maximum(Left);Crack when the vertical crustal stress is maximum(Right)

When the vertical crustal stress is maximum, the crack shape is higher and shorter in horizontal extension. When the horizontal crustal stress is maximum, the crack shape is lower and longer in horizontal extension.

5. Conclusion

Based on the results of numerical simulation, we conclude the following conclusions:

- The high pore pressure regions are mainly concentrated in the first fracture zone(100m from well bottom), and as the velocity increases, the volume of high pore pressure velocity increases.
- When the viscosity of is reduced from 50cP to 10cP, the affected area of the formation increases, but the maximum pore pressure decreases. 200cP has the worst influence on formation.
- When the injection volume of fracturing fluid is between 0 and 1440m^3, the effect of fracture propagation increases gradually. When the injection volume of fracturing fluid is larger than 1440m^3, the contribution of continuous injection become weak.
- The maximum vertical crustal stress results in the fracture extending to the depth. The maximum horizontal crustal stress results in the horizontal propagation of fractured fractures.

6. Reference

[1] Su Yuan 2018 J. Brief introduction of carbonate reservoir. *Petrochemical Technology*. **25** 193.

[2] Chengcheng Xiang, Wei Gao 2018 J. Analysis of research status and development trend of hydraulic fracturing technology . *Chemical Industry Management*.**02** 241.

[3] Chuanqiu Liu,Shifeng Xue 2016 J.Numerical simulation of crack propagation in hydraulic fracturing. *Chinese Journal of Computational Mechanics*. **33** 760-66.

[4] Diguang Gong,Zhanqing Qu,Jianxiong Li,Guanzheng Qu 2016 J.Finite element simulation of hydraulic fracture propagation based on ABAQUS platform.*Rock and Soil Mechanics*.**37** 1512-20.

[5] Mao Sheng,Gensheng Li 2014 J. Extended finite element numerical simulation method for hydraulic fracturing process. *Engineering Mechanics*. **31** 123-28.

IOP Conf. Series: Materials Science and Engineering **744** (2020) 012011 doi:10.1088/1757-899X/744/1/012011

Identification of impregnated activated carbon used as adsorbent in protective equipment for chemical warfare agent

K Won-in[1], J Chaiprapa[2], P Chirawatkul[2] and P Dararutana[3]

[1] Department of Earth Sciences, Faculty of Science, Kasetsart University, Bangkok, Thailand
[2] Synchrotron Light Research Institute, Nakhon Ratchasima, Thailand
[3] Royal Thai Army, Bangkok, Thailand
E-mail: kritwonin@gmail.com

Abstract. It is known that X-rays spectroscopy has been developed to apply to many advantages. In this work, Scanning electron microscope coupled with energy dispersive X-ray spectrometer (SEM-EDS) and synchrotron radiation facilities, such as micro-beam X-ray fluorescent spectroscopy (μ-XRF) and X-ray absorption spectroscopy (XAS) have been carried out to study on impregnated activated carbon samples that have been used as a chemical warfare agent adsorbent in military protective equipment. The elemental composition and distribution of sample surface have been analyzed. Their microstructures were highly porous. The results showed a detection of many kinds of metals, especially chromium (Cr), iron (Fe) and copper (Cu) with different from elemental distribution. It was also detected Cr(VI) which was a carcinogen in some samples. It is proposed that these methods can be used as fingerprint to identify various types of adsorbent.

1. Introduction

It is well known that military conflicts and terrorist incidences in counting from the present highlighted the importance of armed or emergency response operations for dealing with the military and terrorist related events, involving the CBRN (Chemical, Biological, Radiological and Nuclear) agents. It is defined that the chemical warfare is a type of warfare which is operated by using the chemical agents that are incapability, damaging, toxic and lethal components. They are cheap and toxic compounds which are easily made and stored. Meanwhile, the development of chemical industry around the world in nowadays has contributed to a higher risk of exposure to the TICs (Toxic Industrial Chemicals). They are some considered as the CWA (Chemical Warfare Agent) and can be also a consequence of a terrorist attack. The undertaking for the first response remains with individual nations to build on their resources to respond and mitigate the consequence of an emergency, especially to the human being and the environmental contamination. Protective equipment for the chemical warfare is designed to provide protection from serious injuries or illnesses affected from action with the CWA which effect on the respiratory system and parts of the body, such as eyes and skin, etc. The primary protection against the CWA is the respiratory protection by using a protective mask. While the body protection with a protective clothing is also significant. A properly fitted protective mask combined with a protective suit covering the body can offer an excellent protection.

It is also noted that activated carbon is the most common material used to remove toxic gases and vapors from breathable air. It is usually applied as an adsorbent for air filtration both in the military and civilian activities, because of its distinction adsorbing properties, availability durability, lifespan and low cost [1-10]. Activated carbon impregnated with metal salts or metals (such as Fe, Cr, Cu, Ag, and

IOP Conf. Series: Materials Science and Engineering **744** (2020) 012011 doi:10.1088/1757-899X/744/1/012011

Ni) is being used in the military filtration system. Cr which contained Cr(VI) which is specifically added to blood agent destruction is toxic as a carcinogen [11-12]. However, it is shown that the granular and powdered activated carbons are less expensive, while the fibrous and fabric forms are very highly expensive.

Figure 1 shows the military gas mask equipped with a canister which separates into internal and external canisters. The internal canister is a slightly curved triangular in shape which composed of light-weight filter material that fits into the cheek pouches of the mask, as shown in figure. 2a(3). While the external canister is normally a cylindrical in shape which constituted of a multilayer component of different filters: an aerosol filter which consists of metal grids and fold cellulose sheets or glass fibers and a layer of absorbent, as shown in figure 2b(3). The permeable protective suit as shown in figure 2 is composed of a layer of finely distributed activated carbon, either bound in polyurethane foam or particles, which bound between two layers of textile. The activated carbon consisting layer permits water vapour released from the body to pass through. While, the activated carbon adsorbs the CWA and thereby prevents them from passing through the skin.

(3) Filter canister component
(a) Internal filter canister (b) External filter canister

Figure 1. The military protective mask couple with internal canister (a) and external canister (b)

In this work, the impregnated activated carbon samples which were selected from different sources were aimed to characterize the elemental composition and distribution, structure and oxidation state of chromium by using SEM-EDS, μ-XRF and XAS.

(a) Types of suit

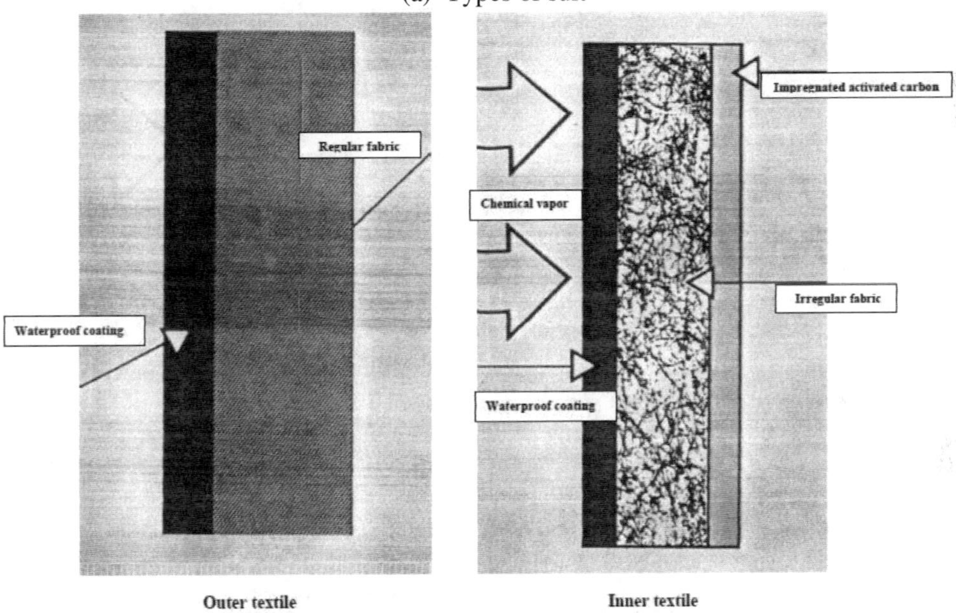

(b) Multilayer textile

Figure 2. The permeable protective suit

2. Experimental

2.1 Sample
Fourteen impregnated activated carbon samples were selected from different sources, as illustrated in Table 1.

2.2 Methods
2.2.1 SEM-EDS. The samples were gold-coated using SC7620 sputter coater (Quorum, UK). Morphologies and elemental compositions of the samples were analyzed using QUANTA 450 scanning electron microscope (SEM) (FEI Company, USA) coupled with X-Max^n 50 energy dispersive X-ray spectrometer (EDS) (Oxford Instruments, UK) at the Synchrotron Light Research Institute, Thailand

(SLRI). The SEM was operated at 20 kV accelerating voltages. Quantitative analysis of XRF spectra was performed using the AZtech program (Oxford Instruments, UK).

Table 1. The selected impregnated activated carbon samples

Sample	Manufactured			Form
	Year	Country	Used	
A	WW II era	Germany	External canister/CWA[a]	broken GAC[b]
C	2017	Thailand	Commercial/TICs[a]	broken GAC
D	2017	Thailand	Commercial/TICs	broken GAC
E	1990	USA	Internal canister/Training[a]	broken GAC
F	1990	USA	Internal canister/CWA	broken GAC
G	2007	USA	External canister/ Training	broken GAC
H	2007	USA	External canister/CWA	broken GAC
J	2004	USA	External canister/CWA	broken GAC
K	2004	USA	External canister/CWA	broken GAC
L	2008	UK	External canister/CWA	shaped GAC[b]
M	2012	UK	External canister/CWA	shaped GAC
P	2000	USA	Protective suit/CWA	broken GAC
Q	2000	USA	Protective suit/CWA	broken GAC
R	2018	Czech	External canister/CWA	broken GAC

[a] CWA, TICs and Training denoted to canister used for Chemical warfare agents, Toxic industrial chemicals and Training chemicals, respectively

[b] broken GAC and shaped GAC denoted to broken and shaped granular activated carbon, respectively

2.2.2 FESEM. The surface topography and porosity of samples were investigated using MIRA3 FE SEM (TESCAN, Czech Republic) at SLRI. FE SEM was operated with 20 kV accelerating voltages.
2.2.3 μ-XRF. SR μ- XRF measurement which was used to analyze elemental composition and distribution were performed at the BL6b of the SLRI, Thailand. Samples were wrapped with PP film, mounted on acrylic plate, and placed on the sample holder. The 2-10 keV white beam with a size of 5 x 2 mm (V x H) was magnetic focused by a polycapillary half lens in order to deliver X-ray beam with a size of 30 μm onto the sample. SR μ- XRF mapping was performed with 30 μm step size in x- and y directions with dwell time of 30 seconds per point. The area of 300 x 300 μm was selected for mapping using a camera with 10X macro lens prior to the SR-beam exposure. The Vortex Si (drift) detector (with resolution of 160 eV) was used for fluorescence yield detection. The X-ray fluorescence data were processed by PYMCA software for creation of elemental mapping in the samples.
2.2.4 XAS. For XAS measurements, the samples were sealed in polypropylene bags and measured as received. Due to the expected low concentration of Cr in the samples. All samples were measured in the fluorescence mode using the 19-element Ge detector available at the BL1.1W. Cr foil was used for energy calibration of the Cr K-edge. The XANES spectra of chemical standards Cr2O3, CrO2 and CrO3 were also measured along with the samples as reference for oxidation states 3+, 4+ and 6+. The spectra were processed and normalized using Athena.

3. Results and discussions

Table 2 showed the elemental composition of impregnated activated carbon samples investigated using SEM-EDS and SR μ-XRF. It was evident that the samples were mainly composed of C and O. Minor amounts of Cu and Cr were all present. Amount of C contented for approximately 51.0 and 90.9 wt%, while O was between 7.1 – 34.7 wt%.

The presence of Cl and P was a result of the chemical activation process that used $ZnCl_3$, and $ZnCl_3$ and H_3PO_4 as activating agents, respectively [2]. The contents of Al (0.1-1.1 wt%), Si (0.1-1.0 w%), Mg (up to 0.2 wt%), Ca (up to 0.5 wt%), Fe (up to 0.5 wt%), and Ti (up to 0.3 wt%) in the samples corresponded to their contents from the original raw materials which were naturally gas adsorbents for CO and NH_3 [3, 9]. Fe content can be originated from impregnation process. In previous research, activated carbon impregnated with iron salt was utilized as adsorbent for S (II) [3] and As (V)

IOP Conf. Series: Materials Science and Engineering **744** (2020) 012011 doi:10.1088/1757-899X/744/1/012011

compounds [13] which were the main compositions of the blister agents that classified as sulfur mustard; 1-chloro-2-[(chloroethyl)sulfany]ethane, and lewisite; 2-chloroethylarsonous dichloride, respectively [1-10]. The presence of K (up to 2.9 wt%) was caused by impregnation of K in activated carbon, which enhances the specific adsorption and catalyst properties [5].

Table 2. Elemental composition of impregnated activated carbon samples

Element wt%	Method	Sample A	C	D	E	F	G	H	J	K	L	M	P	Q	R
C	EDS	89.2	90.9	87.0	75.8	75.0	51.0	61.0	72.4	71.7	70.2	69.5	79.0	57.0	71.6
	μ-XRF	N/D	N/D	N/A	N/D	N/D	N/A	N/A	N/D	N/D	N/D	N/D	N/D	N/D	N/A
O	EDS	15.6	7.1	8.7	17.2	14.3	21.5	17.0	17.0	17.6	20.4	18.6	18.5	34.7	15.8
	μ-XRF	N/D	N/D	N/A	N/D	N/D	N/A	N/A	N/D	N/D	N/D	N/D	N/D	N/D	N/A
Ar	EDS	N/D	N/D	N/D	N/D	N/D	N/D	N/D	N/D	N/D	N/D	N/D	N/D	N/D	N/D
	μ-XRF	N/D	N/D	N/A	N/D	N/D	N/A	N/A	√	√	√	√	√	√	N/A
Al	EDS	0.3	0.1	0.1	0.8	0.8	0.6	0.5	0.6	0.6	0.4	0.4	0.2	1.1	0.6
	μ-XRF	√	√	N/A	N/D	N/D	N/A	N/A	√	√	√	√	√	N/D	N/A
Si	EDS	0.2	0.1	0.2	0.6	0.7	0.5	0.5	0.4	0.5	0.2	0.4	0.3	1.0	0.1
	μ-XRF	√	√	N/A	√	√	N/A	N/A	√	√	√	√	N/D	N/D	N/A
P	EDS	N/D	N/D	0.1	0.1	N/D	N/D	N/D	N/D	N/D	0.2	0.2	1.1	N/D	0.1
	μ-XRF	N/D	√	N/A	√	√	N/A	N/A	√	√	√	√	√	√	N/A
S	EDS	0.1	1.1	0.5	0.5	0.5	0.3	0.3	0.3	0.4	0.3	0.5	0.1	0.2	N/D
	μ-XRF	√	√	N/A	√	√	N/A	N/A	√	√	√	√	N/D	N/D	N/A
Cl	EDS	N/D	N/D	0.2	0.3	0.5	N/D	N/D	N/D	N/D	N/D	N/D	N/D	2.4	0.3
	μ-XRF	√	√	N/A	√	√	N/A	N/A	√	√	√	√	√	√	N/A
K	EDS	2.2	0.7	2.9	N/D	N/D	N/D	N/D	0.1	0.1	0.2	0.2	0.1	N/D	1.2
	μ-XRF	√	√	N/A	√	√	N/A	N/A	√	√	√	√	√	√	N/A
Ca	EDS	0.3	0.1	0.3	0.1	0.1	0.1	0.1	0.1	0.1	0.4	0.5	N/D	N/D	N/D
	μ-XRF	√	√	N/A	√	√	N/A	N/A	√	√	√	√	√	N/D	N/A
Ti	EDS	N/D	N/D	N/D	N/D	N/D	N/D	N/D	N/D	N/D	N/D	N/D	N/D	0.3	N/D
	μ-XRF	N/D	N/D	N/A	√	√	N/A	N/A	√	√	√	√	√	√	N/A
Cr	EDS	N/D	N/D	0.3	3.5	6.0	N/D	4.9	2.4	2.5	4.5	3.7	N/D	N/D	1.6
	μ-XRF	√	√	N/A	√	√	N/A	N/A	√	√	√	√	√	N/D	N/A
Fe	EDS	0.1	0.5	N/D	0.1	0.2	0.2	0.2	0.2	0.2	0.2	0.2	N/D	0.2	N/D
	μ-XRF	√	√	N/A	√	√	N/A	N/A	√	√	√	√	√	√	N/A
Mg	EDS	0.1	0.1	0.1	N/D	N/D	N/D	N/D	0.1	0.1	0.1	0.2	N/D	0.1	N/D
	μ-XRF	N/D	N/D	N/A	N/D	N/D	N/A	N/A	N/D	N/D	N/D	N/D	N/D	N/D	N/A
Cu	EDS	0.1	N/D	0.1	2.8	1.8	8.9	15.2	6.7	6.3	5.0	5.9	0.1	N/D	9.1
	μ-XRF	N/D	√	N/A	√	√	N/A	N/A	√	√	√	√	N/D	√	N/A

N/D denotes to no element detected, √ denotes to element detected, and N/A denotes to no analysis
By using EDS: Sample D, Na 0.1; Sample G, Zn 12.8, Mo 2.2, Na 1.4; Sample P, Na 0.3; Sample Q, Na 0.6, F 1.6, Sb 0.5 wt%
By using μ-XRF: Sample E, Zn; Sample F, Zn; Sample G, Zn; Sample Q, Sb, Sc, Mn, Zn

Cu and Cr were effective adsorbents of blood agents such as cyanogen chloride, hydrogen chloride, and cyanogen [11, 14-15]. In military application, it was proposed that large amount of Cr (VI) can promote cyanogen chloride destruction [11, 16]. However, the modern filter canisters were stopped in production and replaced with other types of impregnated activated carbon [11-12, 17] due to Cr^{6+} compounds were toxic and carcinogenic. Ar was detected in samples J, K, L, M, P and Q by using SR μ-XRF technique. The detectable of Zn (in sample E, F, G and Q), Mn (in sample Q), Sb (in sample Q) and Sc (in sample Q) were also found. While, some elements, such as Na (in sample D, G, P, and Q, up to 1.4 wt%), Zn (in sample G, up to 12.8 wt%), Mo (in sample G, up to 2.2 wt%), F (in sample Q, up to 1.6 wt%) and Sb (in sample Q, up to 2.5 wt%) were also analyzed by using the EDS.

It is proposed that the impregnated element used as adsorbents in sample A which used in the external canister during the World War II was mainly Fe. For other ones (which were newer type canisters), not only Fe but also Cr and Cu were impregnated with activated carbon in order to neutralize the effect of CWAs. Sample C and D which produced from the manufacturing factory in Thailand during 2017 were appropriated for environmental and industrial applications. Sample E and F were used in the internal canister between 1960 -1990. Sample G, H, J, K, L and M were the newer ones made between 2004-2012, which used in the external canister. Sample P and Q were coated on the fabric of textile used in the protective suit that produced in 2000. Sample R was contained with the external canister made in 2018 which was the newest and most safety one.

Ag was not found in both EDS and SR μ-XRF results of any sample. It was suggested that it was due to the accuracy limitation of these techniques for analysis of the investigated samples with the elemental composition below 0.1 wt%. The advanced works by using a transmission electron microscope (TEM), it was shown that Ag was presented in the form of the nanoparticle in some samples. In general, the activated carbon which impregnated with 0.05 wt% Ag was classified as ASC-Whetlerite [17-19]. It revealed inhibitory effects against some bacteria, virus, algae and fungi which were classified as biological warfare agents (BWAs). Heavy metals such as lead or mercury also exhibited these effects, but they originated high toxicity on human [18].

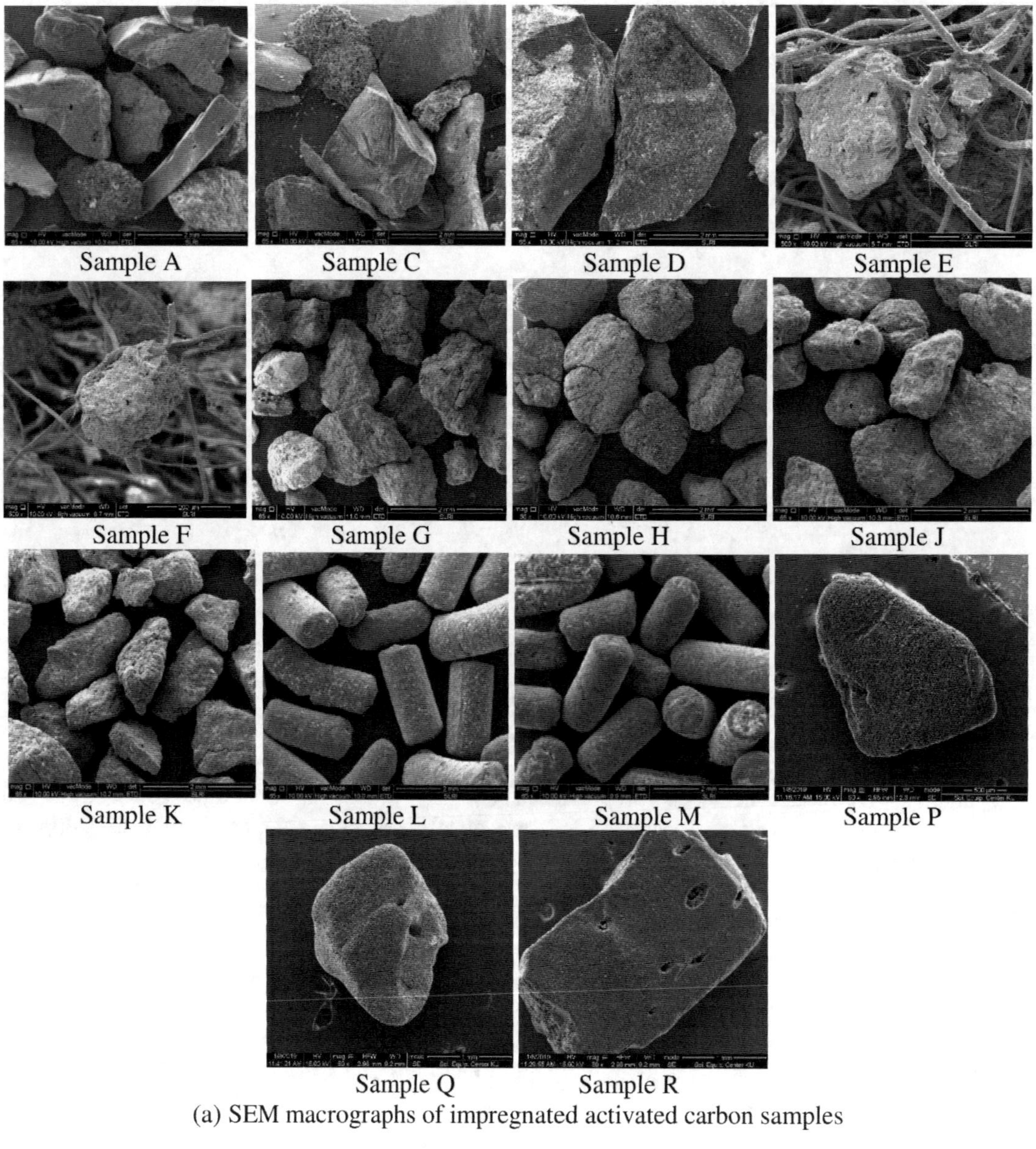

(a) SEM macrographs of impregnated activated carbon samples

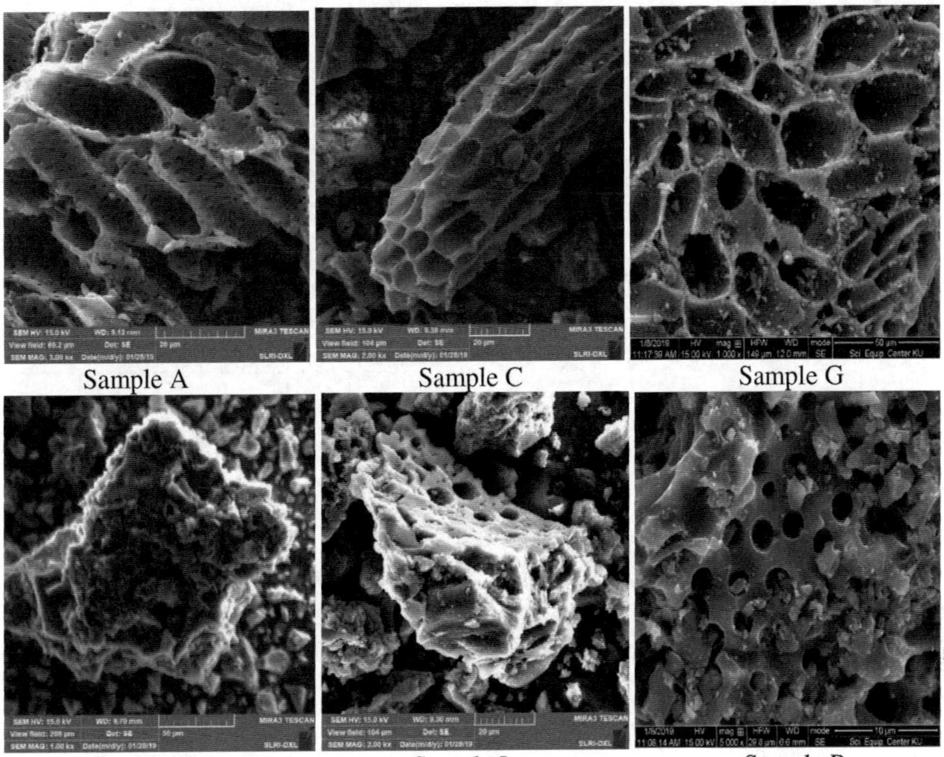

Sample A Sample C Sample G

Sample K Sample L Sample P

(b) FESEM micrographs of impregnated activated carbon samples

Figure 3. SEM macro- and micrographs of impregnated activated carbon samples

By using the SEM (all samples) and FESEM (some samples), the samples possess the highly porous structure of cavities, cracks, and attached fine particles on their surface, as shown in figure 3(b). They showed the macropores, mesopores and micropores with the pore radius (r) of r>50 nm, 50>r>2 nm, and r<2 nm, respectively. It is revealed that these porous structures are used to remove the gases and vaporous components. The macropore and mesopore enable contaminant transportation to the micropore, while the micropore which result in the increasing of the surface energy is able to adsorb a variety of the chemicals [7-9, 20].

SR µ-XRF results provided qualitative information about the elemental compositions of the samples, which were in agreement with the EDS results. However, some elements with low atomic number and/or low concentrations were not detected due to limit of detection of this technique.

Results using the SR XAS which the spectra were processed using Athena [21] and the normalized spectra were shown in figure 4. The standard spectra in figure 4(a) and (b), showed that at the post edge energy of various compounds were different. Although the Cr^{6+} which was non-oxide compounds, it presented the characteristic pre-peak at the 5993 eV [22]. They were classified into 6 groups. It was analysed that Cr^{6+} with the pre-peak at 5993 eV was found in Group 1, 2 and 3, as shown in figure 4(c), (d) and (e), respectively. While the pre-peak for the Cr^{3+} was not appearing at this energy, as shown in figure 4(f), (g) and (h). Furthermore, it was also found that the concentration of the Cr^{6+} contained in Group 1 > Group 2 > Group 3, as shown in figure 4(i).

IOP Conf. Series: Materials Science and Engineering **744** (2020) 012011 doi:10.1088/1757-899X/744/1/012011

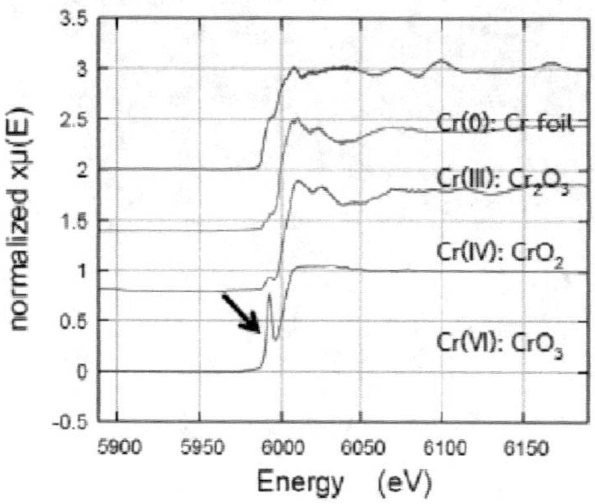

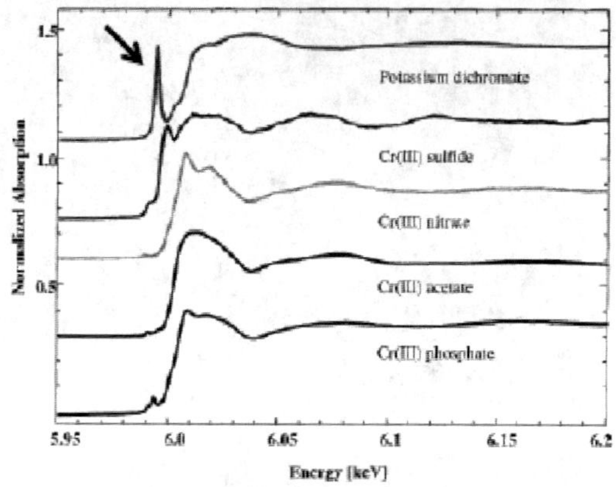

(a) Chemical standards with various oxidation states (b) Chemical standards with various compounds

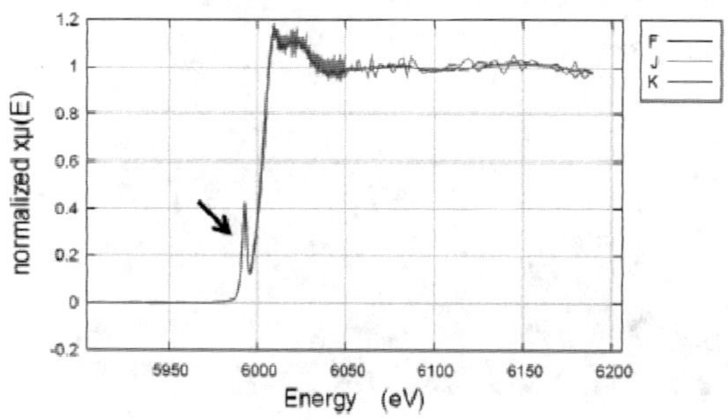

(c) Group 1

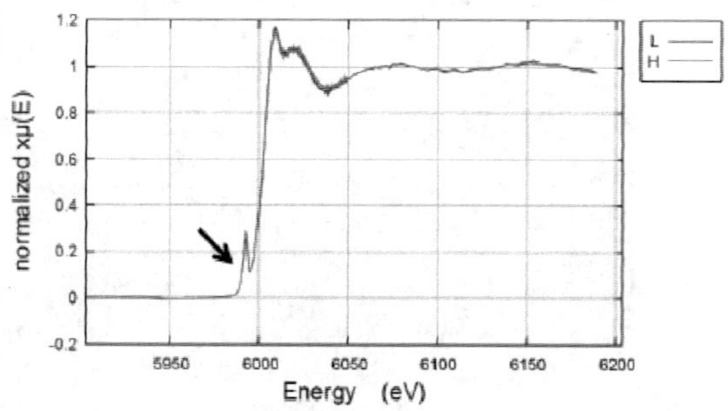

(d) Group 2

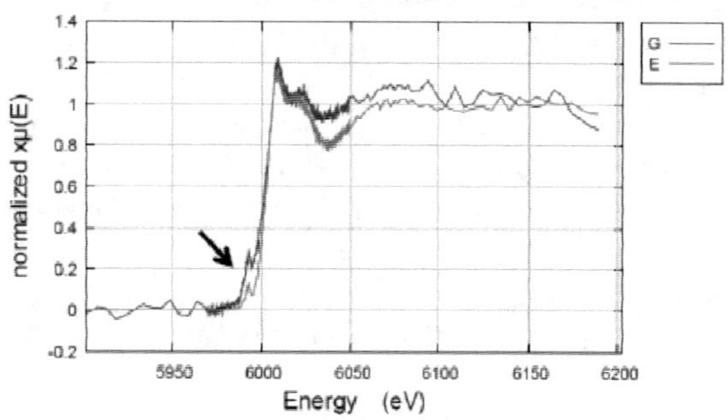

(e) Group 3

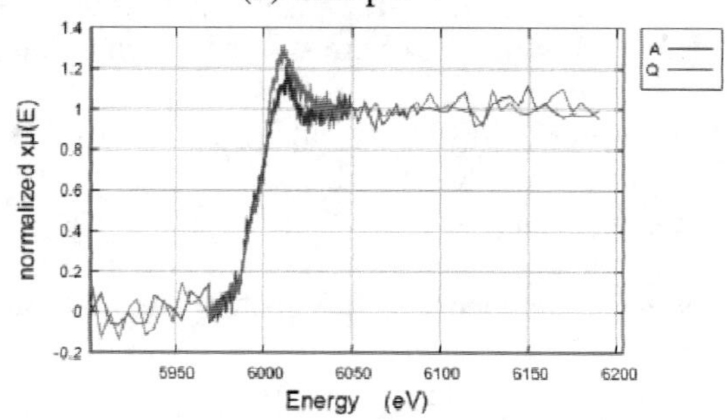

(f) Group 4

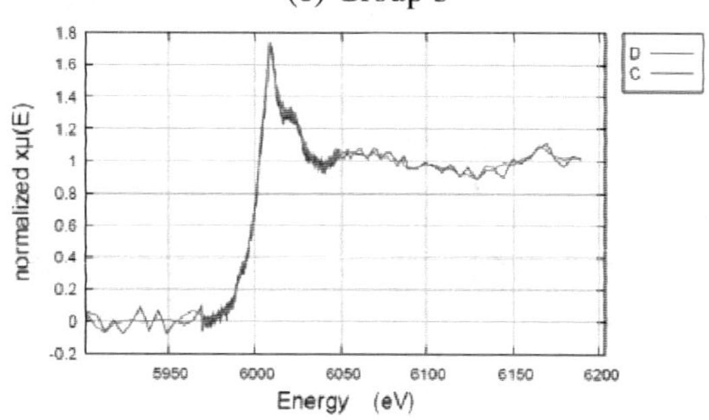

(g) Group 5

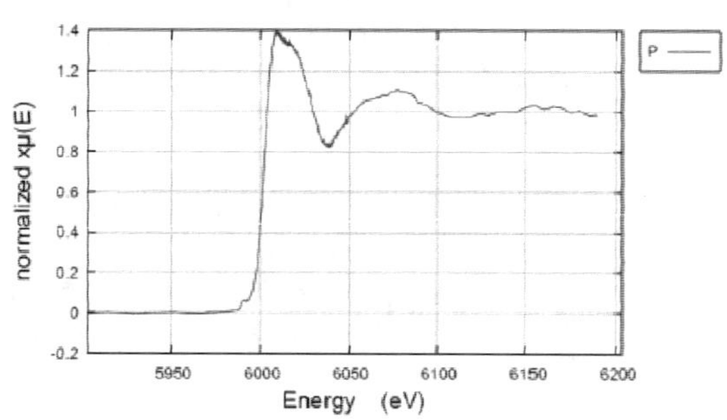

(h) Group 6

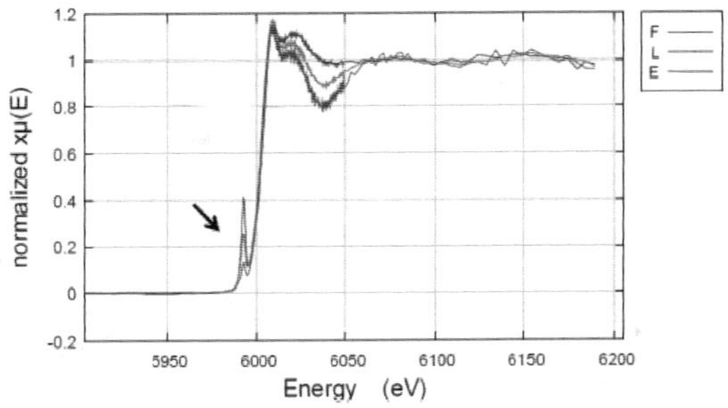

(i) Comparison between Group 1, 2 and 3

Figure 4. The XAS spectra of the selected impregnated activated carbon samples

4. Conclusion

The SEM-EDS results provided the information on the elemental composition of impregnated metals in the carbon-based structure. FE SEM results revealed the pore size and morphology. SR μ-XRF results of elemental composition and distribution were in agreement with the EDS results. The use of adsorbents for impregnated activated carbon in Thailand has been developed from solely Fe (in the World War II era) to Cr, Cu, and Fe (at the present time). It was also found that the Cr detected in some samples was presented as Cr^{6+}.

Acknowledgments

The Royal Thai Army Chemical Department thanks for the samples supporting. This work was partially supported by the Kasetsart University (Department of Earth Science at the Faculty of Science). It has been also facilitated within the activities of the Synchrotron Light Research Institute (BL 1.1W and BL 6b).

References

[1] Karakas SY, Aygun A, Gunes M and Tahtasakal E 2004 *Carbon* **42** 477-484.
[2] Srivastava AK, Saxena A, Singh B and Srivas SK 2007 *Carbon Letters* **8**(4) 274-279.
[3] Prasad GK, Singh B and Vijayaraghavan R 2008 *Defence Science Journal* **58**(5) 686.
[4] Yacob AR, Majid ZA, Dasril SDR and Indaran V 2008 *The Malaysian Journal of Analytical Sciences* **12**(1) 264.
[5] Khah A M and Ansari R 2009 *ChemTech* **1**(4) 859.
[6] Gua Z et al 2014 *Journal of the Taiwan Institute of Chemical Engineers* **45** 2801.
[7] Lee T, Ooi C-H, Othman R and Yeoh F-Y 2014 *Rev.Adv.Mater.Sci.* **36** 118-136.
[8] Coste JBD and Peterson GW 2014 *Chemical Reviews* **114** 5695-5727.
[9] Gulsec II 2016 *Brazilian Journal of Chemical Engineering* **33**(4) 1021.
[10] Tripathi NK et al 2018 *Defence Science Journal* **58**(1) 83.
[11] Rossin JA 1989 ed., Maryland: Chemical Research Development Engineering Center 9.
[12] Mahle JJ et al 2010 *J.Phys.Chem.C* **114**(47) 20083-20090.
[13] Ghanizadeh Gh, Ehrampoush MH and Ghaneian MT 2010 *IranJ.Environ.Health.Sci.Eng.* **7**(2) 145.
[14] Lodewyckx P 2006 in Interface Science and Technology. Activated Carbon Surfaces in Environmental Remediation, vol. **7**, T. J. Bendosz, Ed., Amsterdam: Elsevier 475.
[15] Smith JWH et al 2010 *Journal of Colloid and Interface Science* **34**1 162.
[16] Rossin JA et al 1991 *Carbon* **29**(2) 197.
[17] Morrison RW 2011 ed. Maryland: Guild Associates Inc. 1.
[18] Morrison RW 2011 ed. Maryland: Chemical Research Development Engineering Center 7.
[19] Srivastava AK et al 2012 *Journal of Scientific & Industrial Research* **71** 748.
[20] Kuo X-M et al 2013 *Journal of Aerosol Science* **55** 57-65.
[21] Ravel B and Newville M 2005 Journal of Synchrotron Radiation 12 537.
[22] Aldrich MV et al 2003 *Environmental Science and Technology*, **31**(5) 1573.

IOP Conf. Series: Materials Science and Engineering **744** (2020) 012012 doi:10.1088/1757-899X/744/1/012012

Study of the addition of the hybrid sisal fibers on the dynamic mechanical property of SFs reinforced polylactic acid (PLA) composites

Z H Zhu [1], N Zhang[1], Y Zhang [1], M Y Hao [2] and H W Wu [2]

1 School of advanced manufacturing technology, Guangdong Mechanical & Electrical Polytechnic, Guangzhou, China;
2 The Key Laboratory of Polymer Processing Engineering, Ministry of Education, South China University of Technology, Guangzhou, China
Email: 184823174@qq.com

Abstract. When sisal fiber (SF) is added into polylactide (PLA) matrix, the storage modulus of PLA/SF composites increases greatly which indicate the rigidity of the composite material increases obviously. Comparing PLA composites with pure PLA, it can be found that the glass transition temperature (Tg) of PLA composites is significantly higher than that of PLA which showed the heat resistance of the PLA composites can be improved obviously because of the addition of SFs in PLA matrix. The energy storage modulus of PLA/HSFs is the highest of all PLA composites. HSFs could improve the dynamic mechanical properties of PLA composites more significantly than ASFs and USFs.

1. Introduction

The research of natural fiber reinforced PLA composites has been a hot spot in the field of polymer composites in recent decades. Compared with traditional composites, natural fiber reinforced PLA composites have advantages of environmental friendliness and reproducibility. [1-2] The greatest challenge in preparing natural fibers reinforced PLA composites is their large variation in properties and characteristics. [3] Various kinds of methods for pretreating of fibers had been used to improve the interface bonding of matrix and fibers in the previous research. [4-7] Though these methods improved the interfacial adhesion of fibers and matrix, they eliminated some components and destroyed the construction of fibers, which declined the fiber strength and the mechanical properties of composites couldn't be enhanced as expected. In our previous work, we compounded alkali-treated SFs (ASFs) with untreated SF (USFs) to prepare hybrid SFs (HSFs) which were then filled into PLA matrix to prepare PLA/HSF composites. It was found that HSFs could improve mechanical properties of PLA composites more significantly than ASFs and USFs. In this work, we give a further study of the dynamic mechanical properties analysis (DMA) of the PLA, PLA/ASF and PLA/HSF composite.

2. Experiment

2.1. Materials

PLA 3051D, Nature Works ™, was purchased by Shenzhen Bright China Industrial Company (China). SF bundles (GB/T 15031-94) were provided by Guangxi Sisal Company of Guangxi Province, China. Other chemical reagents were commercial products.

2.2. Samples Preparation

SFs were submerged into 10% NaOH solution (mass concentration) for 3 h. Then, ASFs were washed with distilled water and dried in the open air for 48 h. USFs were mixed with ASFs to fabricate HSFs in a mass ratio of 1:1 and then PLA, ASFs, USFs and HSFs were dried at 100 degrees for 4 h. Then PLA was compounded with SFs with a mass ratio 4:1 using a two-roll plastic mill (SK-160B, Shanghai Rubber Machinery Factory, china) at 180 degrees. Four kinds of samples were made: pure PLA, PLA/USFs, PLA/ASFs, and PLA/HSFs. Then, they were compressed into sheets by a hot press (QBL-350, Wuxi No.1 Rubber Plastics Mechanical Co. Ltd, China). The compression molding temperature and pressure were 190 degrees and 10 MPa, respectively.

2.3. DMA test

Perkin-Elmer DMA-7 was used to test the DMA of PLA and PLA composites. The size of the sample is 60×10×4mm, three-point bending mode is adopted, the vibration frequency is set as 1 HZ, and the heating rate rises from 50 to 250 degrees.

2.4. Polarizing Microscope

Polarizing microscope (POM) investigation was performed with a POM (Axioskop 40 POL, ZEISS, Germany).

3. Results and discussion

3.1. DMA analysis

As one of the test methods to characterize the mechanical properties of materials, dynamic mechanical properties analysis (DMA) is used to measure the strain response of materials under alternating stress. The dynamic energy storage modulus (E) obtained from DMA reflects the rigidity of the material, and the loss factor (tan δ) reflects the micro motion and phase transformation of the polymer chain in the material

Figure 1 shows the relationship between temperature and storage modulus of PLA and PLA composite. It can be observed that the storage modulus of both pure PLA and composite materials decreases slowly, then decreases sharply in the range of glass transition temperature (Tg), and finally tends to be stable as the temperature goes up.

Comparing pure PLA with PLA composite, it can be found that after sisal fiber is added into PLA matrix, the storage modulus of composite materials increases significantly, that is, the rigidity of composite materials is significantly improved For example, the storage modulus of PLA is 3300 MPa, while the storage modulus of PLA composites is above 4500 MPa at 50 degrees. It is mainly because that the motion of molecular chains of PLA matrix is restricted due to the addition of sisal fibers with high specific area. Moreover, the flexural modulus of sisal fiber is about 12GPa which is much higher than 3600MPa of PLA. is blocked because of the interaction force between fillers and matrix.

Comparing PLA/ASFs with PLA/USFs, it can be found that the storage modulus of PLA/ASFs is higher than that of PLA/USFs. It is partly due to the better compatibility of ASFs and PLA matrix. In addition, alkali treatment convert some flexible bonds (cellulose-hemicellulose) of SFs into rigid bonds (cellulose-cellulose) with higher rigidity and thus enhanced the crystallinity of cellulose and ASFs, which can also improve the rigidity of PLA composite.

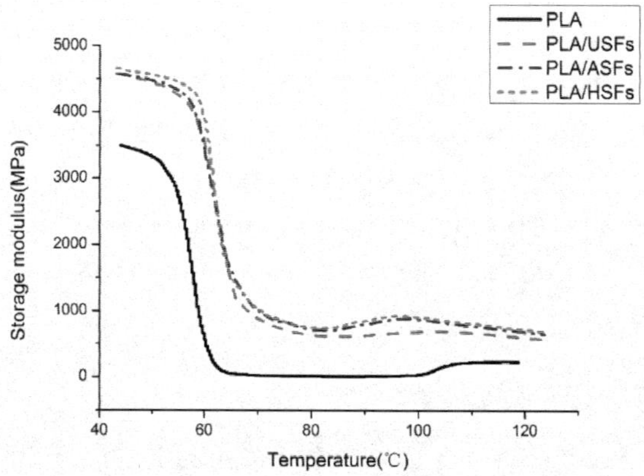

Figure. 1 Curves between storage modulus and temperature of PLA and PLA composites

It can also be observed that the storage modulus of PLA/HSFs is the highest of three composites. It is mainly because that USFs in HSFs have better tensile property and ASFs in HSFs have better interface bonding with PLA matrix. Therefore, they can play a synergistic role and improve the rigidity of PLA/HSFs.

d is the mechanical loss angle, and its tangent is known as the mechanical loss factor, which is commonly used to characterize the magnitude of mechanical loss or internal friction. Figure 2 shows the relationship between temperature and the tand of PLA and PLA composite. The peaks in figure 2 correspond to the α relaxation of the main chain, which are believed to be the glass transition temperature (Tg) of the polymers.

It can be seen from figure 2 that the mechanical loss Angle d is very small when temperature is below Tg, which means the mechanical loss is very small. It is mainly because the rates of the deformation of polymers mainly caused by the deformation of bond lengths and bond angles are quickly under external force and almost completely keep up with the change of the stress.

With the temperature increasing, the polymers change from a glassy state to a highly elastic state. The chain segment starts to move while the viscosity of the system is still very high. Therefore, the friction resistance of the motion of the chain segment is relatively large, which result the high elastic deformation are far behind the change of stress and large internal friction. When tempering temperature further heightens, the deformation gets larger. The chain segment moves freely and thus d and internal friction decreases.

It can also be observed that the d of PLA is much larger than that of PLA composite which indicated that the internal mechanical loss of PLA is much greater than that of PLA composites in the process of deformation. It can be seen from figure 2 that the transition peak of the PLA composites moves to the direction of high temperature obviously. Compared with pure PLA(Tg =63.0 degrees), Tg of PLA/USFs PLA/ASFs and PLA/HSFs is 64.2 degrees, 68.1 degrees and 67.0 degrees respectively, and the height of the transition peak was also reduced. SFs have high rigidity and low density and thus a volume with little mass is still large. In addition, SFs are easier to aggregate. When added SFs to PLA matrix, the homogeneity of the two phases will be worse and the whole molecular chain of the blend system lacks flexibility.

It can also be seen that Tg of PLA/ASFs and PLA/HSFs is obviously higher than PLA/USFs. It is mainly because the interface bonding of the PLA/ASFs and PLA/HSFs is better than the PLA/USFs, which result the chain motion of the matrix molecules of the former two more restricted. In addition, the rigidity of SFs is improved after alkali treatment, which is another reason for the higher glass transition temperatures of PLA/ASFs and PLA/HSFs. Tg is the minimum temperature for the movement of macromolecular segments, which is closely related to the properties of polymers, especially the mechanical properties. Plastics will lose the properties of plastic and translate into rubbers with apparently lower modulus than plastics when temperature is higher than Tg. Therefore, the heat resistance and the operating temperature will be dramatically improved by adding SFs into PLA.

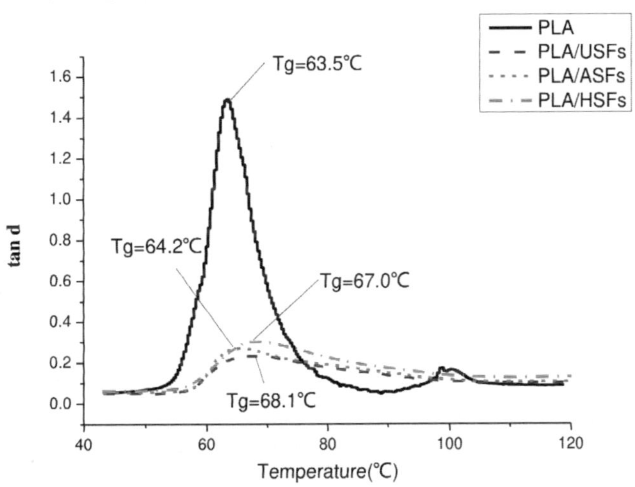

Figure 2 Curves between storage modulus and temperature of PLA and PLA composites

3.2. POM analysis

POM of PLA and PLA composites are presented in figure 3. It can be seen that the crystals of PLA are like branches. However, a layer of crystals perpendicular to the axial direction of sisal fibers appears around SFs which indicated that a new crystal morphology namely transcrystallization appeared because of the addition of SFs in PLA matrix.

The crystals are oriented around the fibers in the direction of the fiber axis in transcrystallization which is mainly because some factors make the fiber become the nucleating material that induces the crystallization of the polymer matrix and thus a large number of crystal nuclei and crystals produced on the fiber surface which squeeze against each other and grow in a directed way. The dynamic and young's modulus of transverse crystal are obviously higher than those of spherulite [8-9], which may be one of the reasons why the rigidity of the PLA composite is higher than that of PLA.

4. Conclusion

The storage modulus and Tg of SFs reinforced PLA composite were higher than PLA, which showed good rigidity and heat resistance of PLA/SF composites. As reinforcement, HSFs can improve the dynamic mechanical properties of PLA composites more significantly than ASFs and USFs.

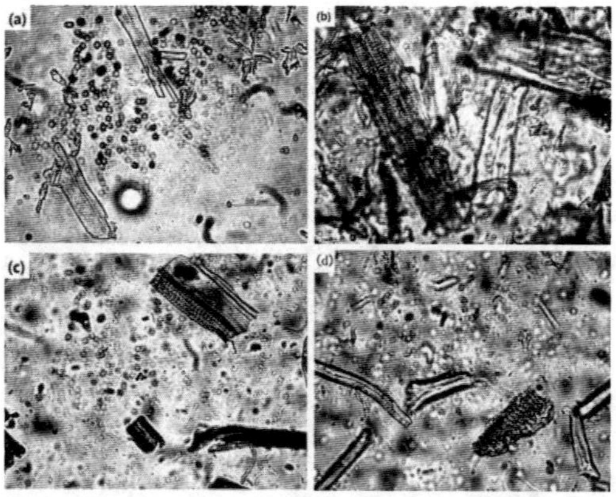

(a) PLA; (b) PLA/USFs; (c) PLA/ASFs; (d) PLA/HSFs

Fig. 3 POM investigation of PLA and composites

Reference

[1]. Auras R, Harte B, Selke S. An overview of polylactides as packaging materials. Macromol. Biosci. 2004, 4, 835-864.

[2]. John M J, Thomas S. Biofibres and biocomposites. Carbohyd. Polym. 2008, 71, 343-364.

[3]. Zhu, Z.Z., Hao, M.Y. Influence of contents of chemical compositions on the mechanical property of sisal fibers and sisal fibers reinforced PLA composites. J. Nat. Fiber. DOI: 10.1080/15440478.2018.1469452.

[4]. Paul S A, Piasta D and Spange S. Solvatochromic and electrokinetic studies of banana fibrils prepared from steam-exploded banana fiber. Biomacromol. 2008, 9, 1802-1810.

[5]. Thakur, V. K., and M. K. Thakur. Processing and characterization of natural cellulose fibers/thermoset polymer composites. Carbohydr. Polym. 2014, 8:102–117.

[6]. Petrissans M, Gerardin P, El Bakali I, et al. Wettability of heat-treated wood. Holzforschung, 2003, 57, 301-307

[7]. Kan C W, Yuen C. Low temperature plasma treatment for wool fabric. Text. Res. J. 2006, 76, 309-314.

[8] Wang C, Liu C R. Transcrystallization of polypropylene composites: nucleating ability of fibres[J]. Polym. 1999, 40(2):289-298

[9] Wu C M, Chen M, Karger-Kocsis J. Effect of micromorphologic features on the interfacial strength of iPP/Kevlar fiber microcomposites[J]. Polym. 2001, 42(1):199-208

Acknowledge

This research was supported by Key Research Platforms and Projects of Universities of Guangdong Province (grant no. 2018GKTSCX111) and Funding Project for Scientific Research of High-level Talents of Guangdong Mechanical & Electrical Polytechnic (grant no. Gccrcxm-201903).

Effects Analysis of Super Surfacing Pavement Performance Using Epoxy Modified Emulsified Asphalt as Bonding Layer

Wen Xuejun[1] and Ming Huang[2]

1 Vice-Chief Engineer, Shanghai Municipal Engineering Design Institute (Group) Co., Ltd., Shanghai 200092, China
2 Postdoctoral, Senior Engineer, Shanghai Municipal Engineering Design Institute (Group) Co., Ltd., Shanghai 200092, China
E-mail: huangming@tongji.edu.cn

Abstract. A large number of studies have shown that emulsified epoxy asphalt has good cohesion and is an ideal choice for road surface treatment technology. However, after emulsification, epoxy resin needs to be emulsified and modified in a very special way to form a good combination with other structural layers in different binders. Based on super surfacing treatment technology, the combination of epoxy resin emulsification and different emulsification modification methods was discussed. Referring to the latest precision design, three bonding methods were used to bond the surface and the underlying layer. That is, no bonding layer, emulsified asphalt bonding layer and epoxy emulsified asphalt bonding layer. The pavement performance comparison test of super surface asphalt mixture was carried out. SBR-A modified emulsified asphalt and SBR-B modified emulsified asphalt were selected as the most commonly used fine surface binder. After comprehensive mix proportion design, three key indicators of wear resistance, water damage resistance and skid resistance are selected to evaluate road performance. The test process and research results show that the epoxy emulsified asphalt binder has excellent pavement performance under the design of test gradation, and has good compatibility with SBR-A modified emulsified asphalt.

1. Introduction

Super surfacing is one of pavement preventive maintenance with the layer thickness below 1 cm, mostly at 0.5-0.8 cm (structure as shown in Figure 1). It is composed of multi-component high-performance binder and fine aggregate[1]. It spreads to the original pavement through special construction equipment. Due to the addition of high-performance bonding layer, compared with the existing preventive maintenance technology such as fog seal and micro-surface, the super surfacing performs a stronger adhesion, more stable structure, better wear resistance and skid resistance[2], and it has no change in elevation and light weight, it therefore could be paved several times in maintenance.

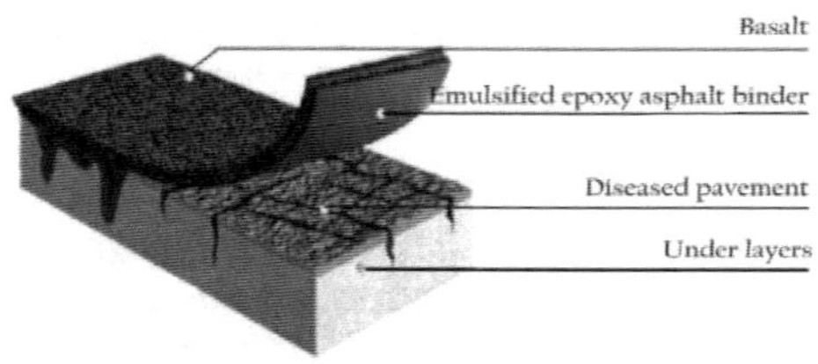

Figure 1. Typical Pavement Structure at Super surfacing.

Super surfacing is a highly efficient maintenance technology in developed countries rather than in China. Because of its high price of emulsified epoxy asphalt, as a high-quality material, it is less used in pavement structure, only used in steel bridge deck bonding layer. With the improvement of road maintenance and the thermosetting resin chemical techniques in China, the research and application of super surfacing have been developed[2-3]. This research pioneered the use of epoxy emulsified asphalt in the bonding layer of cold pavement maintenance, in order to solve the problem of insufficient durability of super surfacing before[4]. At present, the most researchers are focused on gradation analysis and application introduction on super surfacing, and ignored the importance of its bonding layer. The purpose of this study is to analyze the pavement performance changes of 2 conventional refined surfaces after using epoxy emulsified asphalt as bonding layer through laboratory tests.

2. Materials and Tests

2.1. Materials

Epoxy asphalt contains Part A and B. Part A is epoxy resin, Bisphenol A epoxy resin typed E-51 accords to researchers' prophase study[5]. Part B is a mixture of emulsified asphalt and the curing system. The basic properties of E-51 epoxy resin as Part A is shown in Table 1. Table 2 presents the basic properties of Part B and another bonding layer material (common emulsified asphalt) according to the ASTM specification[6]. Part B comprises amine curing agent, compatibilizer and additives[7].

Table 1. Main chemical properties of E-51 type epoxy resin(Bart A).

Chemical composition	density (23℃)(g/cm^3)	viscosity (mPa·s)	epoxy equivalent(g/eq)
2,2-bis(4-(2,3-epoxpropyloxy)phenyl)propane	≤1.10	11000~14000	211~290

There are many kinds of binder innovated in pavement surfacing technology. At present, the most commonly used modified emulsified asphalt is the SBR modification technology[8,9]. In this study, the epoxy emulsified asphalt mixed with thermosetting resin and hardener is used as the bonding layer for the bonding between the lower bearing layer and the upper mixture on the super surfacing, and it actually will merge with the binder in upper mixtures and contribute to the pavement performance. Hence force, it is worth exploring which method can better integrate with the bonding layer (emulsified epoxy) system[10]. Therefore, two representative and innovative binders are selected: domestic SBR latex modified emulsified asphalt (SBR-A for short) and imported SBR latex modified emulsified asphalt (SBR-B for short) for fusion degree verification test.

Table 2. Performance Indicators of the 2 emulsified Asphalts as bonding layer

Items	Specification Requirements	Unit	Epoxy Emulsified Asphalt Part B	Emulsified Asphalt
Screen residual	≤0.1	%	0.04	0.02
Standard Viscosity of Asphalt	12~60	S	18.4	13
Evaporative residue content	≥60	%	62.1	62.1
Penetration (25 ℃)	40~100	0.1mm	67.6	71.4
Softening point (R/B)	≥57	℃	63.2	52.6
Ductility (5 ℃)	≥20	cm	35.8	37.1
1days storage stability	≤1.0	%	0.43	0.21
5 days storage stability	≤5.0	%	2.32	1.32

The test grade is JB-5[1], as shown in Table 4. The oil-stone ratio of the two mixes are that: SBR-A: 9.5%, SBR-B: 8.0%, water content: SBR-A: 2.5%, SBR-B: 4.0%. On the basis of ultra-high performance and cost considerations, the amount of emulsified epoxy is uniformly 1.5%. Wet track abrasion test and pendulum tester friction pendulum were carried out to evaluate the pavement performance of different modified emulsified asphalt. The aggregate are selected in strict accordance with the norms[1].

2.2. Test scheme and procedure

In the study tests, different bonding layers will be planned to undertake the two most common surface layers(SBR-A and SBR-B). Non-bonding layer, common emulsified asphalt bonding layer and epoxy emulsified asphalt bonding layer, respectively. Abbreviated as N/A, EA and EMEA respectively. The test orthogonal combinations are shown in Table 3.

Table 3. Combination scheme of test

Bonding layer / Types of Super surfacing	N/A	Emulsified asphalt	Epoxy modified Emulsified asphalt
SBR-A modified emulsified asphalt	1	2	3
SBR-B modified emulsified asphalt	4	5	6

There will be six groups of experiments and parallel experiments are conducted three times to remove data with coefficient of variation greater than 5%. The test specimens are shown in Fig. 2.

Figure 2. Comparison of Anti-Water Resistan.

The forming steps of laboratory test specimens are as follows: 1. Clean the lower bearing layer (cement concrete surface or asphalt concrete surface); 2. Uniformly coat the bonding layer (0.8L/); 3. Spread out the mixtures on the fine surface and measure their average thickness by vernier caliper; 4. Stand at room temperature for 2 hours before testing. It is noteworthy that epoxy emulsified asphalt, as a bonding layer, must be sprayed almost simultaneously with the fine surface mixture, and the time interval is short. Therefore, synchronous construction should be carried out in the actual project.

3. Results and analysis

3.1. Abrasion resistance

The value of 1h wet track abrasion and loaded wheel test reflect the anti-wear property of the mixture to a certain extent[4]. Three kinds of bonding layer (including none bonding layer) in super surfacing are used to tack the super surfacing mixture. Then the 1h Wet track abrasion test and the loaded wheel bonding test are carried out. The test results are shown in Fig. 2 and 3.

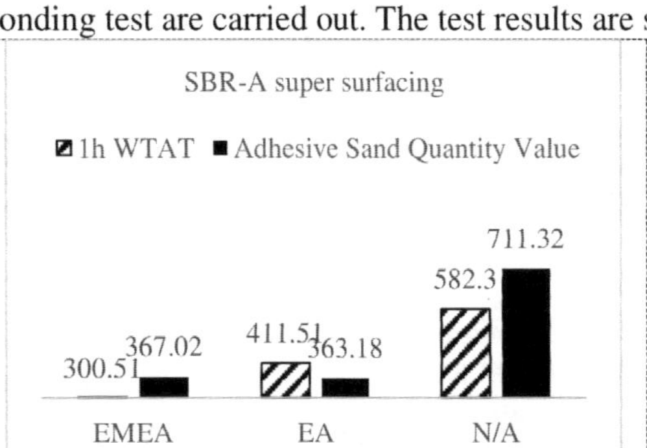

Figure 3. Abrasion Resistance of three kinds

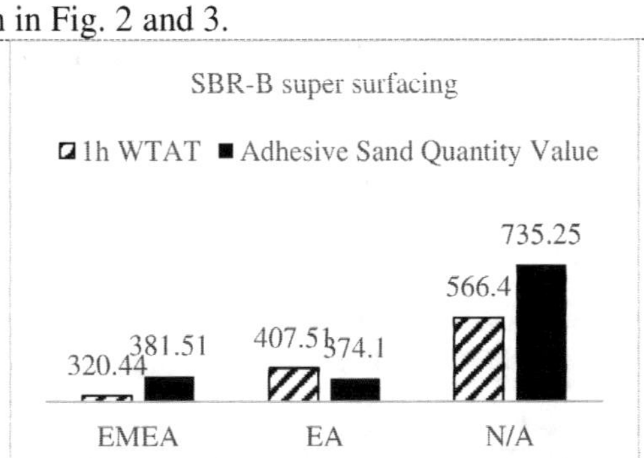

Figure 4. Abrasion Resistance of three kinds of

of bonding layer in Super surfacing(SBR-A). bonding layer in Super surfacing(SBR-B).

As shown in Figure 3 and 4, compared with the wear value of wet wheels for one hour, the performance of EMEA > EA > N/A shows that the wear resistance of SBR-A modified emulsified asphalt is better than that of SBR-B. This is mainly due to the use of rapid setting technology in this kind of domestic SBR, while SBR-B performs high viscous properties, but the strength increases slowly. The amount of adhering sand can reflect the adhesiveness of asphalt and aggregate after the strength of super surfacing mixtures is formed to a certain extent. However, it should be noted that the higher the amount of adhering sand is, the more likely the phenomenon of "sticking wheel" and "dust collection" will appear after open traffic. The order of the loaded wheel valuue is EMEA > EA > N/A.

3.2. Anti-water resistance

The results of 6-day wet track abrasion test (WTAT) on the precise surface of three modified emulsified asphalts are shown in Fig. 5 and 6.

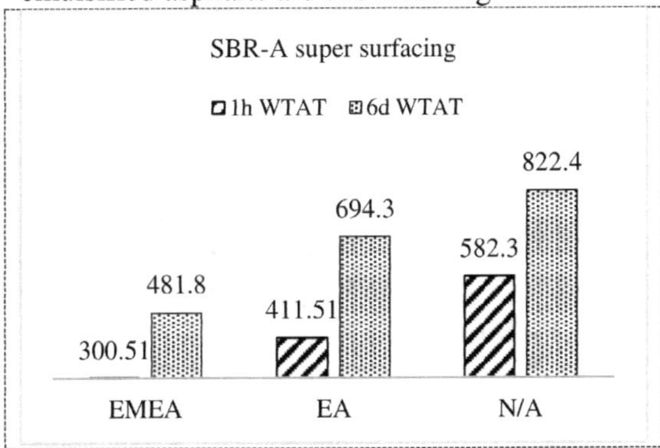

Figure 5. Anti-Water Resistance of three kinds of bonding layer in Super surfacing(SBR-A).

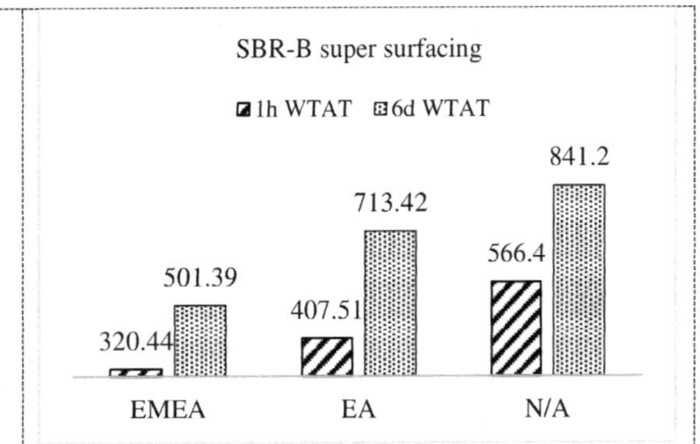

Figure 6. Anti-Water Resistance of three kinds of bonding layer in Super surfacing(SBR-B).

In Figure 5 and 6, the data gap between 6d wet track abrasion value and 1h wet track abrasion value is more obvious. The order of 6d wet track abrasion value is EMEA> EA > N/A. Overall, the performance of SBR-A fine surface is better than SBR-B.

3.3. Skid resistance

As a friction layer directly serving the wheel load, the skid resistance of the finished surface is of particular concern. Pendulum BPN value is the most common and universal method to test the anti-skid ability of pavement. There are differences in the compatibility of different cements with gradation and aggregate at the finish surface, so it is necessary to analyze the anti-skid performance before implementation. Because different emulsified asphalt may have a great influence on the final strength formation of Super surfacing mixtures, this comparative test set up 60 for 2, 4 and 6 hours to analyze and compare the skid resistance of Super surfacing mixtures with different emulsified asphalt under different curing time. The experimental results are shown in Fig. 7 and 8:

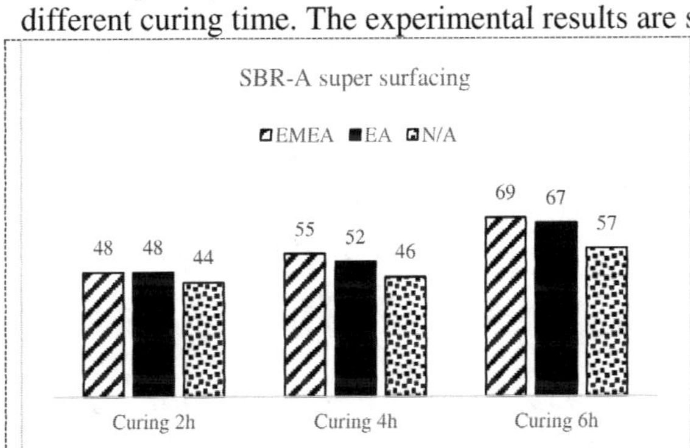

Figure 7. Skid resistance of three kinds of

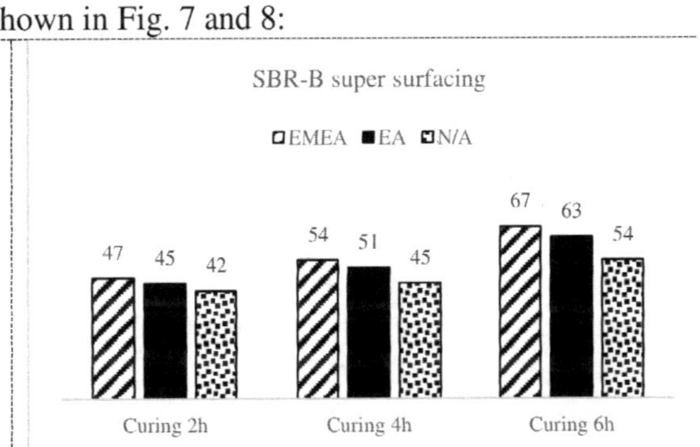

Figure 8. skid resistance of three kinds of

bonding layer in Super surfacing(SBR-A).	bonding layer in Super surfacing(SBR-B).

From the test process and results shown in Fig. 7 and 8, it can be seen that combination of EMEA and SBR-B modified emulsified asphalt super surfacing shows the best anti-sliding performance. The reason is that its imported SBR polymer has poor compatibility with emulsified epoxy, and there is a lot of segregation, which results in the uneven distribution of modified substances in the binder, the uneven distribution of liquid and solid substances in the binder, and the waste of modifier. Meanwhile, it has brought negative impact on the anti-skid performance.

4. Conclusions

The pavement performance of four different emulsified asphalts was compared in order to study the advantages and disadvantages and evaluation effects of the currently used in integration with epoxy modified emulsified asphalt in super surfacing, and to compare the gradation and ring used in the test. The specific conclusions are as follows:

(1) The wear resistance of super surfacing used EMEA bonding layer is the best way. Bonding layer of epoxy modified emulsified asphalt is more completely wrapped between asphalt and aggregate than the other two ways, but it is easier to peel in blocks due to vehicle loads. It needs to prolong the sealing time properly and dry and clean the surface in time to improve its use effect.

(2) EMEA bonding layer greatly enhanced cohesion between asphalt and aggregate after demulsification in super surface, and in areas with complex water action, the combination of EMEA and SBR-A super surface is the best choice.

(3) With the passage of curing time, the anti-sliding ability of three bonding layer ways of super surfacing abrasive layers is increasing. Three kinds of emulsified asphalt surface can meet the requirements of open traffic within 6 hours of curing time (BPN > 60). After any curing time, the order of anti-skid performance is: EMEA>EA>N/A.

5. References

[1] *Technical Guide for Super Surfacing of Asphalt Pavement*. Communications Press: Beijing, China, 2017.

[2] Qian Weimin. *Production Technology of SBS Modified Emulsified Asphalt [J]*.Petroleum Asphalt, 2005, 19(3): 51-53.

[3] Luo Sang; Qian Zhendong; Wang Hui. *Condition survey and analysis of epoxy asphalt concrete pavement on Second Nanjing Yangtze River Bridge: a ten-year review*. Journal of Southeast University (English Edition), v 27, n 4, p 417-22, Dec. 2011.

[4] Kamiya S, Tasaka S, Zhang X, et al. *Compatibilizer role of styrene-butadiene-styrene triblock copolymer in asphalt*. Polymer journal, 2001, 33(3): 209-213.

[5] Ming Huang, Xuejun Wen. *Experimental study on photocatalytic effect of nano TiO_2 epoxy emulsified asphalt mixture*. Applied Science. 2019, 9, 2464

[6] ASTM D1763 - 00(2013) *Standard Specification for Epoxy Resins.*

[7] Xiong Yuanfan, Xia Jin Kui, Xu Anjun, Xiao Zhe. *A preparation method of SBS water emulsion*. Invention Patent: CN15547002004-12-15.

[8] Qian Zhang, Yi-heng Xu, Zhi-guang Wen. *Influence of water-borne epoxy resin content on performance of waterborne epoxy resin compound SBR modified emulsified asphalt for tack coat.* Construction and Building Materials 153 (2017) 774–782

[9] H.T. Li, B. Yu, *Fatigue performance and prediction model of multilayer deck pavement with different tack coat materials*, J. Mater. Civ. Eng. 26 (5) (2014) 872–877.

[10] Yildirim Y. *Polymer modified asphalt binders*. Construction and Building Materials, 2007, 21(1): 66-72.

Acknowledgments

The research was supported by Technology Program of Shanghai Municipal Engineering Design Institute (Group) Co., Ltd. No. K2015K024 and K2018K081.

IOP Conf. Series: Materials Science and Engineering **744** (2020) 012014 doi:10.1088/1757-899X/744/1/012014

New Application Area for Nitride Based Ceramics as Optical Materials

S Kurama[1,2] and S Avcioglu[3]

1 Eskisehir Technical University, Faculty of Engineering, Department of Materials Science and Engineering, Iki Eylul Campus, 26000 Eskisehir, TURKEY
2 Eskisehir Technical University, Faculty of Aeronautics and Astronautics, Iki Eylul Campus, 26000 Eskisehir, TURKEY
3 Ondokuz Mayıs University, Faculty of Engineering, Department of Metallurgy and Materials Engineering, Kurupelit Campus, 55139 Atakum / Samsun, TURKEY
E-mail: skurama@eskisehir.edu.tr

Abstract. This study is focused on the production of translucent nitride based ceramics changing different parameters such as additive, starting powder size and sintering conditions. Therefore, two different processes were followed in this research. In first one translucent SiAlON ceramics were improved by controlling specific parameters which were mentioned above. In the second stage, translucent Si_3N_4 ceramic was obtained also the controlling same parameters. This is a very important result for aerospace and defence industry applications due to these materials high temperature resistivity and mechanical properties.

1. Introduction

One of the most required properties in engineering ceramics is good mechanical properties at high temperature applications. Therefore, SiAlON ceramics are preferred because of the most attractive high-temperature properties. It is a good candidate for high temperature application due to its unique characteristics, such as high hardness and the potential of cleaning the grain boundary phase. However, the application areas of SiAlON ceramics which are very interested with high temperature resistance and mechanical properties are limited with structural applications. On the contrary, it will be an important contribution expanding the use of the functional application area of this material, which has excellent features for defence industry applications which are of great importance today [1, 2]. Many research works have been carried out for transparent/translucent oxide ceramic. Most of the oxide ceramics like Y_2O_3 and YAG are transparent due to their cubic crystal structure and have no birefringence [3]. However, unlike SiAlON ceramics, these oxide ceramics degrades their properties in severe conditions. There is always a demand for a transparent/translucent material that can resist extreme conditions. New technologies need optical materials with the best mechanical properties. Therefore, in recent years, researchers have paid more attention to developing functional application area for SiAlON ceramics. There are some studies related to the effect of starting raw materials and sintering process on the production of transparent SiAlON ceramics.
Transparent ceramics for armour, hard coating and IR window/dome applications should have superior mechanical and thermal properties. Therefore, nitride ceramics have attracted much attention, especially for this kind of application since the 1980s. Kuramoto et al. report that it was possible to produce translucent AlN polycrystalline ceramics in 1989 [4]. Cheng et al. attempt to improve optical properties

IOP Conf. Series: Materials Science and Engineering **744** (2020) 012014 doi:10.1088/1757-899X/744/1/012014

of aluminium nitride ceramics using a microwave sintering process [5]. As well as aluminium nitride, silicon nitride and boron nitride ceramics optical properties have also been investigated by various research groups. SiAlONs are solid solutions of Si_3N_4 and Al_2O_3, which are also a well-known class of structural ceramics. Due to the substitution of Si by Al, with a corresponding atomic replacement of N by O, SiAlON production is easy by liquid phase sintering. These materials, generally preferred due to its high temperature stability and mechanical resistivity in structural application areas. However, improving in defence and aerospace industry also required some new materials which have high temperature resistivity, high mechanical and friction properties also optically properties in functional application areas. Therefore, starting in this point translucent SiAlON ceramics researched were started in 1981 by Mitomo et.al [6]. In 1999, Mandal, et al. showed that translucent α-SiAlON ceramics doped with Nd_2O_3-CaO, Nd_2O_3, CeO_2-CaO and Nd_2O_3-Yb_2O_3 can be obtained and its properties were improved by heat treatment at 1450°C [7].

SiAlON ceramic is a solid solution and it has different types of phases in its structure such as; α-SiAlON, β-SiAlON, grain boundary phases and porosity etc. All these phases have an import role in the transparency of materials. It was also showed that all different types of phases have an important effect on α/β-SiAlON ceramics. As the amount of these phases increases, the translucency of SiAlON materials worsens. At this point, the refractive index differences emerge. It is known that SiAlON has higher refractive index (2) than Al_2O_3 (1.76) and YAG (1.84) materials. Therefore, increasing of secondary phase such as alpha or beta SiAlON or grain boundary phase is not requested in this process. However, this is not easy in SiAlON ceramic sintering. From this point, decreasing grain boundary and increasing translucency is the first aim of this research.

In the present study, two different research were carried out. In first one, it was investigated that all effects of sintering, dopants and particle size parameters on optical properties of SiAlONs. In the second one, similar researches were also carried out for Si_3N_4 ceramics. Si_3N_4 material is an alternative candidate to nitride based optical materials. Their processing is not much common as an optical materials production and application in literature due to the sintering process difficulty of this material. In practice, sintering of silicon nitride based ceramics is a challenge due to the covalent bond between N and Si atoms. Furthermore, a difference in refractive index between grains and grain boundaries must be minimized in order to obtain a transparent polycrystalline ceramic. To overcome these problems Eu_2O_3 is used as a sintering additive to promote densification of Si_3N_4 ceramics and improve optical properties where they dissolve in the solid solution with the main phase. In this research, process controlling was helped to improve translucent Si_3N_4 ceramic production.

2. Experimental Procedure

2.1. Composition and powder preparation of SiAlON ceramic

Two different -SiAlON compositions were prepared using the following starting powders: -Si_3N_4 (UBE-10, containing 1.6%oxygen); AlN (Tokuyama, containing 1% oxygen); Al_2O_3 (99.99%, AKP50); Dy_2O_3 and Y_2O_3 (99.99%HC Starck). Samples were doped with single and dual cation and they were called as D, Y and DY, respectively. To investigate the effect of particle size, Dy^{3+}- doped composition was milled by Pulverisette 6 Fritsch, Germany in isopropanol alcohol and the sample was named as $D_{(i)}$. The solvent ratios were determined as given in our earlier research [8-10].

2.2. Sintering process and densification of SiAlON ceramic

Sintering of samples was carried out by gas pressure sintering process at 1850–1900°C under a gas pressure of 100 bars for a 2 h holding time. The entire process was performed in N_2 atmosphere. Surface grinding processes were applied to the sintered products in order to thin out the samples down to the desired thickness and to improve the surface quality.

3. Results and Discussion

3.1. Effect of dopant and starting powder size on translucent properties of SiAlONs

It is well known that type of dopant stability is an important effect on grain boundary formation of SiAlON ceramics. In beta-SiAlON ceramic structure, stabilization of dopants is not possible due to its structure whereas this can be tolerated in alpha-SiAlON. Therefore, it is important to selection of suitable atoms which can be stabilized in alpha-SiAlON structure. On the other side, the second important parameter is the amount of liquid phase. SiAlON ceramics can be sintered via liquid phase sintering method and it required enough amount of liquid phase for the process. This liquid phase can be supplied by dopants such as MgO, Y_2O_3, rare earth, etc. Therefore, in this researched it was focused on decreasing amount of grain boundary, prevention secondary phase formation and increasing densification in the microstructure of SiAlON ceramics. To reach these aims two different dopants Y_2O_3 and Dy_2O_3 were used as a single or dual cation in SiAlON compositions. After sintering process of these samples, it was observed that Dy_2O_3 or Y_2O_3–doped samples have lower density than dual cation (Dy/Y-doped) doped sample, 3.29 g/cm^3, 3.30 g/cm^3 and 3.34 g/cm^3 respectively [11]. The main phase was observed as an alpha-SiAlON in all samples whereas some amount of AlN-polytpe phase has been seen in Y-doped sample (Fig. 1-2). Their transparency was changed related to their microstructure.

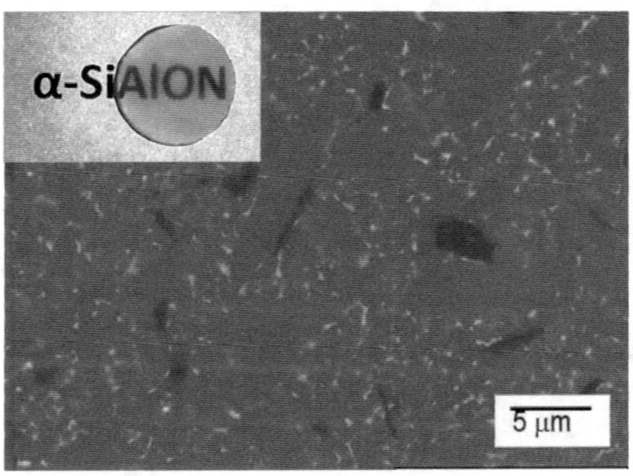

Figure 1.a Microstrucure of Y_2O_3-doped sample at 1900°C.

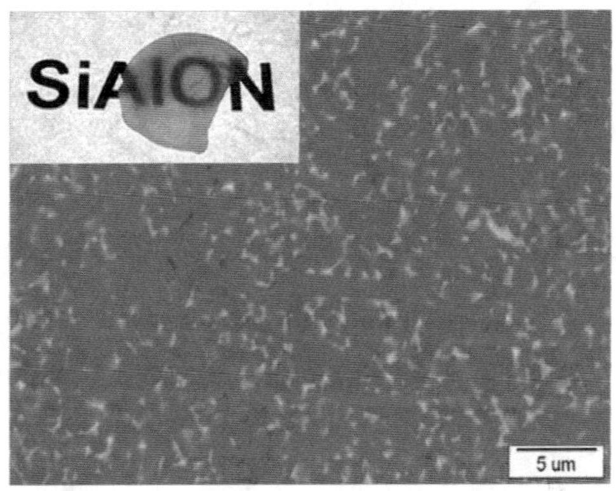

Figure 1.b Microstrucure of Dy/Y -doped sample at 1850°C.

Y_2O_3-doped sample has more elongated grains and higher amount of glassy phase than other samples. Additionally, it was observed that some amount of AlN-polytpe in its structure. All these parameters have not deteriorated the transmittance of the sample which is about 45% (Figure 1.a). Comparing with Y_2O_3-doped sample, Dy_2O_3-doped one has higher amount of grain boundary phase whereas its grain size distribution is narrower than that of Y-doped. The transmittance of this sample is about 33% even though has higher amount of grain boundary phase (Figure 1.b). Dual cation doped SiAlON composition (Dy/Y-doped) were prepared to increase stabilization of Dy-cation in alpha-SiAlON structure and decrease the amount of grain boundary phase to improve transparency. As it is seen from microstructure, grain size growth was observed whereas the grain boundary phase was still higher than Y-doped one. Therefore, its transmittance is decreased to 25%.

In this point, increasing of the amount of dopant would result with higher amount of grain boundary in samples. Therefore, a novel process was applied to the system to increase density whereas decrease grain boundary phase in the structure of samples. To reach this aim, composition $\mathbf{D}_{(i)}$ was prepared by milling high energy ball milling system. The powder size of composition was decreased to 137.9 nm from 263.6 nm. This progress was resulted with lowering sintering temperature and decreasing grain boundary phase as it is seen in Fig. 2. Even though decreasing starting powder size the amount of grain boundary phase was not much reduced, as we reported in our previous studies [12]. Additionally, grains were formed in elongated morphology which negatively effects transparency of the sample. On the other hand, the amount of AlN-polytype was reduced by decreasing powder size. This result has a positive effect on the transmittance percentage of sample as 35%.

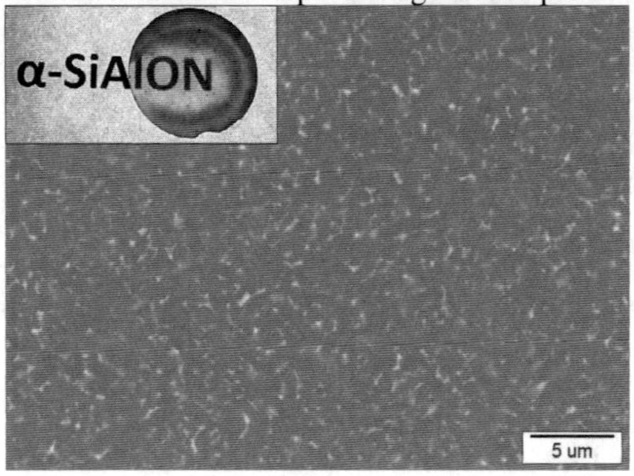

Figure 2.a Microstrucure of Dy_2O_3-doped sample at 1850°C.

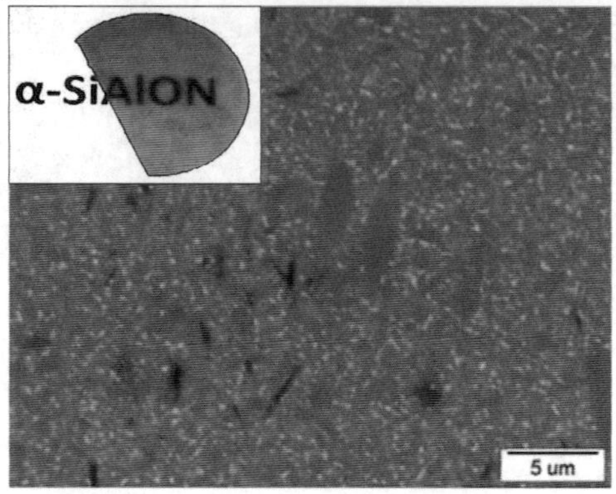

Figure 2.b Microstrucure of $\mathbf{D}_{(i)}$ sample at 1850°C.

3.2. An alternative material to SiAlON

Structure of SiAlON materials is very complicated and its composition stability is not much easy. Therefore, sintered Si_3N_4 has a much attraction as an alternative optical material for SiAlON. However, there is no more research in this area. In this research the effect of dopant and sintering process was investigated for Si_3N_4 ceramic materials. Sintering of these materials was carried out by spark plasma sintering. The sintering additives and relative density of Si_3N_4 ceramics are given in Table 1.

Table 1. Sintering additives and density of samples.

Sample	Type of dopant	Density (g/cm^3)
Eu-S@LT	Eu_2O_3	3.09
Eu-S@HT	Eu_2O_3	3.29

Results showed that sintering conditions have an important role in densification and transparency of the sample. The results were given in figure 3a and b. As it is seen in Figure 3a, the microstructure of Eu-S@LT has a very high amount of porosity and low density. Its theoretical density is about 88%. Therefore, transparency of this sample could not be observed due to porosity and low densification. This is related with lower sintering temperature of samples. Eu-S@HT sample was prepared with the same amount of Eu_2O_3. This sample was sintered about 55°C higher than the first one. This result showed that without changing type and amount of additive higher densification can be obtained by sintering at higher temperature. This sample reached 94 % density, as 3.29 g/cm^3. This result was also supported by microstructural analyses result of sample (Fig. 3b). Even though both samples have alpha and beta-Si_3N_4 phases, their amount has an important role related with sintering temperature. Alpha-Si_3N_4 phase shows low temperature stability whereas beta-Si_3N_4 is stable at higher temperature. Therefore, alpha-Si_3N_4 phase transforms to beta phase above 1500-1650°C. This temperature range is not constant for all systems, it changeable with amount of grain boundary phase in its structure. All these parameters, such as the amount of alpha and/or beta –Si_3N_4, porosity and microstructure have an important role on transparency of Si_3N_4 material. This prior study showed that if sample has high densification microstructure, the amount of grain boundary phase positively affects the translucent properties of samples.

IOP Conf. Series: Materials Science and Engineering **744** (2020) 012014 doi:10.1088/1757-899X/744/1/012014

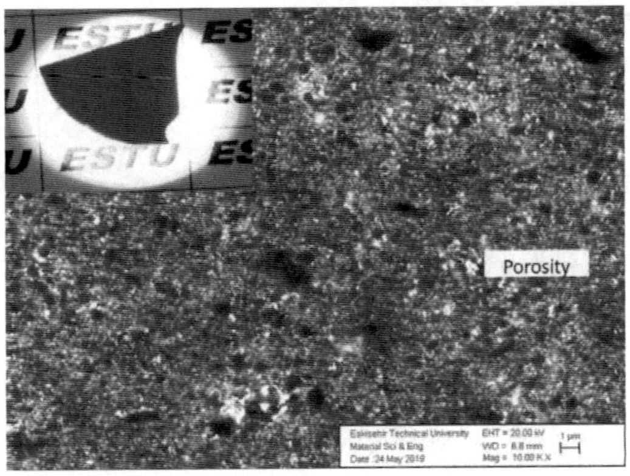

Figure 3.a Microstructure of Eu-S sample sintered at low temperature

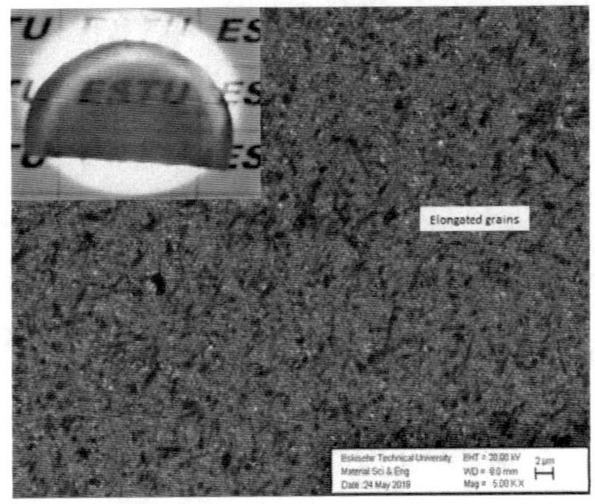

Figure 3.b Microstructure of Eu-S sample sintered at high temperature.

4. Conclusion

All results showed that sintering conditions and type of additive have an important role on microstructural and optical properties on nitride based materials. Reducing particle size of starting powder is one of the dominant factors to determine the final morphology and phase assemblage of the nitride based ceramics. Doping SiAlON structure with Dy^{+3} and Y^{+3} also improves the IR transmission. Additionally, this research showed that it is also possible to gained optical property to Si_3N_4 which is a strong candidate as translucent polycrystalline ceramic for military applications.

5. References

[1] Karunaratne B S B, Lumby R J, Lewis M H 1996 Rare earth-doped -SiAlON ceramics with novel optical properties, *Journal of Materials Research*, **11** 2790-4.
[2] Mandal H 1999 New developments in -SiAlON ceramics, *Journal of the European Ceramic Society* **19**: 2349-57.
[3] Krell A, Hutzler T, Klimke J 2009 Transmission physics and consequences for materials selection, manufacturing, and applications *J. Eur. Ceram. Soc.* 29 2 p.207-221.
[4] Kuramoto N, Taniguchi H, Aso I 1989 Development of translucent aluminum nitride ceramics *Am. Ceram. Soc. Bull.* 68 p.886.
[5] Cheng J, Agrawal D, Zhang Y, Roy R 2002 Development of translucent aluminum nitride (AlN) using microwave sintering process, *J. Electroceram.* 9 p.67-71.
[6] Mitomo M, Moriyoshi Y, Sakai T, Ohsaka T, Kobayashi M 1981 Translucent -SiAlON ceramics *Journal of Materials Science Letter* 1 p.25.
[7] Mandal H 1999 *Journal of the European Ceramic Society* 19 p.2349-2357.
[8] Eser O, Kurama S 2012 A compression of sintering techniques using different particle sized -SiAlON powders, *Journal of the European Ceramic Society* **32**: 1343-47.
[9] Eser O, Kurama S 2010 The effect of the wet milling process on sintering temperature and the amount of additive of SiAlON ceramics, *Ceramic International* **36**, 1283-88.
[10] Eser O, Kurama S, Gunkaya G 2010 The production of β-SiAlON ceramics with low amounts of additive at low sintering temperature, *Journal of the European Ceramic Society*, **30**: 14, 2985-2990.
[11] Avcioglu S, Kurama S 2017 Investigation of the influence of initial powder size on the optical properties of Dy- -SiAlON ceramics fabricated by gas pressure sintering *Ceram. Int.* 43 p. 8449-8454.
[12] Kurama S, Avcioglu S, Ayas E 2015 Optimization of the optical properties of α-SiAlON through a decrease in starting powder size, *Journal of the European Ceramic Society* **35** 3229–3235

IOP Conf. Series: Materials Science and Engineering **744** (2020) 012015 doi:10.1088/1757-899X/744/1/012015

Computational Simulation about the Impact of High Voltage Electrostatic Field on Crystallization of Calcium Carbonate Aqueous Solution

D F Wang[1, a], D D Meng[1, 2, b *], X D Cui[1, c], F J Wang[1, d] and Y Liu[1, e]

[1]College of Arts and Science, Qingdao Binhai University, Qingdao 266555, China
[2]Key Laboratory of New Metallic Functional Materials and Advanced Surface Engineering in Universities of Shandong, Qingdao Binhai University, Qingdao 266555, China
Shandong Provincial Natural Science Fund, ZR2019BA025, China
E-mail: [a]wdfcg2006@126.com, [b*]mengdongmail@163.com, [c]cuixindong@163.com, [d]wangfengjuna@126.com, [e]lycser0311@163.com
Corresponding Author: mengdongmail@163.com

Abstract. Computational simulation method was used in two models of calcium carbonate aqueous solution to explore the mechanism of anti-scaling under electrostatic field on the molecular level. The binding energy between the surface and ions indicated that the surface (104) was more favorable for the growth of the new crystal, while it was much easier to decompose under field strength 1300 V/m. The diffusion coefficients of Ca^{2+} and CO_3^{2-} revealed that the chemical reaction of Ca^{2+} and CO_3^{2-} ions in Model-II was much difficult to form calcium carbonate crystal under electrostatic field. The hydration degree of Ca^{2+} with O atoms from water molecules showed that the electric field could enhance the hydration degree of Ca^{2+} under certain field length, hence preventing the chemical reaction with CO_3^{2-} to produce crystal. In short, the scale inhibition effect under electrostatic field in Model-II was much better.

1. Introduction

In the process of industrial production and daily life, the occurrence of scaling not only caused enormous economic loss but also lots of serious security incidents. Over the past few decades, researchers had developed many kinds of methods for anti-scaling, one of which was using high voltage electrostatic field that originated in the United States [1]. This technique was developed in Japan in 1970s, which was firstly utilized in China in 1975.

In the past few years, the study of anti-scaling was almost about experimental method on the macro-quantitative level. However, the complexity of water system caused the great chanciness of experimental results [2]. In order to study the mechanism of anti-scaling under electrostatic field on the molecular level, and provide academic support for the application of anti-scaling using electrostatic field, computational simulation (molecular dynamics, MD) method was used in this paper to study the impact of high voltage electrostatic field on the crystallization of calcium carbonate aqueous solution.

In this paper, two types of aqueous solution of calcium carbonate that containing calcite crystal $(1\bar{1}0)$ surface and (104) surface (labeled as Model-I and II, respectively) were utilized to simulate the impact of crystallization under high voltage electrostatic field, and then to explore the mechanism of anti-scaling. Through computational simulation, we analyzed the binding energy, diffusion value and hydration degree of two models to explore the impact of electrostatic field on crystallization of calcium carbonate aqueous solution, and found that the Model-II was much sensitive to the electric field so that the scale inhibition effect under high electrostatic field in Model-II was much better.

2. The Method of Computational Simulation

Molecular Dynamics Simulation is a computing physics method that widely used to simulate the classical multi-particle system. Materials Studio (MS) is an analog-computing software that developed by the Accelrys Company, which can be directly run on the PC for the simulation. The computing module—DISCOVER is used for simulation in this work, which is running on the server side, the data generated can be remotely shared with other PC in time. Discover module contains multiple force fields, among which the powerful force field—COMPASS will be used in the present work, that can be used to accurately predict the structure, conformation, vibration, and thermodynamic properties of various molecules in an isolated or condensed system under physical conditions. Complex potential energy functions are adopted by COMPASS, among which the bond stretching energy function adopting three stage series expansion and the non-bonding energy term adopting Lennard-Jones-9-6 function, as detailedly reported in our other works [3].

3. Parameters Setting and System Balanced Judgement

The crystals of $CaCO_3$ studied in this paper were composed of calcite structure with two main growth surfaces $(1\bar{1}0)$ and (104), respectively.

The bottom Layer of Model-I contained surface $(1\bar{1}0)$, the size was 1.996×1.706×1.008 nm^3, including 7 layers of Ca^{2+} and CO_3^{2-} ions. The volume of calcium carbonate solution was 1.996×1.706×2.602 nm^3, which was constructed by 300 water molecules, $2Ca^{2+}$ and $2CO_3^{2-}$ ions. The bottom Layer of Model-II was calcite surface (104) that composed of 4 layers of Ca^{2+} and CO_3^{2-} ions. The middle layer of the solution system was constructed by 300 water molecules, $2Ca^{2+}$ and $2CO_3^{2-}$ ions, and the volume was 2.429×1.996×1.828 nm^3.

The high voltage electrostatic field was added to the 2 models by computer programming language BTCL, of which the field strength varying from 1000 V/m to 2000 V/m .

System balance conditions were $\Delta E_{converge} \le 0.001$(convergence parameter) and $R \le 0.001$(the ratio of total energy fluctuation to the kinetic energy fluctuation). If the system was not accordance with the two conditions, the calculating system would be optimized automatically and calculated repeatedly until satisfying the balance conditions.

4. Results and Discussion

4.1. The influence of electrostatic field on binding energy between the surface and Ca^{2+}/CO_3^{2-} ions

The binding energy between surface $(1\bar{1}0)/(104)$ and Ca^{2+}, CO_3^{2-} ions under the electrostatic fields were shown in Figure 1, respectively. The results showed that the variations of electrostatic field strength had no obvious effect on the binding energy in Model-I. While in Model-II, the binding energy dropped steeply to 3630 kcal/mol under the field strength 1300 V/m, revealing that the binding energy between surface (104) and ions was weakened by electric field. Viewed as a whole, the binding energy between ions and surface (104) (Model-II) was much stronger than that combining with $(1\bar{1}0)$ (Model-I). This result demonstrated that the ions were much easier to be adsorbed by surface (104), hence the surface (104) was more favorable for the growth of the new crystal. This result is the same as our previously reported temperature-dependent binding energies [3]. As the way of energy definition was different, the binding energies that we calculated were larger than the reality. According to solid state physics theories, the influence of electrostatic field on atom's energy mainly due to it's impact on energy band. The present simulating results proved that electrostatic fields have no obvious effects on binding energies, in other words, the variation of field strength from 1000 V/m to 2000 V/m have no effect on the energy bands of Ca^{2+}, CO_3^{2-} ions.

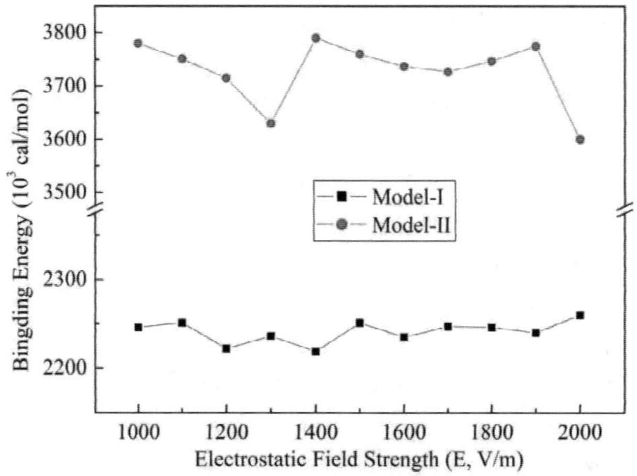

Figure 1. The influence of electrostatic field on the binding energies between surface $(1\bar{1}0)$ / (104) and Ca^{2+}, CO_3^{2-} ions.

4.2.The diffusion of Ca^{2+} and CO_3^{2-} under the influence of electrostatic field

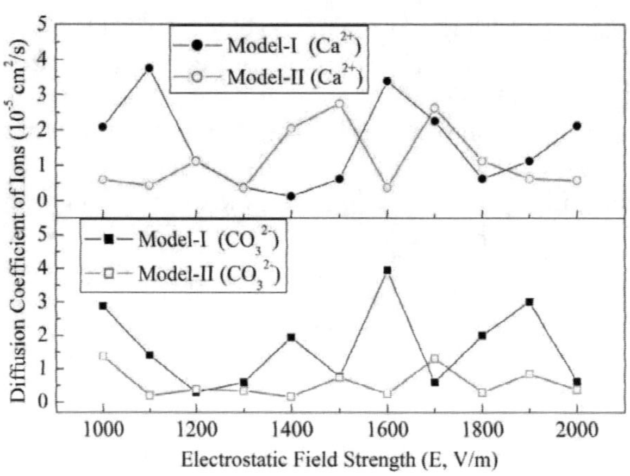

Figure 2. The diffusion coefficients of Ca^{2+} and CO_3^{2-} ions under different electrostatic fields in Model-I and II, respectively.

The influence on the diffusion coefficients of Ca^{2+} and CO_3^{2-} ions under different electrostatic field in Model-I and II were shown in Figure 2, respectively. As a whole, the diffusion of Ca^{2+} in Model-I was affected more obviously than that in Model-II, field strengths 1100 V/m and 1600 V/m in particular, the diffusion values reached $3.750\times10^{-5}cm^2/s$ and 3.385×10^{-5} cm^2/s, respectively. On the contrary, the diffusion of Ca^{2+} in Model-II showed two pronounced anomalies under field strengths 1500 V/m and 1700 V/m that the values reached 2.750×10^{-5} cm^2/s and 2.625×10^{-5} cm^2/s, respectively, as shown in Figure 2 (up).

As shown in Figure 2 (down), CO_3^{2-} in Model-I showed more diffusive on the whole than that in Model-II under different field strengths. The diffusion of CO_3^{2-} in Model-I exhibited three pronounced anomalies under field strengths 1400 V/m, 1600 V/m and 1900 V/m, respectively. While in Model-II, the diffusion showed faint variations under 1500 V/m, 1700 V/m and 1900 V/m, respectively.

On the whole, the Ca^{2+} and CO_3^{2-} ions in Model-I were more active than that in Model-II under the influence of electrostatic field, and the anormalies' driving electrostatic field strengths were universally lower than that in Model-II. The present results revealed that the chemical reaction of Ca^{2+} and CO_3^{2-} ions in Model-I was much easier, especially under the field strength 1600 V/m.

4.3. The influence of electrostatic field on the hydration degree of Ca^{2+} with O atoms from water molecules in the two models

The Radial-Distribution-Function (RDF) can provide a lot of information about the motional characteristics of particles in the system, which can be used to understand the "structure" of liquids. The RDF can be interpreted as the ratio of the regional density to the average density of the system [4]. The density of the region near the reference molecule is different from the average density of the system, but it should be the same as the average density when the distance from the reference molecule is long, this means that the value of the RDF should be close to 1 when the value is large [5].

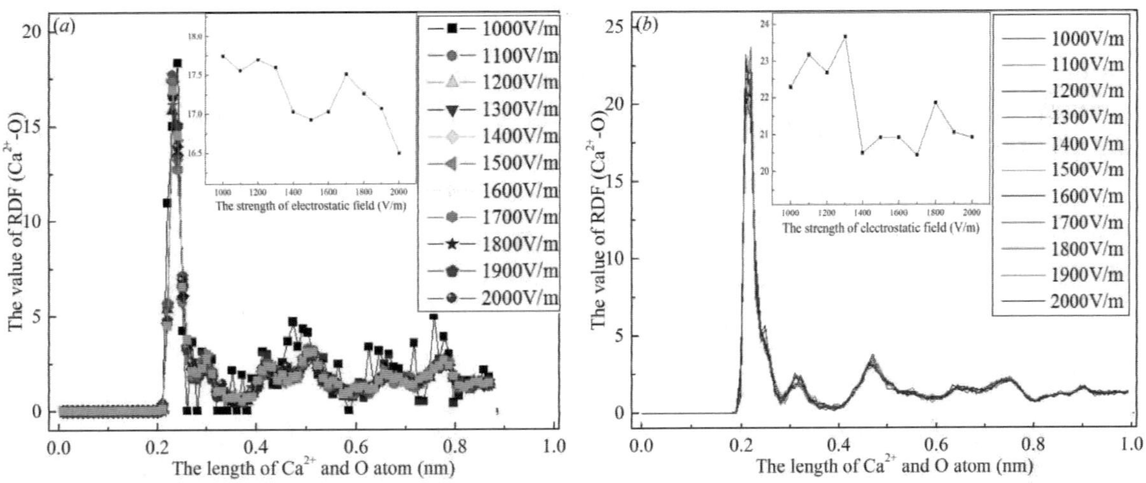

Figure 3. The influence of electrostatic field on RDF values (Ca^{2+}-O) in (*a*) Model-I and (*b*) Model-II.

In the solution, a certain number of water molecules are arranged around Ca^{2+} and CO_3^{2-} to form hydrated ions, in which water molecules act as the ligand and bind with ions through coordination bonds. Obviously, the higher the hydration degree of Ca^{2+} and O atom is, the more difficult to form calcium carbonate crystals.

The RDF about hydration degree with particle length of Ca^{2+} in aqueous solution and O atoms from water molecules in the two models were shown in Figure 3 (a) and (b), respectively. One can see from the figures that the spike of RDF in Model-I showed variations under the influence of field length ~1200 V/m and 1700 V/m, which in Model-II exhibited obvious variations under field length 1300 V/m and 1800 V/m, as shown in the insets of Figure 3 (a) and (b). The present results indicated that the electric field could enhance the hydration degree of Ca^{2+} under certain field length, hence preventing the chemical reaction with CO_3^{2-} to produce calcium carbonate crystal.

On the other hand, the RDF spike values of Ca^{2+}-O in Model-II was significantly larger than that in Model-I. It demonstrated that the hydration degree of Ca^{2+} in Model-II was higher than that in Model-I. From this point of view, the scale inhibition effect of high voltage electric field in Model-II was much better.

5. Summary

Computational simulation (molecular dynamics) method was used to study the mechanism of anti-scaling under electrostatic field on the molecular level.

By analyzing the calculated binding energy data, one can know that the electrostatic field did not have an obvious influence on the binding energy between ions in aqueous solutions and crystal growth surface $(1\bar{1}0)$ (Model-I). The binding energy between ions and surface (104) (Model-II) was much stronger which demonstrated the ions were much easier to be adsorbed by surface (104), hence it was more favorable for the growth of the new crystal in Model-II. On the other hand, the binding energy dropped steeply under the field strength 1300 V/m in Model-II, revealing the Model-II was much sensitive to electric field.

The diffusion values of Ca^{2+} and CO_3^{2-} under electrostatic field revealed that the Ca^{2+} and CO_3^{2-} ions in Model-I were more active than that in Model-II under the influence of electrostatic field, hence the chemical reaction of Ca^{2+} and CO_3^{2-} ions in Model-I is much easier, especially under the specific field strengths.

The RDF about hydration degree of Ca^{2+} in aqueous solution and O atoms from water molecules in the two models indicated that the electric field could enhance the hydration degree of Ca^{2+} under certain field length and prevent the chemical reaction with CO_3^{2-} to produce calcium carbonate crystal. Because the hydration degree of Ca^{2+} in Model-II was higher than that in Model-I, the scale inhibition effect under high electrostatic field in Model-II was much better.

Acknowledgements

Thanks to Prof. J. Zhang and L. B. Huang (China University of Petroleum) for providing the software <Materials Studio> in the process of calculation. This work is supported by Shandong Provincial Natural Science Foundation (ZR2019BA025).

Reference

[1] Cheng L X and Yang J H 2000 *Petro-Chemical Equipment* **29** 22
[2] Yao H L, Gong J T, Guo Y, Yao Z M and Wang C 2001 *Industrial Water Treatment* **21** 24
[3] Wang D F, Meng D D, Wang F J and Cui X D 2019 *Materials Science: Properties and Technologies*. **394** 73
[4] Chen Z L, Xu W R and Tang L D 2007 *Theory and Practice of Molecular Simulation* (Beijing: Chemical industry Press) p 112
[5] Frenkel D and Smit B 1996 *Understanding Molecular Simulation: From Algorithms to Applications* (New York: Academic Press) p 230

Comparisons of Conventional, SRI, EAS and Hybrid-Stress Plane Q4 and Q9 Finite Element Models in Displacement and Energy Norms

H B Xiao[1], H C Hu[2], H Y Zhang[3,4], Q Xie[5] and Y X Zhou[4,3]

1 Aircraft Aerodynamics & Energy System Division, Powerplant and Fuel System Department, Shanghai Aircraft Design and Research Institute, Shanghai, China
2 Department of Mechanical Engineering, University of Hong Kong, Hong Kong SAR, China
3 Faculty of Civil Engineering and Mechanics, Jiangsu University, Jiangsu, China
4 School of Science, Harbin Institute of Technology, Shenzhen, Guangdong, China
5 School of Highway, Chang'an University, ShaanXi, China
E-mail: zhouyexin@hit.edu.cn, xiaohaibin@comac.cc

Abstract. Advanced finite element models have been being pursued for several decades. For plane models, the popularly employed formulations include the selectively reduced integration, enhanced assumed strain, hybrid/mixed methods, stabilization methods, among others. The accuracies of the advanced finite element models with respect to the standard element models are often demonstrated by some popular numerical tests in which displacement and stress components at a few selected points are benchmarked. The more comprehensive but laborious means of assessing the accuracy by computing the relative errors in displacement norm (Sobolev norm of order zero) and energy norm are less popularly employed. In this paper, these norms of the standard and some advanced 4-node and 9-node quadrilateral finite element models for a number of tests are computed and reported. In contrast to the common perception, the advanced models are not always more accurate than the standard ones. The discussions in this paper can be helpful for the understanding and usage of the numerical benchmark tests.

1. Introduction

Tremendous efforts have been put on developing advanced finite element models which are expected to be more accurate than the standard finite element models. For plane elements, the popularly employed formulations include but not restrict to selectively reduced integration (SRI) [1], enhanced assumed strain (EAS) [2], hybrid/mixed [3,4] and stabilization methods [5]. Note worthily, the accuracy of the advanced finite element models with respect to the standard element ones are traditionally demonstrated by popular numerical tests in which displacement and stress at a few selected points are compared [6]. On the other hand, the more comprehensive but laborious means of assessing the accuracy in terms of the following relative error in energy norm e_σ, which is recommended for error estimate in some well-known textbooks such as [7,8], is less popularly employed to assess the advanced finite element models:.

$$e_\sigma = \frac{\| \mathbf{u}^h - \mathbf{u}^{\text{exact}} \|_E}{\| \mathbf{u}^{\text{exact}} \|_E} = \frac{\left(\int_\Omega (\boldsymbol{\sigma}^h - \boldsymbol{\sigma}^{\text{exact}})^T \mathbf{C}^{-1} (\boldsymbol{\sigma}^h - \boldsymbol{\sigma}^{\text{exact}}) d\Omega \right)^{1/2}}{\left(\int_\Omega (\boldsymbol{\sigma}^{\text{exact}})^T \mathbf{C}^{-1} \boldsymbol{\sigma}^{\text{exact}} d\Omega \right)^{1/2}} \tag{1}$$

IOP Conf. Series: Materials Science and Engineering **744** (2020) 012016 doi:10.1088/1757-899X/744/1/012016

in which Ω, $\mathbf{u}$, $\boldsymbol{\sigma}$ and $\mathbf{C}$ denote the problem domain, displacement vector, vector of stress components and the material stiffness matrix, respectively. The superscripts "h" and "*exact*" refer to the numerical and exact solutions, respectively. One would naturally expect that the advanced models should be more accurate than the standard ones in terms of e_σ.

What trigger us to prepare this paper is that the advanced models do not often meet the expectation. To achieve more understanding on the accuracy, we employ further the following relative error in displacement (Sobolev norm of order zero) norm e_u:

$$e_u = \frac{\| \mathbf{u}^h - \mathbf{u}^{\text{exact}} \|_0}{\| \mathbf{u}^{\text{exact}} \|_0} = \frac{\left(\int_\Omega (\mathbf{u}^h - \mathbf{u}^{\text{exact}})^T (\mathbf{u}^h - \mathbf{u}^{\text{exact}}) d\Omega \right)^{1/2}}{\left(\int_\Omega (\mathbf{u}^{\text{exact}})^T \mathbf{u}^{\text{exact}} d\Omega \right)^{1/2}} . \quad (2)$$

Intuitively speaking, e_σ and e_u reflect the accuracies of the stress and displacement predictions, respectively.

In this paper, four-node quadrilateral (Q4) and 9-node quadrilateral (Q9) standard, SRI, EAS and HS (hybrid stress) models will be briefly reviewed first. Then, the numerical tests to be conducted and the convergence plots for the computed e_σ and e_u will be presented. From the results, it is found that while advanced models are more accurate than the standard model in terms of e_u in most cases, however, there are tests in which advanced models are less accurate than the standard ones in terms of both e_σ and e_u.

2. The Standard Model

2.1 Four-Node Quadrilateral (Q4) Element

For Q4 element, its coordinate and displacement interpolations are:

$$\mathbf{x} = \begin{Bmatrix} x \\ y \end{Bmatrix} = \sum_{i=1}^{4} N_i \mathbf{x}_i , \quad \mathbf{u} = \begin{Bmatrix} u \\ v \end{Bmatrix} = \sum_{i=1}^{4} N_i \mathbf{u}_i = \mathbf{N}\mathbf{d} \quad (3)$$

where $\mathbf{x}_i$ is the nodal coordinate, $\mathbf{u}_i$ is the nodal displacement, $\mathbf{d}$ is the element displacement vector that contains all $\mathbf{u}_i$s, $\mathbf{N}$ is the standard Q4 displacement interpolation matrix [7,8].The interpolated displacement leads to the strain:

$$\boldsymbol{\varepsilon} = \begin{Bmatrix} \varepsilon_x \\ \varepsilon_y \\ \gamma_{xy} \end{Bmatrix} = \begin{Bmatrix} u,_x \\ v,_y \\ u,_y + v,_x \end{Bmatrix} = \mathbf{L}\mathbf{u} = \mathbf{L}\mathbf{N}\mathbf{d} \quad (4)$$

where $\mathbf{L} = \begin{bmatrix} \partial/\partial x & 0 & \partial/\partial y \\ 0 & \partial/\partial y & \partial/\partial x \end{bmatrix}^T$ is the strain-displacement operator. The total potential of the element is:

$$\Pi^e = \frac{1}{2}\int_{\Omega^e} (\boldsymbol{\varepsilon}^T \mathbf{C} \boldsymbol{\varepsilon}) d\Omega - P^e = \frac{1}{2}\mathbf{d}^T \left(\int_{\Omega^e} (\mathbf{L}\mathbf{N}^T \mathbf{C} \mathbf{L}\mathbf{N}) d\Omega \right) \mathbf{d} - P^e = \frac{1}{2}\mathbf{d}^T \mathbf{k}^e \mathbf{d} - P^e \quad (5)$$

in which

$$\mathbf{k}^e = \int_{\Omega^e} (\mathbf{L}\mathbf{N})^T \mathbf{C} (\mathbf{L}\mathbf{N}) d\Omega$$

is the stiffness matrix and P^e is the work done by the external force. The standard Q4 element (Q4Std) is evaluated by the 2nd order quadrature and is said to be fully integrated.

2.2 Nine-Node Quadrilateral (Q9) Element

For Q9 element, its coordinate and displacement interpolations are

$$\mathbf{x} = \begin{Bmatrix} x \\ y \end{Bmatrix} = \sum_{i=1}^{9} N_i \mathbf{x}_i , \quad \mathbf{u} = \begin{Bmatrix} u \\ v \end{Bmatrix} = \sum_{i=1}^{9} N_i \mathbf{u}_i = \mathbf{N}\mathbf{d} \quad (6)$$

in which $\mathbf{N}$ is the standard Q9 displacement interpolation matrix. The subsequent procedures to obtain the strain, the total potential and the element stiffness matrix are the same as those for Q4, Eqs. (4)-(5). The standard Q9 element (Q9Std) is evaluated by the 3rd order quadrature and is said to be fully integrated.

3. Selectively Reduced Integration (SRI)

For plane strain isotropic materials, the elastic material stiffness matrix and the stress can be split into

$$\mathbf{C}=\mu\begin{bmatrix}2&0&0\\0&2&0\\0&0&1\end{bmatrix}+\lambda\begin{bmatrix}1&1&0\\1&1&0\\0&0&0\end{bmatrix}=\mathbf{C}_\mu+\mathbf{C}_\lambda \qquad \boldsymbol{\sigma}=\mathbf{C}\boldsymbol{\varepsilon}=\mu\begin{bmatrix}2&0&0\\0&2&0\\0&0&1\end{bmatrix}\boldsymbol{\varepsilon}+\lambda\begin{bmatrix}1&1&0\\1&1&0\\0&0&0\end{bmatrix}\boldsymbol{\varepsilon} \tag{7}$$

where μ and λ are Lame's constants, E is the elastic modulus, and υ is the Poisson's ratio. Thus, the total potential can be expressed as

$$\Pi^e=\frac{1}{2}\int_{\Omega^e}\boldsymbol{\varepsilon}^T(\mathbf{C}_\mu+\mathbf{C}_\lambda)\boldsymbol{\varepsilon}d\Omega-P^e=\frac{1}{2}\int_{\Omega^e}(\boldsymbol{\varepsilon}^T\mathbf{C}_\mu\boldsymbol{\varepsilon}+\boldsymbol{\varepsilon}^T\mathbf{C}_\lambda\boldsymbol{\varepsilon})d\Omega-P^e. \tag{8}$$

In the above expression, the terms associated with $\mathbf{C}_\mu$ and $\mathbf{C}_\lambda$ correspond to the shear and volumetric energies, respectively. The corresponding stiffness matrix of the element can be written as:

$$\mathbf{k}^e=\int_{\Omega^e}(\mathbf{LN})^T\mathbf{C}_\mu(\mathbf{LN})d\Omega+\int_{\Omega^e}(\mathbf{LN})^T\mathbf{C}_\lambda(\mathbf{LN})d\Omega=\mathbf{k}^e_\mu+\mathbf{k}^e_\lambda \tag{9}$$

in which $\mathbf{k}^e_\mu$ and $\mathbf{k}^e_\lambda$ are the stiffness matrices associated with $\mathbf{C}_\mu$ and $\mathbf{C}_\lambda$ respectively.

For nearly incompressible materials, the Poisson's ratio υ approaches to 0.5, λ becomes extremely large and the well-known dilatational locking may be induced. SRI alleviates the locking by evaluating $\mathbf{k}^e_\mu$ with full integration (the 2nd and 3rd order quadratures for Q4 and Q9, respectively), and $\mathbf{k}^e_\lambda$ with reduced integration (the 1st and 2nd order quadratures for Q4 and Q9, respectively). The SRI Q4 and Q9 elements are here denoted as Q4SRI and Q9SRI, respectively.

4. Enhanced Assumed Strain (EAS) Formulation

EAS is another commonly used method to alleviate the locking phenomenon. Following the idea of Simo & Rifai [2], the total strain in an element is split into the compatible strain $\boldsymbol{\varepsilon}^u$ and the enhanced strain $\tilde{\boldsymbol{\varepsilon}}$ as:

$$\boldsymbol{\varepsilon}=\boldsymbol{\varepsilon}^u+\tilde{\boldsymbol{\varepsilon}} \tag{10}$$

where $\boldsymbol{\varepsilon}^u=\mathbf{Lu}$ is the compatible part and $\tilde{\boldsymbol{\varepsilon}}$ is directly assumed in the EAS method as $\tilde{\boldsymbol{\varepsilon}}=\mathbf{E}\boldsymbol{\alpha}$ in which the matrix $\mathbf{E}$ to be defined for various elements contains the assumed shape functions and the vector $\boldsymbol{\alpha}$ contains the element internal parameters. Substituting Eq. (10) into the total potential yields:

$$\Pi^e=\frac{1}{2}\begin{Bmatrix}\mathbf{d}\\\boldsymbol{\alpha}\end{Bmatrix}^T\begin{bmatrix}\mathbf{k}_{dd}&\mathbf{k}_{d\alpha}\\\mathbf{k}_{\alpha d}&\mathbf{k}_{\alpha\alpha}\end{bmatrix}\begin{Bmatrix}\mathbf{d}\\\boldsymbol{\alpha}\end{Bmatrix}-P^e \tag{11}$$

in which $\mathbf{k}_{dd}=\int_{\Omega^e}(\mathbf{LN})^T\mathbf{C}(\mathbf{LN})d\Omega$, $\mathbf{k}_{d\alpha}=\int_{\Omega^e}(\mathbf{LN})^T\mathbf{CE}d\Omega$, $\mathbf{k}_{\alpha d}=\int_{\Omega^e}\mathbf{E}^T\mathbf{C}(\mathbf{LN})d\Omega$ and $\mathbf{k}_{\alpha\alpha}=\int_{\Omega^e}\mathbf{E}^T\mathbf{CE}d\Omega$.
Variation of Eq.(13) with respect to $\boldsymbol{\alpha}$ leads to

$$\mathbf{k}_{\alpha d}\mathbf{d}+\mathbf{k}_{\alpha\alpha}\boldsymbol{\alpha}=\mathbf{0} \text{ and } \boldsymbol{\alpha}=-\mathbf{k}_{\alpha\alpha}^{-1}\mathbf{k}_{\alpha d}\mathbf{d}. \tag{12}$$

Substituting the last expression into Eq.(13), the total potential can be written as:

$$\Pi^e=\frac{1}{2}\mathbf{d}^T(\mathbf{k}_{dd}-\mathbf{k}_{d\alpha}\mathbf{k}_{\alpha\alpha}^{-1}\mathbf{k}_{\alpha d})\mathbf{d}-P^e \tag{13}$$

4.1 EAS Q4 Models

For Q4 models, the following sets of four and seven incompatible EAS modes are commonly used [9]:

$$\tilde{\boldsymbol{\varepsilon}}=\begin{Bmatrix}\tilde{\varepsilon}_{xx}\\\tilde{\varepsilon}_{yy}\\2\tilde{\varepsilon}_{xy}\end{Bmatrix}=\frac{1}{J}\bar{\mathbf{T}}_\varepsilon\begin{bmatrix}\xi&0&0&0\\0&\eta&0&0\\0&0&\xi&\eta\end{bmatrix}\begin{Bmatrix}\alpha_1\\\mathrm{M}\\\alpha_4\end{Bmatrix},\quad \tilde{\boldsymbol{\varepsilon}}=\begin{Bmatrix}\tilde{\varepsilon}_{xx}\\\tilde{\varepsilon}_{yy}\\2\tilde{\varepsilon}_{xy}\end{Bmatrix}=\frac{1}{J}\bar{\mathbf{T}}_\varepsilon\begin{bmatrix}\xi&0&0&0&\xi\eta&0&0\\0&\eta&0&0&0&\xi\eta&0\\0&0&\xi&\eta&0&0&\xi\eta\end{bmatrix}\begin{Bmatrix}\alpha_1\\\mathrm{M}\\\alpha_7\end{Bmatrix} \tag{14}$$

where $\bar{\mathbf{T}}_\varepsilon=\mathbf{T}_\varepsilon\big|_{\xi=\eta=0}$, $\mathbf{T}_\varepsilon=\mathbf{T}_\sigma^{-T}=\begin{bmatrix}x,_\xi x,_\xi & x,_\eta x,_\eta & 2x,_\xi x,_\eta\\ y,_\xi y,_\xi & y,_\eta y,_\eta & 2y,_\xi y,_\eta\\ x,_\xi y,_\xi & x,_\eta y,_\eta & x,_\xi y,_\eta+y,_\xi x,_\eta\end{bmatrix}^{-T}$

and $\mathbf{T}_\sigma$ is the stress transformation matrix. In the subsequent numerical examples, the Q4 elements using the four and seven EAS modes will be abbreviated as Q4E4 and Q4E7, respectively.

4.2 EAS Q9 Model

For Q9 element, the following set of eleven EAS modes were proposed by Bischoff & Ramm [10]:

$$\tilde{\boldsymbol{\varepsilon}}=\begin{Bmatrix}\tilde{\varepsilon}_{xx}\\ \tilde{\varepsilon}_{yy}\\ 2\tilde{\varepsilon}_{xy}\end{Bmatrix}=\frac{1}{J}\bar{\mathbf{T}}_\varepsilon\begin{bmatrix} f_\xi & \eta f_\xi & f_\xi f_\eta & 0 & 0 & 0 & 0 & 0 & 0 & 0 & 0\\ 0 & 0 & 0 & f_\eta & \xi f_\eta & f_\xi f_\eta & 0 & 0 & 0 & 0 & 0\\ 0 & 0 & 0 & 0 & 0 & 0 & f_\xi & \eta f_\xi & f_\eta & \xi f_\eta & f_\xi f_\eta\end{bmatrix}\begin{Bmatrix}\alpha_1\\ \vdots\\ \alpha_{11}\end{Bmatrix} \tag{15}$$

where $f_\xi = 3\xi^2-1$ and $f_\eta = 3\eta^2-1$. This model shall be abbreviated as Q9E11 in the subsequent numerical examples.

5. Hybrid Stress (HS) Formulation

In HS elements, the element-wise Hellinger-Reissner functional is expressed as:

$$\Pi^e = \int_{\Omega^e}(-\tfrac{1}{2}\boldsymbol{\sigma}^T\mathbf{C}^{-1}\boldsymbol{\sigma}+\boldsymbol{\sigma}^T(\mathbf{Lu})-\mathbf{b}^T\mathbf{u})d\Omega - P^e \tag{16}$$

in which the displacement **u** is again interpolated as in Eq.(3) or Eq.(6). The stress $\boldsymbol{\sigma}$ is independently assumed from **u** and written as $\boldsymbol{\sigma}=\mathbf{P}\boldsymbol{\beta}$. With this stress and the interpolated **u**, the functional Eq.(16) can be further manipulated to be

$$\Pi^e = -\frac{1}{2}\boldsymbol{\beta}^T\mathbf{H}\boldsymbol{\beta}+\boldsymbol{\beta}^T\mathbf{Gd}-P^e \tag{17}$$

where $\mathbf{H}=\int_{\Omega^e}\mathbf{P}^T\mathbf{C}^{-1}\mathbf{P}d\Omega$ and $\mathbf{G}=\int_{\Omega^e}\mathbf{P}^T\mathbf{C}^{-1}\mathbf{LN}d\Omega$. Variation of Eq.(17) with respect to $\boldsymbol{\beta}$ enforces:

$$\mathbf{H}\boldsymbol{\beta}=\mathbf{Gd} \text{ or } \boldsymbol{\beta}=\mathbf{H}^{-1}\mathbf{Gd}. \tag{18}$$

Substituting it into Eq.(17) yields:

$$\Pi^e = \frac{1}{2}\mathbf{d}^T(\mathbf{G}^T\mathbf{H}^{-1}\mathbf{G})\mathbf{d}-P^e. \tag{19}$$

5.1 HS Q4 Model

For the HS Q4, the well-known 5β element of Pian & Sumihara [3] is considered in this paper. In their element, the assumed stress is:

$$\boldsymbol{\sigma}=\begin{Bmatrix}\sigma_{xx}\\ \sigma_{yy}\\ \sigma_{xy}\end{Bmatrix}=\begin{Bmatrix}\beta_1\\ \vdots\\ \beta_3\end{Bmatrix}+\bar{\mathbf{T}}_\sigma\begin{bmatrix}\eta & 0\\ 0 & \xi\\ 0 & 0\end{bmatrix}\begin{Bmatrix}\beta_4\\ \beta_5\end{Bmatrix} \tag{20}$$

where $\bar{\mathbf{T}}_\sigma=\mathbf{T}_\sigma\big|_{\xi=\eta=0}$. In the subsequent examples, the HS Q4 element shall be abbreviated as Q4H5.

5.2 HS Q9 Model

For the HS Q9, the hybrid stress model of Sze [4] is considered in this study. The assumed hybrid stress in the model is:

$$\begin{Bmatrix}\sigma_{xx}\\ \sigma_{yy}\\ \sigma_{xy}\end{Bmatrix}=[\mathbf{I}_3,\xi\mathbf{I}_3,\eta\mathbf{I}_3,\xi\eta\mathbf{I}_3]\begin{Bmatrix}\beta_1\\ \vdots\\ \beta_{12}\end{Bmatrix}+\bar{\mathbf{T}}_\sigma\begin{bmatrix}\xi f_\eta & 0\\ 0 & \eta f_\xi\\ 0 & 0\end{bmatrix}\begin{Bmatrix}\beta_{13}\\ \beta_{14}\end{Bmatrix} \tag{21}$$

In the subsequent examples, this model shall be abbreviated as Q9H14.

6. Numerical Tests

In this section, numerical tests will be conducted by using the afore-reviewed quadrilateral elements whereas the relative errors e_σ and e_u will be computed and compared. Both the full integration (FI) and the uniform reduced integration (URI) by using Gaussian quadrature are employed for computing e_σ while nodal integrations (compounded trapezoidal rule for Q4 and compounded Simpson's rule for Q9) are used for computing e_u. It is well-known that the SRI element can also be viewed as a hybrid element with an assumed pressure or volumetric strain field. In this regard, the strain will also be interpolated/extrapolated at the reduced order integration points in the course of computing the stress e_σ for the SRI elements. For elements of the same order, the relative computational efficiency is trivial among various formulations when compared to the computational cost of solving the global linear equation. To obtain the convergence plots, each element in the initial mesh shall be sub-divided into $n \times n$ elements where n = 1, 2, 4, 8 and 16 for Q4 and n = 1, 2, 4 and 8 for Q9. Thus, the maximum number of nodes used by Q4 and Q9 are the same.

6.1 Cantilever modelled by 2n×n rectangular elements

Figure 1 shows a cantilever modelled by rectangular elements under plane strain condition and subjected to an end shear force P. For υ = 0.25, the results predicted by Q4s and Q9s are plotted in **Figure 2**(a-c) and (d-f), respectively. The standard and SRI elements provide almost the same results for e_σ(FI), e_σ(URI) and e_u. Q4E4, Q4E7 and Q4H5 yield identical predictions in this case [11, 12] and they are significantly more accurate than Q4Std and Q4SRI except for e_σ (URI) with only 2×1 elements. Q9E11 and Q9H14 produce the same e_σ (FI) while their e_σ (URI) and e_u are also very close to each other. Q9E11 and Q9H14 are markedly more accurate than Q9Std and Q9SRI.

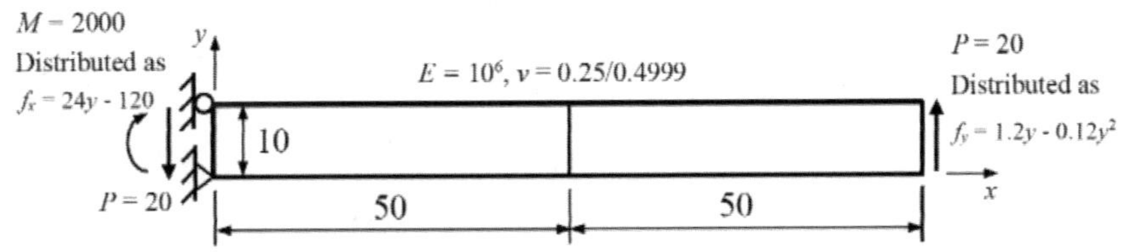

Figure 1. A cantilever modelled by 2×1 rectangular elements subjected to shear force.

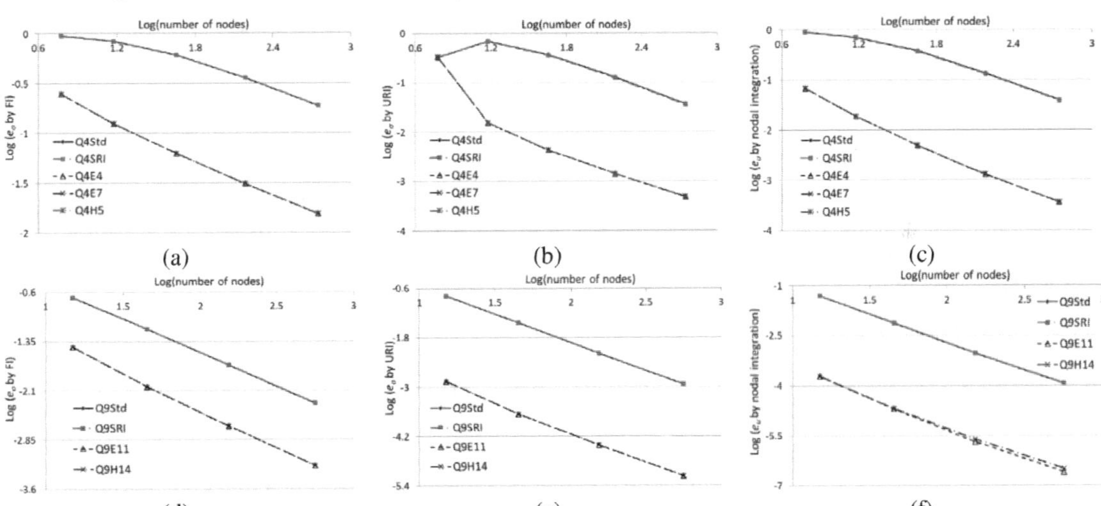

Figure 2. The relative errors e_σ and e_u for the cantilever beam modelled by $2n \times n$ rectangular elements with Poisson's ratio υ = 0.25. (a)-(c): e_σ(FI), e_σ(URI) and e_u for the Q4 models; (d)-(f): e_σ(FI), e_σ (URI) and e_u for the Q9 models.

For υ = 0.4999, the results are plotted in **Figure 3**. Q4Std seriously suffers from the dilatational locking while the Q4SRI, Q4E4, Q4E7 and Q4H5 are locking-free. It should be remarked that since SRI integrates $\mathbf{k}_e^\lambda$ only at the reduced order Gaussian points, using the full integration to compute e_σ appears to be unfair and the relative error e_σ (URI) may be more pertinent. For Q9s, the dilatational locking is not noted in Q9Std whilst the SRI does not improve the predictions. Again, EAS and HS elements are the accurate elements in terms of both the displacement and energy norms.

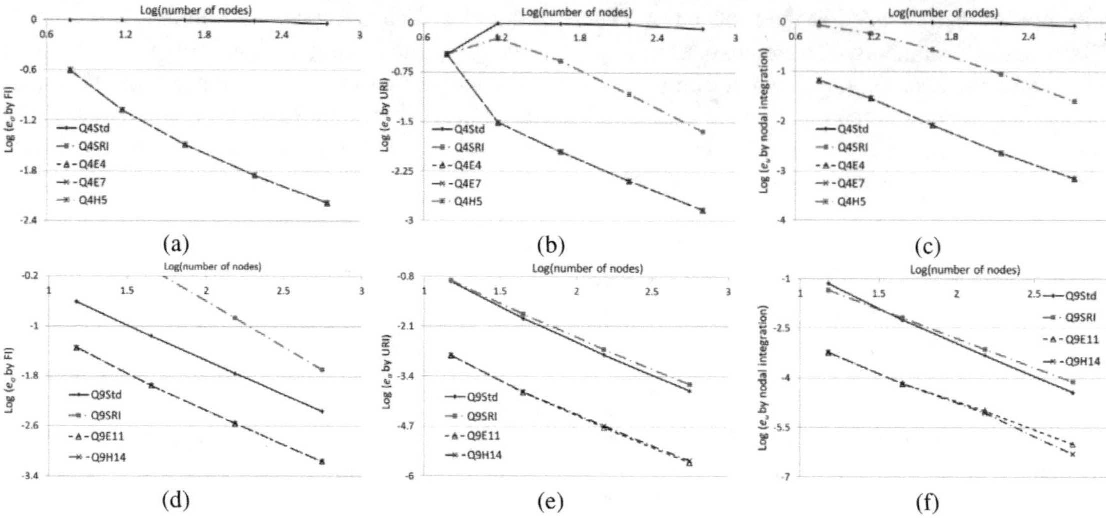

Figure 3. The relative errors e_σ and e_u for the cantilever beam modelled by $2n\times n$ rectangular elements with Poisson's ratio $\upsilon = 0.4999$. (a)-(c): e_σ (FI), e_σ (URI) and e_u for the Q4 models; (d)-(f): e_σ (FI), e_σ (URI) and e_u for the Q9 models.

6.2 Cantilever modelled by $5n\times n$ trapezoidal elements

Figure 4 considers again the cantilever problem but the beam is initially modelled by 5 trapezoidal elements. The computed error norms for $\upsilon = 0.25$ are plotted in **Figure 5**. The Q4 EAS and HS elements are the most accurate Q4s followed by Q4SRI which is more accurate than Q4Std in terms of e_σ (URI) and e_u. However, it is interesting to note that Q9Std and Q9SRI are the most accurate Q9s in terms of e_u. All the Q9s yield close e_σ and Q9E11 is a bit better.

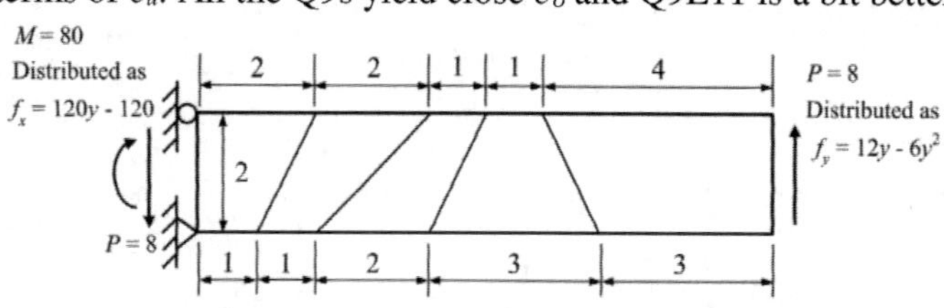

Figure 4. A cantilever modelled by 5 trapezoidal elements subjected to shear force.

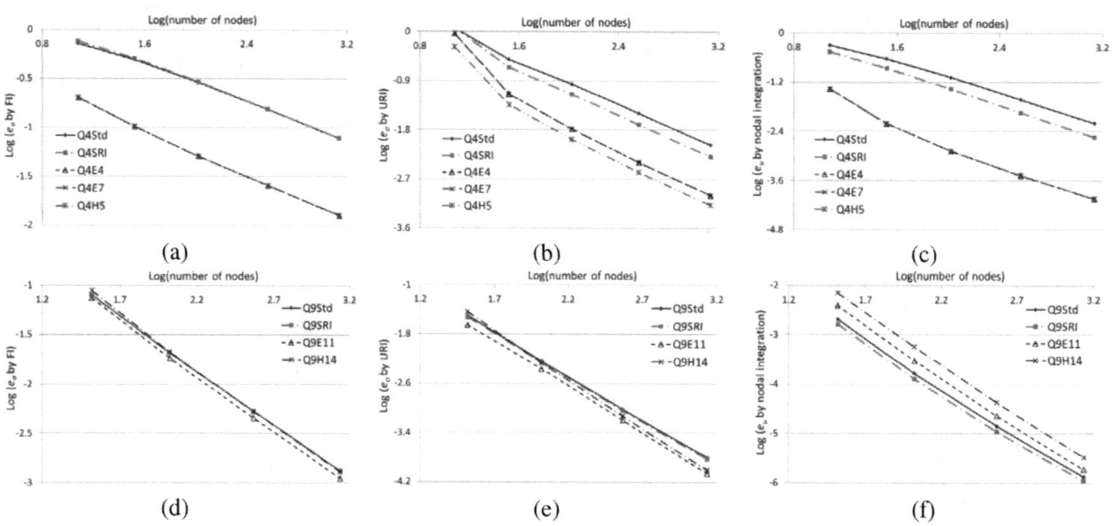

Figure 5. The relative errors e_σ and e_u for the cantilever beam modelled by $5n\times n$ trapezoidal elements with Poisson's ratio $\upsilon = 0.25$. (a)-(c): e_σ (FI), e_σ (URI) and e_u for the Q4 models; (d)-(f): e_σ (FI), e_σ (URI) and e_u for the Q9 models.

When υ = 0.4999, the standard elements are the least accurate ones due to the locking and the Q4 EAS and HS elements are again the most accurate Q4s. Thus, the results are not plotted here. However, it should be noted that Q9SRI is more accurate than the Q9E11 and Q9H14 in terms of e_u.

6.3 Square panel with a circular cutout

This example considers a plane strain panel with a circular hole at the center. Owing to the symmetry, only one-quarter of the panel is modelled and the initial mesh contains 2 elements, see **Figure 6**. While the conditions of symmetry are applied along the x- and y- axes, the tractions along x = 4 and y = 4 are prescribed according to the analytical stress solution [13].

$$\begin{Bmatrix} \sigma_{xx} \\ \sigma_{yy} \\ \tau_{xy} \end{Bmatrix} = \begin{Bmatrix} 1-\frac{1}{r^2}(\frac{3}{2}\cos 2\theta+\cos 4\theta)+\frac{3}{2r^4}\cos 4\theta \\ -\frac{1}{r^2}(\frac{1}{2}\cos 2\theta-\cos 4\theta)-\frac{3}{2r^4}\cos 4\theta \\ -\frac{1}{r^2}(\frac{1}{2}\sin 2\theta+\sin 4\theta)+\frac{3}{2r^4}\sin 4\theta \end{Bmatrix} \tag{22}$$

The analytical displacement solution for this example can also be found in Ref.[13]:

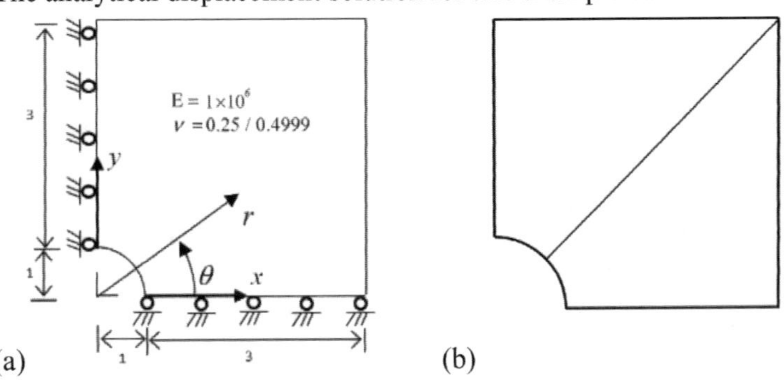

Figure 6. Square panel with a circular cutout. (a) Geometry, (b) initial mesh.

When υ = 0.25, the predictions of all Q4s are close in both the displacement and energy norms, see **Figure 7**(a-c). The SRI model is marginally better than the others in e_u and the EAS and HS elements are marginally better in e_σ (FI). It is interesting to note that Q9Std and Q9H14 are the most and least accurate elements in e_σ, respectively, yet they are least and most accurate in e_u, respectively, as shown in **Figure 7**(d-f).

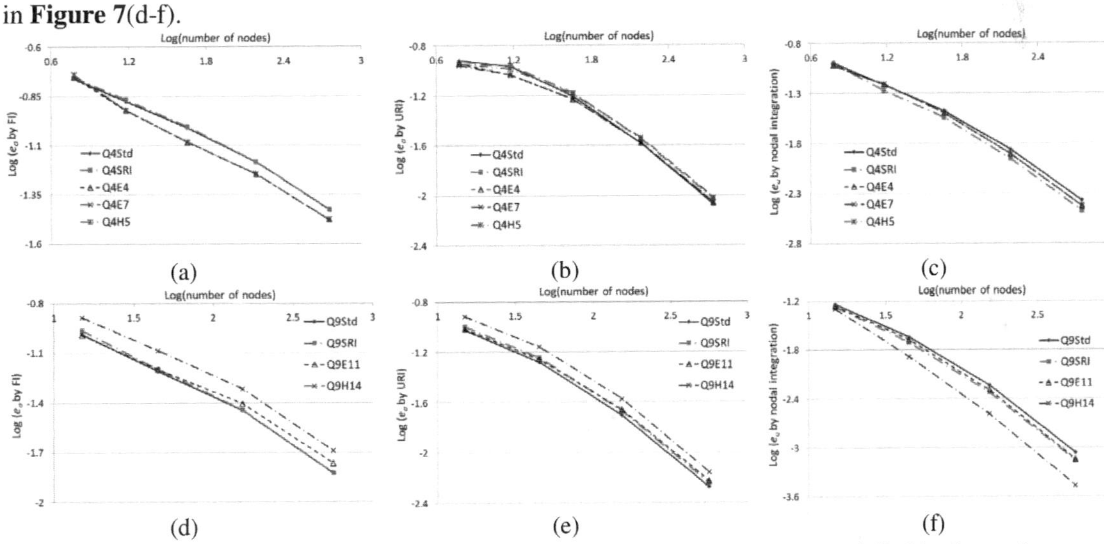

Figure 7. The relative errors e_σ and e_u for the square panel with circular cutout modelled by $2n$xn elements with Poisson's ratio υ= 0.25. (a)-(c): e_σ (FI), e_σ (URI) and e_u for the Q4 models; (d)-(f): e_σ (FI), e_σ (URI) and e_u for the Q9 models.

For υ = 0.4999, the standard models again suffer from the dilatational locking and the predictions are the poorest. Q4SRI is the most accurate Q4s in terms of e_u and its e_σ (URI) is also close to that of Q4H5. Q9H14 is the most accurate Q9 element in terms of e_u while Q9E11 and Q9SRI are the most accurate ones in terms of e_σ.

6.4 Thick-wall cylinder

This example considers a thick-wall cylinder subjected to internal pressure p = 1. Owing to symmetry, only one-quarter of the cylinder is modelled. The computational domain is initially divided into 2 elements as shown in **Figure 8**. The analytical stress solution can be found in Ref.[13].

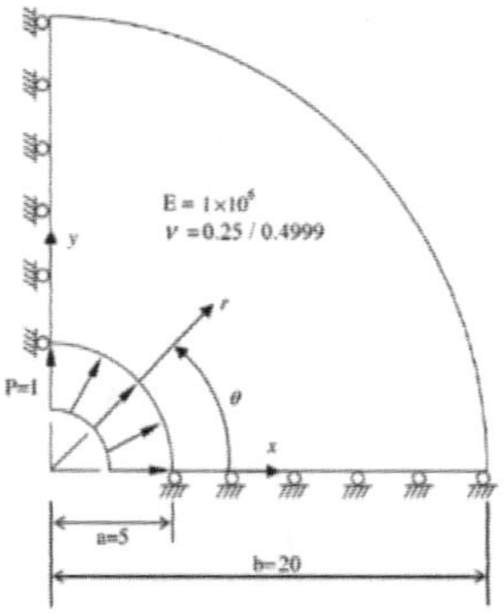

(a) (b)

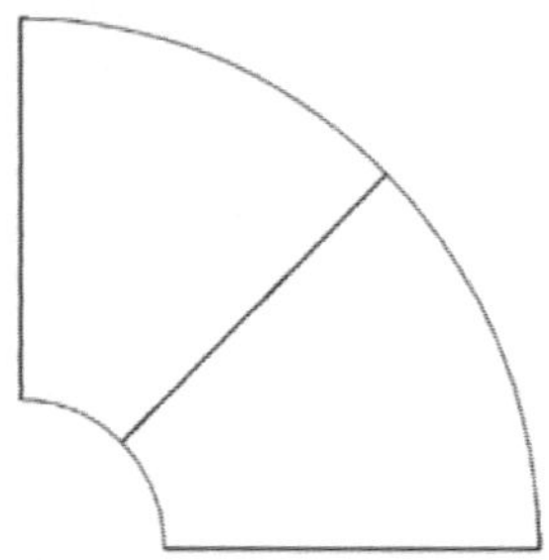

Figure 8. A thick-wall cylinder under inner pressure. (a) Geometry and load, (b) initial.

When v = 0.25, Q4SRI and Q4Std are the most accurate Q4 elements in terms of e_u and e_σ (URI), respectively, see **Figure 9**. The EAS and HS elements are only more accurate in terms of e_σ (FI). Q9H14 is the most but least accurate Q9 element in terms of e_u and e_σ. Q9E11 and Q9Std are the most accurate models in terms of e_σ (FI) and e_σ (URI), respectively.

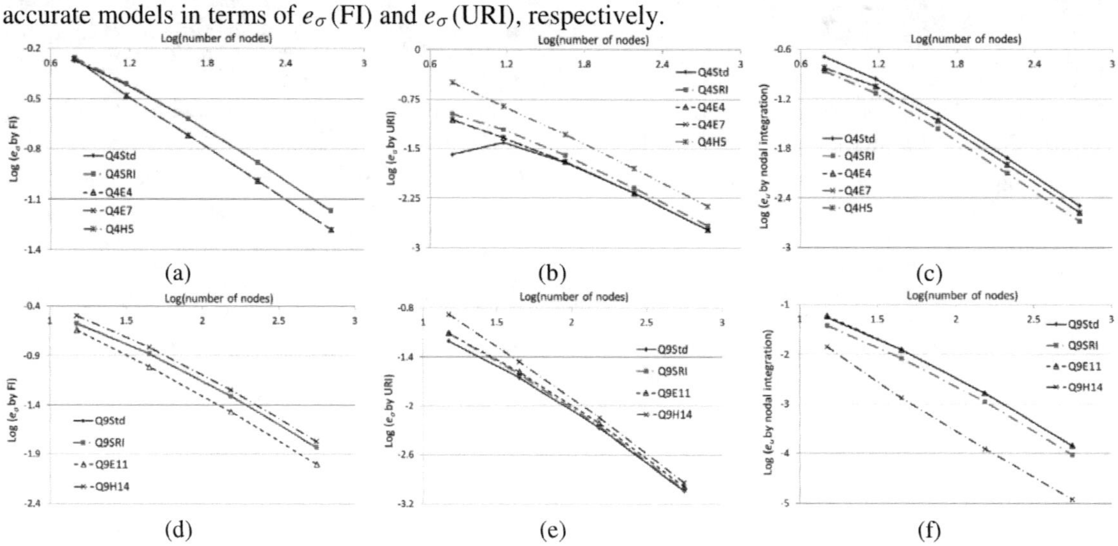

Figure 9. The relative errors e_σ and e_u for the thick-wall cylinder modelled by $2n$xn elements with Poisson's ratio υ= 0.25. (a)-(c): e_σ (FI), e_σ (URI) and e_u for the Q4 models; (d)-(f): e_σ (FI), e_σ (URI) and e_u for the Q9 models.

For v = 0.4999, the standard models are again locked. Among the advanced Q4s, Q4SRI is the most accurate one in terms of e_u and e_σ (URI). Among the advanced Q9s, Q9H14 yields the best displacement prediction but is worse than Q9E11 in terms of e_σ (FI). Q9SRI, Q9E11 and Q9H14 give almost the same e_σ (URI).

6.5 Square panel loaded by a double-sinusoidal body force

This example considers a unit plane strain square panel subjected to the double-sinusoudal body force

$$f = \begin{Bmatrix} f_x \\ f_y \end{Bmatrix} = \begin{Bmatrix} \frac{9}{2}\pi^2(2\mu+\lambda)sin(\frac{3}{2}\pi x)cos(\frac{3}{2}\pi y) \\ \frac{9}{2}\pi^2(2\mu+\lambda)\cos(\frac{3}{2}\pi x)\sin(\frac{3}{2}\pi y) \end{Bmatrix}. \quad (23)$$

The boundary conditions for the square panel are:

At $x = 0$: $u_x = 0; f_y = 0$; At $y = 0$: $u_y = 0; f_x = 0$; At $x = 1$: $u_y = 0; f_x = 0$; At $y = 1$: $u_x = 0; f_y = 0$.
When the Poisson's ratio approaches 0.5, λ and thus the body force would be very large. To avoid the large force, only $v = 0.25$ is considered. The analytical displacement solution for this problem is:

$$\begin{Bmatrix} u_x \\ u_y \end{Bmatrix} = \begin{Bmatrix} \sin(\frac{3}{2}\pi x)\cos(\frac{3}{2}\pi y) \\ \cos(\frac{3}{2}\pi x)\sin(\frac{3}{2}\pi y) \end{Bmatrix}. \quad (24)$$

The square panel is initially modelled by 1 element and then is divided into $n \times n$ regular elements. The results are plotted in **Figure 10**. Q4Std is the most accurate Q4 element in terms of e_u and e_σ (FI) while the EAS and HS elements are better in terms of e_σ (URI). For Q9s, Q9Std is the most accurate one in terms of e_u and e_σ (FI). Noticeably, Q9Std is only marginally less accurate than Q9E11 in e_σ (URI).

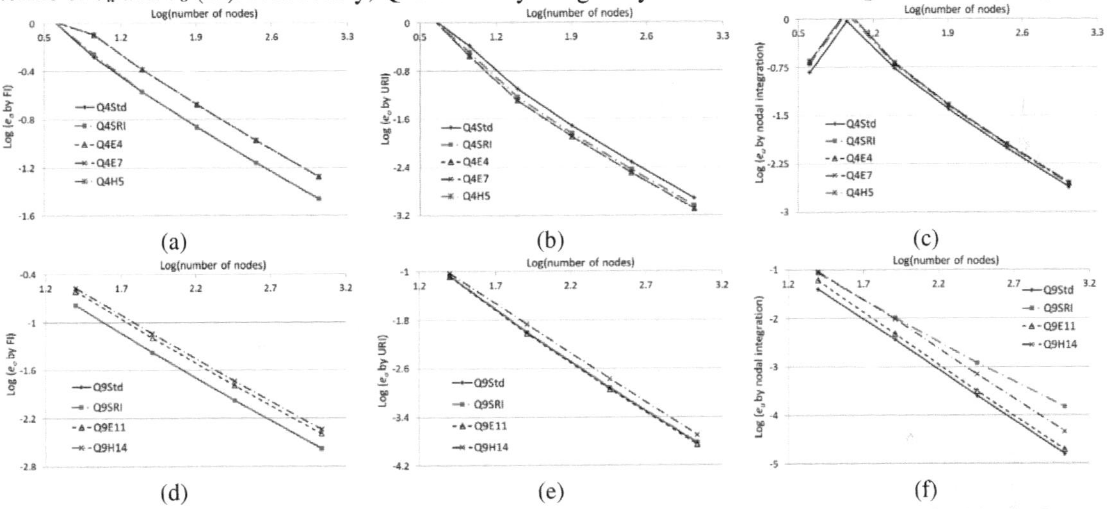

Figure 10. The relative errors e_σ and e_u for the square panel loaded by a double-sinusoidal body force modelled by $n \times n$ elements with with Poisson's ratio v= 0.25. (a)-(c): e_σ (FI), e_σ (URI) and e_u for the Q4 models; (d)-(f): e_σ (FI), e_σ (URI) and e_u for the Q9 models.

7. Discussion

From the above examples, the following features are remarked:

1) Since the EAS and hybrid models are devised with due consideration to the bending modes, they behave well in bending examples (Section 6.1- 6.2) and are more accurate than the standard ones in these cases in terms of both e_u and e_σ.
2) In examples where the bending modes are not dominating, the advantages of the advanced models are not so obvious. For an example, the HS models usually give accurate predictions in terms of e_u but its predictions are not so good in terms of e_σ. In Section 6.5, the standard elements even give more accurate predictions than those of the advanced elements in terms of both e_u and e_σ.
3) The standard elements suffer from both the dilatational and trapezoidal locking. Thus, the superiority of the advanced elements is most obvious in problems prone to locking.
4) SRI is a simple scheme to improve the standard element. When bending deformation dominates, the present SRI scheme are less effective than EAS and HS. Otherwise, SRI are competitive among the advanced formulations. The bending response of the considered SRI can be further improved by computing the element strain in a local Cartesian coordinate and applying reduced integration to the shear energy defined with respect to the local coordinates.

5) The advanced formulations are more effective for improving the lower order elements. The accuracy improvements in the Q4 elements are more prominent than those in the Q9 elements.

8. Closure

In developing the advanced finite elements, they are often demonstrated to be more accurate than their standard counterparts by popular numerical tests in which displacement and stress predictions at a few selected points are compared. In this paper, a series of the numerical tests are conducted to assess not only the standard elements but also the advanced elements including those formulated by SRI, EAS and hybrid-stress methods in terms of the relative errors in displacement and energy norms. The results reveal that the common perception is not always true. While the advanced elements show significant improvement in accuracy over the standard ones in problems where the bending deformation and/or locking dominate, there are also examples in which the standard elements give better predictions than the advanced counterparts. Thus, more careful assessment may be necessary during developing novel advanced finite element models, especially when practical application scenarios need to be considered.

References

[1] Malkus D S & Hughes T J. Mixed finite element methods-reduced and selective integration techniques: a unification of concepts. Computer Methods in Applied Mechanics and Engineering, 1978, 15, 63-81.
[2] Simo J C & Rifai M S. A class of mixed assumed strain methods and the method of incompatible modes. International journal for numerical methods in engineering, 1990, 29, 1595-1638.
[3] Pian T H H & Sumihara K. Rational approach for assumed stress finite elements. International Journal for Numerical Methods in Engineering, 1984, 20, 1685-1695.
[4] Sze K Y. A novel approach for devising higher-order hybrid elements. International journal for numerical methods in engineering, 1993, 36, 3303-3316.
[5] Belytschko T & Bachrach W E. Efficient implementation of quadrilaterals with high coarse-mesh accuracy. Computer Methods in Applied Mechanics and Engineering, 1986, 54, 279-301.
[6] Macneal R H & Harder R L. A proposed standard set of problems to test finite element accuracy. Finite elements in analysis and design, 1985, 1, 3-20.
[7] Zienkiewicz O C, Taylar R L, Zhu J Z. The Finite Element Method: its basis and fundamentals (6th Edition), Elsevier Butterworth-Heinemann, 2005.
[8] Bathe K J. Finite element procedures (2nd Edition), Watertown, 2014 and Higher Eduacation Press, China, 2016
[9] Andelfinger U & Ramm E. EAS-elements for two-dimensional, three-dimensional, plate and shell structures and their equivalence to HR-elements. International Journal for Numerical Methods in Engineering, 1993, 36, 1311-1337.
[10] Bischoff M & Ramm E. Shear deformable shell elements for large strains and rotations. International Journal for Numerical Methods in Engineering, 1997, 40, 4427-4449.
[11] Bischoff M, Ramm E, Braess D. A class of equivalent enhanced assumed strain and hybrid stress finite elements. Computational Mechanics, 1999, 22, 443-449.
[12] Yeo S T & Lee B C. Equivalence between enhanced assumed strain method and assumed stress hybrid method based on the Hellinger–Reissner principle. International Journal for Numerical Methods in Engineering, 1996, 39, 3083-3099.
[13] Timoshenko S P & Goodier J N. Theory of elasticity, McGrawHill, New York, 1987.

Acknowledgments

The authors acknowledge the support of Natural Science Foundation of Jiangsu Province (BK20150479, BK20170521), Jiangsu Natural Science Foundation for Universities and Colleges (17KJB130006), the support of Entrepreneurship and Innovation Doctor Program in Jiangsu Province and Start-up Foundation of Harbin Institute of Technology, Shenzhen.

Absorptive protective padding with electrospinned polyester fibers with hydrogel matrix

C. B. Soh[1], X. H. Lee[1], Samuel W. Y. Lim[1], N. F. B. Sipalah[2], Paul Ho[3]

[1]Singapore Institute of Technology, 10 Dover Drive, Singapore 138683
[2] Newcastle University Upon Tyne, Newcastle upon Tyne NE1 7RU, United Kingdom
[3]G4 Pte Ltd 5 Elias Green #10-03 Singapore 519961

E-mail: Chewbeng.Soh@SingaporeTech.edu.sg

Abstract. Hip fractures has been prevalent especially among aged elderly. This can lead to permanent disability of the elderly as they are wheel-chaired bound after suffering from hip fractures. In this work, we explore the design of a protective hip pad which encapsulate the hydrogel and fibres composite in a laminated sheet. The hydrogel was fabricated using Polyvinyl alcohol (PVA) with distilled water. Two types of electrospinned fibres, Nylon-6 fibres and polyester fibres were used as filler. The fabrication of hydrogel matrix composite was varied by the weight percent of nylon 6,6 pellets dissolving in formic acid as compared to Hydrolysed Poly(Vinyl alcohol) to form crosslinked PVA with water. The impact absorption concept was to allow the gel-like hydrogel to flow through the fibres pores and during this process, part of the energy would be dissipated in the matrix composite. This lower the impact force being transferred to the user's hip and prevent the facture of the elderly's hips during fall. The protective padding with 20 layers of polyester fibres with 2.5wt % of PVA had the lowest impact force of 0.8 kN as compared to conventional protective padding with measured 1.07kN impact force for impact study.

1. Introduction

Falls are considered a major public health problem in the world. The total number of hip fractures are expected to be more common among elderly woman due to a disease known as osteoporosis which weakens the bone structure and caused them to fracture more easily [1]. As such, the World Health Organisation (WHO) has positioned falls as the second leading cause of accidental or unintentional injury deaths worldwide. Typically a hip fracture occurs after a lateral fall. Majority of the elderly is unable to break the fall and this result in the transfer of the primary impact during a fall onto the great trochanter of the proximal femur. [2]

There are various strategies to target those who are at risk of a hip fracture due to a fall which included wearing of hip protectors, floor mats and other former of early detection of fall and sensory devices. [3] Hip protectors can be widely found and there are many different designs available commercially. Each design had also claimed its own advantage. However, one commonality amongst the typical hip protector design was that it is incorporated with a pair of protective padding that prevents all the fall impact to be directly transferred to the hip. The protective padding generally comes in the form of hard-shell or soft-shell where the hard-shell type is typically used to shunt energy and the soft-shell type substantially reduced the total impact force during simulated sideway falls via energy absorption. [4] However, there has been report of inadequate validation of existing protector due to the lack of standardization of test system. [5]

Polyvinyl alcohol has been utilized for preparation of hydrogel by physical cross-linking mechanism due to its non-toxicity and biocompatibility. [6] Varying freeze/ thaw cycle can be applied to attain different viscosity level based on the degree of cross-linking [7]. Hence in this work, we will like to explore the use of hydrogel based matrix with electrospinned fibers as a filler to strengthen the polymeric composite for use in soft protector pad. Impact test will be carried out to test the absorbance nature of the hip protector pad using the drop tower configuration. [5] The rationale for using the following material combination is to benefit from both energy absorbing nature of the cross-linked PVA composite matrix and high tensile properties of the porous electrospinned nylon fibers.

2. Experimental Details

2.1 Materials used

Hydrogels are polymer materials which is made up of three-dimensional network that exhibits high moisture retention property with proper encapsulation. In this work, we have explored the use of Hydrolyzed Poly(Vinyl alcohol) from Sigma Aldrich for making of hydrogel. Different weight ratio of PVA is mixed with deionized water which serve as solvent. The PVA will have to undergo heat treatment at 95°C for 1hr and metallic rolling for another 15 min before undergoing the freeze-thaw cycle.[8] As most hydrogels are extremely soft material with a modulus of 0.1MPa, fillers can be added to strengthen the composite. [9] For this reason, in our study two types of materials are explored, (i) commercial continuous filament polyester and (ii) electrospinned nylon 6 fibers. The fabrics of polyester has a thickness of 0.1mm and size of 22.5 cm by 22.5 cm. The electrospinned nylon 6 fibers on the otherhand is obtained by dissolving the Nylon6, 6 pellets with 1,1,1,3,3,3-Hexafluoro-2-propanol, 99% (HFP) from Matrix Scientific. Nylon 6,6 is a semi-crystalline polyamide material which exhibits moisture retention property of up to 8.5% of its mass. Water serves as a solvent to alter the nylon 6,6 tensile properties.

2.2 Sample Preparation

Figure 1a shows the Electrospinning system used to spin nylon 6 fibers at a high potential with source solution prepared with Nylon 6,6 pellets and HFP. The fibres were collected on a planar Aluminium surface collector in the system. The flow rate, distance of separation and potential between the needle tip of the syringe head with the collector stage can be adjusted. The heat treated gel-like PVA solution is prepared in a metal tray as shown in Figure 1b before placing it into the fridge for freeze-thaw cycles. The freeze-thaw cycles and duration was varied so as to verify the effectiveness of cross-linked formation between adjacent porous polymeric chains in the viscous hydrogel on the overall composite

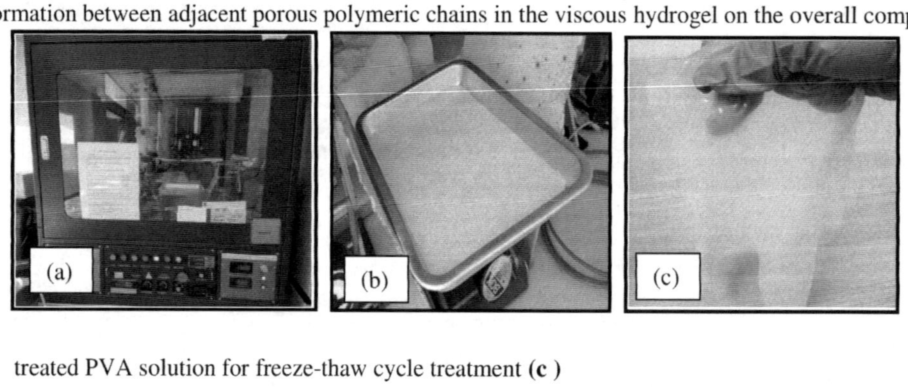

treated PVA solution for freeze-thaw cycle treatment **(c)** performance as a protector pad. [8]. Figure 1c shows an image of the gel-like viscous hydrogel obtained after 2 freeze-thaw processing steps.

Once the fibers and fabric are prepared, they are infused into the gel-like hydrogel by layering process. Part of the viscous hydrogel is first poured into the tray before placing the nylon fibers or the polyester fabric. The remaining of the hydrogel will then be poured in the tray once the fibers are completely immersed in the hydrogel. Figure 2a shows the step to fold up layers of the hydrogel infused nylon fibers or polyester fabric. The hydrogel matrix composite with the fibers are then heat sealed in thermoplastic polyurethane film to form the protector pad.

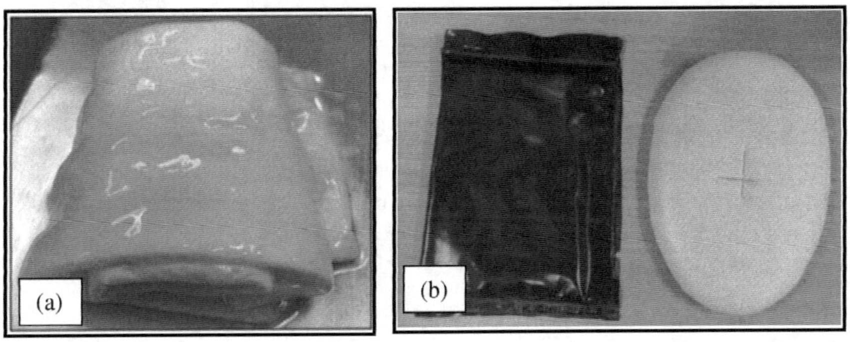

Figure 2(a) Folded layers of polyester matrix infused in hydrogel matrix (b) customized composite heated sealed in thermoplastic polyurethane film (Left) and commercial hip protector pad (Right).

Air void formation can be avoided in the process by ensuring that layers of polyester fabric is completely wetted with gel-like viscous hydrogel layer (which has been thawed for 30min). The layers of gel-like hydrogel during the thaw cycle can be sliced out in accordance with the dimension of the hip pad to be heated sealed in a stretchable thermoplastic polyurethane film. An image of the sealed composite in polyurethane film (left) with the commercial protector pad (right) is as shown in Fig. 2b. Thermoplastic polyurethane film is chosen as the outer layer of the protector pad as it is elastic, and resistance to oil, grease and abrasion

3. Results and Discussion

3.1 Morphology of the electrospinned fibers and commercial polyester fabrics

To have a better understanding of the structure and the dimension of these fibres, SEM imaging is carried out for the electrospinned fibers generated at different condition as well as fibers embedded in the hydrogel.

3.1.1 Different concentration used for the Nylon 6,6 pellets with formic acid

Figure 3 shows the comparison of the electrospinned fibers using formic acid with varying flow rates. When the fibers are spunned at a 10 wt % of Nylon 6 and formic acid, the electrospinned fibers are very fine (from 5nm to 200nm in dimension as shown inset of Fig. 3a). The network of fibers are also prone to beads formation as illustrated in Fig. 3a. This is undesirable as it can impedes the flow of the hydrogel through the fibers during impact and they are too fine for ease of the gel to flow through, resulting in poor energy absorption. Figure 3b shows the electrospinned fibers at 15 wt % of formic acid which results in thicker bead-less fibers network. It could be seen that the fibres thickness of the fibers are much thicker (between 0.5μm to 2μm) but with poor uniformity. At some of the regions, the fibers are flatten which can result in poor fibre integrity and prone to fiber fracture during load impact.

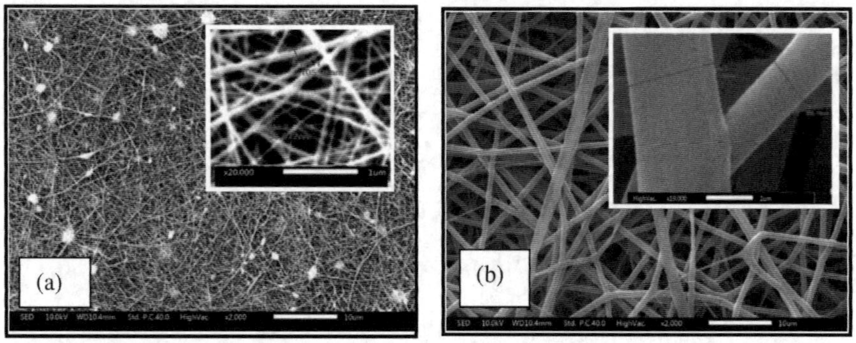

Figure 3 Electrospinned Nylon 6 fibers in formic acid with (a) 10 wt % of nylon 6,6 pellet which results in fine fibers (b) 15 wt % of nylon 6,6 pellet which results in network of fibers with thickness varying from 500nm to 2μm

3.1.2 Solution of Nylon 6,6 pellets with HFP for electrospinning as compared to commercial polyester fibres fabric

The positive effect of using Formic acid and mixtures of formic acid/dichloromethane, such as clear dissolution of polyamide-6 has allowed for subsequent electrospinning to obtain fibers of high aspect ratio [10]. However, the network of main fibers structures with varying thickness and existence of flatten fibers is not ideal for our work. The existence of a finer Nylon 6 mesh with high aspect ratio structure may hinder the flow of hydrogel into the mesh. There is also the risk of incomplete evaporation of formic acid, leading to defective fibers [11]. Due to the challenge in obtaining well-aligned, uniform thickness of electrospinned nylon 6 fibers with formic acid as the solvent, we have attempt to use Hexafluoro 2 propanol (HFP) instead [12,13].

Figure 4 SEM image of **(a)** electrospinned nylon with 15 wt % nylon 6,6 in Hexafluoro 2 propanol solvent (with inset of showing the fibers with thickness of ~800 nm and fragment of the nylon 6 fabric) and **(b)** Commercial polyester fibers with fiber thickness of approximately 2μm to 3 μm with inset of its fabric liner.

Figure 4(a) shows the electrospinned fibres using Nylon-6 with HFP solvent. The dimension of the fiber ranges from 0.7 μm to 0.9 μm which is fairly well-controlled in terms of its physical dimension. By comparing it with the polyester fibres, it could be seen that the polyester fibres were more spaced apart with a thickness of about 2 to 3 μm as shown in Fig. 4(b). As such, there were be greater amount of voids and gaps for the hydrogel to pass through the network of fibers of the polyester fabric. Since there was more bond stretching for polyester, more energy was absorbed and hence the impact force transferred to the user would be much lower.

3.1.3 Embedded structure of the fibers and fabrics in hydrogel

Figure 5 shows the embedded structure of both types of fibers with hydrogel. Hydrogel has been known to have autonomous self-healing properties with diverse reversible non-convalent bonding [14], hydrophobic interaction [15] and π-π stacking [16]. In our stack embedded structure as shared in Figure 2a under sample preparation, the fibers strengthened the hydrogel matrix and prevent them from undergoing matrix deformation. Figure 5a shows that the infused hydrogel matrix are coated into patches of the fiber. The strand of the fiber appears to have broaden due to its interaction with the hydrogel matrix. It is likely that the hydrophilic Nylon 6 fiber absorbed the moisture content and interact with the gel to form a lump of composite at some of the regions. On the otherhand, for the polyester fibre, they appear to have retained its structure which is affirmed by their hydrophobic nature [17]. The hydrogel matrix forms a web-like structure sticking onto some parts of the hydrophobic polyester fiber. Due to the bigger voids between the polyester fibers, this enables the hydrogel cross-linked matrix to be compressed in the fabric and force to the side of the encapsulated thermoplastic polyurethane film. This mechanism differ from the nylon fiber interaction with hydrogel where the matrix soaked nylon 6 fibers will be squeeze out from the nylon 6 fabric under the dynamic impact.

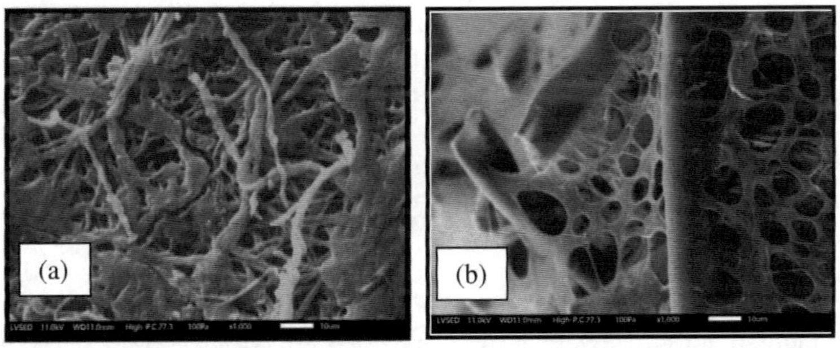

Figure 5 SEM image of the embedded **(a)** Nylon 6 fibers **(b)** polyester fabric in hydrogel matrix

3.2 Dynamic Impact study

The fabric of embedded nylon 6 fiber and the polyester fiber are then encapsulated in the thermoplastic polyurethane film. Dynamic impact test is then done on both samples as well as the commercial hip pad. Figure 6a shows the impact test system which is used for our study. The impact load is 15kg dropped from a height of 20 cm. The commercial hip pad is placed on the stage as shown and the impact force is determined by the load cell. The dwell time is measured from the fast speed camera connected at its side which triggered and measure the relative position of the load cell to the stage.

IOP Conf. Series: Materials Science and Engineering **744** (2020) 012017 doi:10.1088/1757-899X/744/1/012017

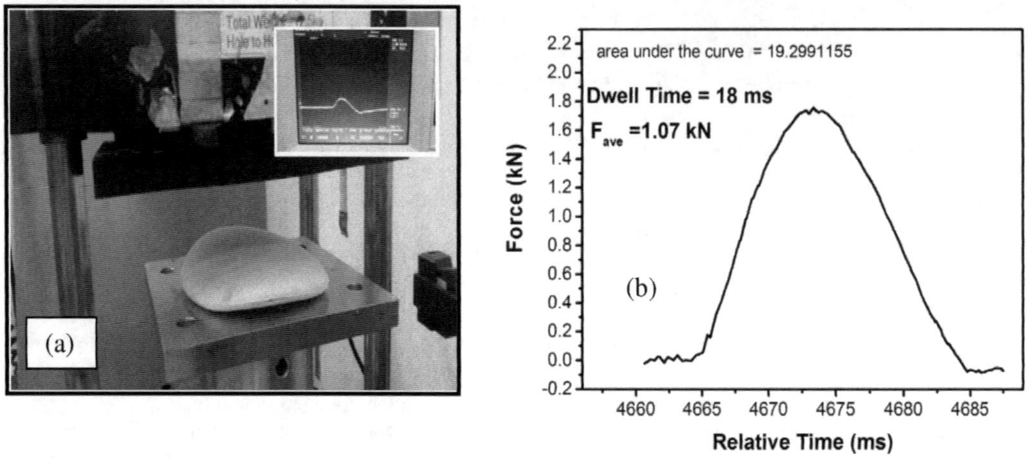

Figure 6(a) Mechanical impact drop test on the commercial protective padding (b) Plot of the the force versus the relative time to determine the average impact force with the dwell time

The samples were subjected to a mechanical impact drop test to find out its impact force and dwell time. This test was done as to simulate the amount of force being absorbed by the protective padding samples and the remaining force being transferred to the user during a fall. The drop test consists of a 15kg load cell that would be dropped from a height of 0.2m onto the test samples to obtain the results. Figure 6(a) shows the drop test being conducted on the commercial protective padding. The load cell (in black) covers the test area of all the test samples' sufficiently so that full area of the pad can be used for absorbance. The maximum amount of gravitational potential energy would be converted to work done as the load cell drops from a height and comes in contact to the test sample. Work done is related to force and distance. Hence, the remaining amount of force after it has been absorbed by the test pad samples would be transferred to the load cell and hence be detected on the monitoring system. The dwell time indicates the period of time when the system load remains in physical contact with the sample. The area under the force time graph was determined to obtain the average force of impact, F_{ave} by dividing the total integrated area over the dwell time.

Table I shows the tabulation for the average force and dwell time for the commercial pad, Nylon 6 fiber with hydrogel and lastly polyester fiber with hydrogel. The results show that the conventional

Table 1. Tabulation of the average force and dwell time

	Materials	Average force, F_{ave}	Dwell Time
Sample A	commercial pad with foams	1.07 kN	18 ms
Sample B	Nylon 6 fiber filler in hydrogel matrix	0.85 kN	20.5 ms
Sample C	polyester fiber filler with hydrogel matrix	0.80 kN	18 ms

Commercial pad has a highest average force of 1.07kN as compared to the two samples with the hydrogel matrix. The polyester fiber with hydrogel matrix, Sample C gives the lowest value of average impact force. This could be attributed to the larger gap between the lower density polyester fiber which allows the greater ease of viscous hydrogel to infuse into it. With the impact force, it is harder for the more dense Nylon 6 fiber to allow the hydrogel to diffuse through. For this reason, Sample B has higher average impact force as compared to Sample pad C. However, the advantage of having B is that the dwell time is much longer which allows the impact force to be better dissipated. This is possible due to its hydrophilic nature of the nylon 6 fiber which allows the absorbed water and hydrogel to be gradually squeeze out from the fiber. This also account for Sample B to be having the highest dwelling time of 20.5 ms.

4. Conclusion

The viscosity of the hydrogel, the nature and the thickness of the fibers can affect the impact absorbent properties of the pad. The generation of the nylon 6 fiber by electrospinning allows the user to gather the fibers to be infused with hydrogel. The hydrophilic nature of the nylon 6 fibers has the advantage over the polyester fiber in that it can soaked up more hydrogel which dispense the gel when a higher load is applied. This called for further work on optimizing the nature of the fiber by varying the freeze-thaw cycle and its duration and possibly different weight composition of the Nylon 6,6 is a semi-crystalline polyamide with solvent water. The impact test affirmed the effectiveness of the hydrophilic nature of nylon 6 fiber to absorb the hydrogel so as to infuse into the fabric for effective impact absorbance.

5. References

[1] Kannus P, Parkkari J, Sievänen H, Heinonen A, Vuori I, Järvinen M, 1996 Epidemiology of hip fractures. Bone, 18(1), 57S-63S

[2] Parkkari J, Kannus P, Palvanen M, Natri A, Vainio J, Aho H, and Järvinen M 1999 Majority of hip fractures occur as a result of a fall and impact on the greater trochanter of the femur: A prospective controlled hip fracture study with 206 consecutive patients. *Calcified Tissue International*, 65(3), 183–187.

[3] Li N, Tsuushima E, Tsuushima H, 2013 Comparison of impact force attenuation by various combinations of hip protector and flooring material using a simplified fall-impact simulation device *Journal of Biomechanics,* 46, 1140-1146.

[4] Laing Andrew C, Robinovitch S, 2008 Effect of soft shell hip protectors on pressure distribution to the hip during sideways falls *Osteoporosis International* 19(7), 1067-75.

[5] Yahaya S A, Ripin Z M, Ridzwan M J Z, 2019 Test Systems for biomechanical evaluation of hip portectors: a systematic review, *Osteoporosis Intern*ational, https://doi.org/10.1007/s00198-019-05128-x

[6] Lu J, Wang T, Drzal L T, 2008 Preparation and properties of microfibrillated cellulose polyvinyl alcohol composite materials Compos Part A: Appl Sci Manuf 39(5), 738-46

[7] Mahanta N, Teow Y, Valiyaveettil S, 2013 Viscoelastic hydrogel from (polyvinyl)alcohol –Fe (III) complex , Biomater Sci i(5), 519-27.

[8] Buldwa S J, Boere K W M, Dijkstra P J, Vernmonden T, Hennink W E, 2014 Hydrogels in historical perspective: form simple network to smart materials, J. Control, Release 190, 254-73

[9] Gupta G, Gupta A, Dhanola A, Raturi A, 2016 Mechnaical behaviour of glass fiber polyester hybrid composite filled with natural fillers IOP Conf Sereis: Mat Sci and Eng 149, 012091.

[10] Bazbouz M B, Stylios G K, 2008 Alignment and Optimization of Nylon 6 Nanofibers by Electrospinning, J. Appl. Polymer Sci, 107, p3023-3032

[11] Nirmala R, Pant H R, Park S-J, Kim H Y, 2010 Effect of solvents on high aspect ratio polyamide-6 nanofibers via electrospinning, Macromolecular Research 18(8): 759-765.

[12] Hopkins A R, Rasmussen P G, Basheer R A, 1996 Characterization of Solution and Solid State Properties of Undoped and Doped Polyanilines Processed from Hexafluoro-2-propanol, *Macromolecules* 29(24), 7838-7846

[13] Gururajan G, Sullivan SP, Beebe TP, Chase DB, Rabolt JF, 2011 Continuous electrospinning of polymer nanofibers of Nylon-6 using an atomic force microscope tip, Nanoscale 3(8), 3300-8.

[14] Zhu D, Ye Q, Lu X, Lu Q , 2015 Self-healing polymers with PEG oligomer side chains based on multiple H-bonding and adhesion properties, Polym Chem, 6, 5086-92.

[15] Tuncaboylu D C, Sari M, Oppermann W, Okay O, Tough and self-healing hydrogel formed via hydrophobic interaction, Macromolecules 44, 4997-5005.

[16] Xu Y, Wu Q, Sun Y, Bai H, Shi G, 2010 Three dimensional self-assembly of graphene oxide and DNA into multi-funcitonal hydrogel, ACS Nano 4(12), 7358-62

[17] Azeem M, Wiener J, Khan M Z, 2018 Hydrophobic analysis of nano-filament polyester fabric, Fibres and Textiles, (1), 8-12.

[18] Moutos F T, Freed L E, Guilak F, 2007 A biomimetic three dimensional wove composite scaffold for functional tissue engineering of cartilage, Nat Mater, 6(2), 162.

Comparison of Tensile and Impact Absorption properties of Bio-Inspired Helicoidal stacked with Cross-Ply Stacked Carbon Fiber Laminate

Samuel W.Y. Lim[1], Soh Chew Beng[1#], Shawn Lim Jian Hui[2], Paul Ho[3]

1 Singapore Institute of Technology, 10 Dover Drive, Singapore 138683
2 Newcastle University Upon Tyne, Newcastle upon Tyne NE1 7RU, United Kingdom
3 G4 Pte Ltd 5 Elias Green #10-03 Singapore 519961
E-mail: ChewBeng.Soh@Singaporetech.edu.sg

Abstract. Herrigbone structures are identified on the chitins of the Mantis shrimp. Bio-inspired helicoidal arrangement of composite layers can mimic these herringbone structures. First, the carbon composite samples are fabricated using the hand lay-up method, followed by a curing process using a hot press system. The impact absorption properties of these samples will be analyzed through a series of impact tests. The samples will also be tested for its tensile properties to compare the effect of plies arranged in helicoidal [0,45,90], [0,30,60,90] and Cross-ply orientations to form the laminate structure. The results showed that helicoidal samples have higher penetration tolerance and toughness as compared to Cross-ply samples during impact tests. It is found that the specimen with helicoidal stacking sequence has better debonding resistance and improved damage tolerance. The 6-ply helicoidal samples exhibits higher tensile strength as compared to 6-ply Cross-ply samples due to better fiber to matrix adhesion for helicoidal structure in the direction of pulling force.

1. Introduction

Mantis shrimp is a type of crustacean that has a hard-shelled body and a heavily mineralized dactyl club which it uses to attack its preys. [1] Its dactyl club and telson have shown itself to be useful tools for survival. The chitinous organic matrix of its dactyl club can withstand recurring high velocity impact forces that last in milliseconds without catastrophic failure. [2,3,4] Therefore, the mantis shrimp is a formidable hunter and fighter not to be trifled with. [5] Another interesting point to note is the telson of the mantis shrimp. During ritualized fighting, the mantis shrimp uses its telson as a defense mechanism against the extreme impact force inflicted by the appendages of the perpetuator mantis shrimp. [5,6,7] The fact that the dactyl club and telson can take this extreme impact without breaking has inspired researchers into developing and strengthening armor that mimics the structures of probably nature's most efficient force absorbers/dissipaters. [3,6,9] In a research article by Weaver et.al and others, the bio-structure on the dactyl

IOP Conf. Series: Materials Science and Engineering **744** (2020) 012018 doi:10.1088/1757-899X/744/1/012018

club is shown to be fibrillar and helicoidal in nature and consists of multiphase composite of oriented hydroxyapatite and amorphous calcium phosphate and carbonate. [1,3,4] Hence in the process of this study, we aimed to mimic these structures by fabricating helicoidal oriented carbon fiber composites.

2. Materials and Methods

2.1 Materials

The hot press system used for this work is supplied by LMS Technologies Pte Ltd. With this system, it assures the uniformity of the carbon fiber composite sample surface during fabrication. This system also allows for variations of temperature, pressure and duration during hot press process. The helicoidal sample is first layered by hand lay-up method with a mixture of epoxy resin to hardener in a ratio of 5:3. before the hot press process is carried out

Plain weave carbon fiber fabrics and epoxy mix are supplied by Wee Tee Tong Chemicals Pte Ltd. Carbon fiber is a brittle material known to exhibit high tensile properties. [10,11] When infused with epoxy as the matrix, the end-product will be a carbon fiber composite with enhanced material properties.

2.2 Methods for Testing

The tensile test system from Shimadzu as shown in Figure 1a can measure tensile strength of up to 20KN with a measurement accuracy of up to 0.5%. The composite samples are cut to the required dimensions by waterjet cutting which have minimal defect causation compared to other suitable cutting methods. [12] The dimensions of the tab composite samples prepared are with adherence to ASTM D5083 standard. [13] For impact test, the semi-spherical steel impactor made of high strength steel alloy with diameter of 5cm is held by an electromagnet at a height of 1m as shown in figure 1b. The impact test results in insignificant deformation of the impactor relative to the CFRP plate. The 8-ply sample thickness for impact test is compliant to ASTM D7136 standard. [14] Both tensile and impact tests on the samples will be carried out with increasing tensile stress and impact load till they reach their point of failure or when visual defects are observed.

Figure 1 (a) Tensile machine (b) Impact Test System

2.3 Carbon Fiber Composite Fabrication

Three different orientations are explored in the making of the composite samples for this work. Their orientations are (i) Cross-ply and helicoidal stacked arrangement at (ii) $[0^{\circ},45^{\circ}, 90^{\circ}]_2$ and (iii) $[0^{\circ},30^{\circ}, 60^{\circ}, 90^{\circ}]_2$. These samples will be used for tensile tests and impact tests to observe the correlations between fiber orientation, tensile property and durability.

Table 1. Number of plies corresponding to sample thickness

No. of Plies	Thickness(mm)
4	0.73
6	1.04
8	1.46

The number of plies shows a direct proportional increase in sample thickness from 4 to 8 plies.

Table 2. Parameters for Hot Pressing

Parameters	Value &Units
Temperature	80°C
Pressure	30Bar
Duration	3600 Seconds

The parameters of hot press curing were kept constant throughout the study as tabulated in Table 2. For hand lay-up method, the two caul plates applied with mold release wax were used to sandwich the prepared layers before hot pressing as shown in Figure 2.

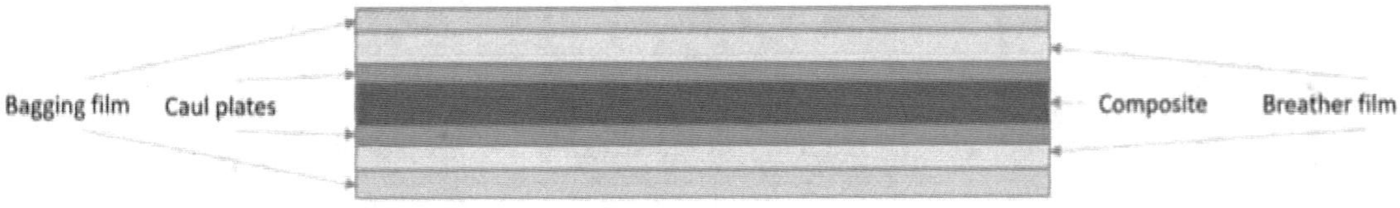

Figure 2 Schematic diagram of prepared layers by Hand Lay-up method

3. Results and Discussion

3.1 Drop test results on the Cross-ply and Helicoidal samples

The impact test results show an increasing trend of resilience to deformation with increase in number of layers. The nature of resilience to deformation is drastic especially when the number of layers is increased from 6 to 8. Helicoidal samples showed better toughness and penetration resistance as compared to 0°/90° samples. The reason for this behavior is due to the different fiber orientations which prevent continuous matrix cracking that has occurred on Cross-ply samples. Therefore, helicoidal orientation allows for optimal dissipation of energy in every fiber direction [15]

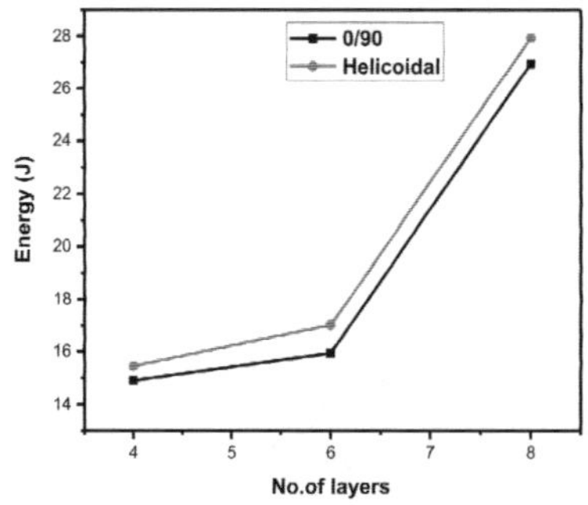

Figure 3 Impact test results for amount of impact energy for sample failure

3.2 Sample Failure Analysis

3.2.1 Impact tested samples

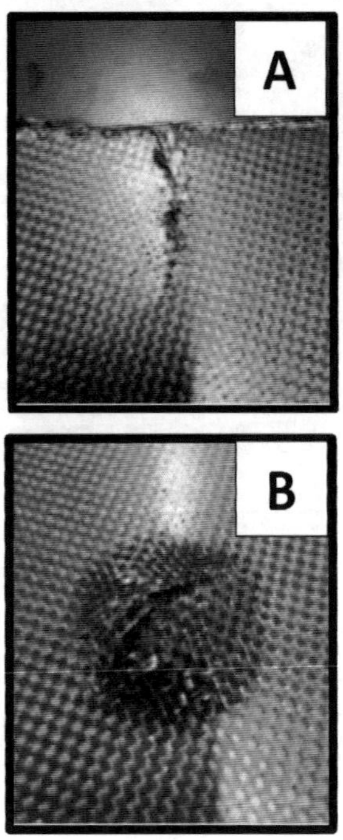

Figure 4 (A) 4 ply failure in 0°/90° orientation and (B) 6 ply

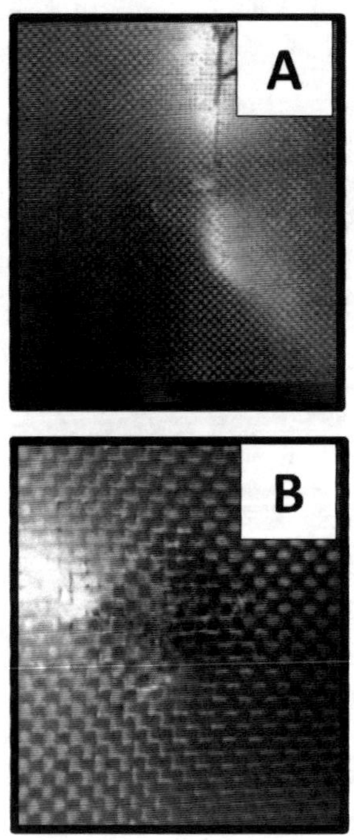

Figure 5 (A) 6 ply failure in 0°/90° orientation and (B) 6 ply helicoidal orientation

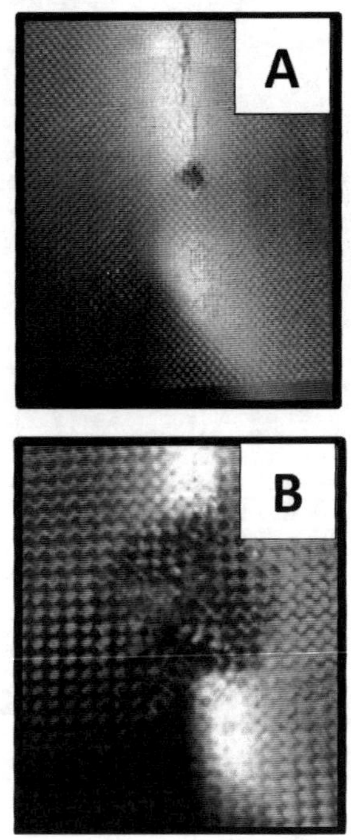

*Figur*e 6 (A)8 ply failure in 0°/90° orientation and (B) 6 ply helicoidal orientation

4 ply and 6 ply samples are used for tensile tests and impact tests in order to find the correlation between number of layers to toughness and different angular orientation. Based on Figure 4 to 6, crack propagation is observed on Cross-ply samples while global indentations are more prominent on helicoidal samples. The reason for the crack propagation to occur on the Cross-ply sample is due to the impact wave being able to readily traverse along the plane. Whereas for helicoidal samples, the orientation of the fibers allows for a larger surface area per unit crack length in the main direction of projectile impact [15]. The high shear resistance from the difference in minute angle between laminas can disrupt the flow of impact wave along a plane and instead evenly distributes the impact force propagation along planar orientations of the helicoidal arrangement. [1,15]

3.2.2 Tensile Test Results

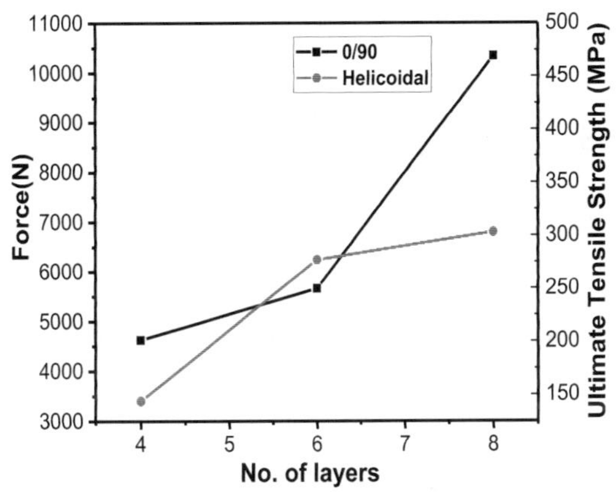

Figure 7 Plot of Tensile force applied and ultimate tensile strength until sample failure

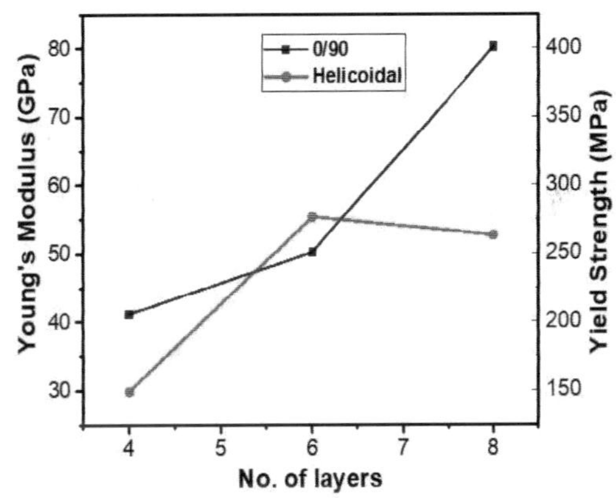

Figure 8 Young's Modulus and Yield Strength until sample failure

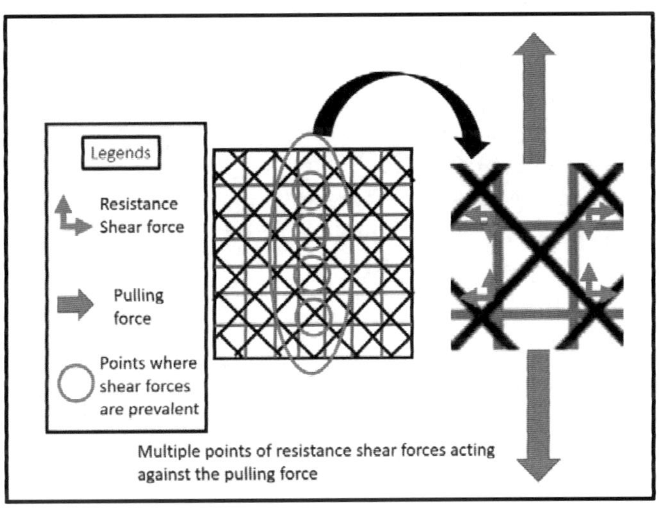

Figure 9 Schematic diagram showing multiple points of resistance

It is clear from the results that Cross-ply samples is better than helicoidal sample in terms of tensile strength. However, it is an exception for 6 ply composites; $[0^{\circ},45^{\circ}, 90^{\circ}]_2$. 2 layers of carbon fiber fabrics are pointing 45° away from the direction of 90°. The two plies of 45° deviation carbon fiber complements the shear strength and tensile properties of the epoxy matrix as the sample is being pulled in the longitudinal direction. It prevents the phenomenon of carbon fibre pullout from the epoxy matrix which can occur in the case when the fibers are arranged in Cross-ply orientation. The helicodial orientation of the fibers samples enables it to exhibit higher shear resistance between laminas [16]. The shear resistance to delamination between the orientated laminas by the epoxy interaction with the 45° orientated fibers give rise to higher ultimate tensile strength as compared to Cross-ply samples as shown in Figure 9. During the tensile test on helicoidal samples, delamination begins to occur first at the mid-plane where there are higher traverse shear stresses. The drastic failure suggested by *J.L.Liu et al.* occurs when the fiber failure meets the delamination point at the mid-plane.[17] From Figures 7 and 8, the increase in the tensile force and the ultimate tensile strength is gradual when the plies were increased from 4 to 6 plies but more significant when it increases to 8 plies.

An observation from the failure analysis of the carbon fiber helicoidal composite that undergoes the tensile test is that there is little delamination, but more signs of drastic fiber breakage. The reason was that the epoxy matrix can adhere to the carbon fiber well, and distribution of force can be consistent throughout the sample during pulling. This optimizes the tensile performance of the helicoidal composite. However, the carbon fiber will reach its breaking point first as it has higher Young's Modulus compared to epoxy. This also suggests that the tensile strength of the sample is based largely on the carbon fiber instead of the epoxy matrix. An observation of failure for cross-ply samples is that there are many signs of delamination traverse to the pulling force after sample failure. Cracking of the epoxy matrix already begins to form at an early stage of the tensile test before the pull out phenomenal of carbon fiber takes place [18,19,20].

More studies can be carried out on the correlations of orientations using different angles in helicoidal samples to better optimize the tensile force transition from one fiber orientation to another. This may allow the elastic properties of both carbon fiber and epoxy matrix to match closely, thereby reducing the chances of delamination between each layer [15].

4. Conclusion

From the results, we can identify that helicoidal orientations of different step angles can further enhance tensile performance of the carbon fiber composite as seen in Figures 7 and 8. Although samples with helicoidal orientations have inferior tensile properties, impact absorption properties are enhanced, and crack propagation is significantly reduced as compared to Cross-ply samples. Helicoidal samples, when compared to Cross-ply samples, have the potential to exhibit higher energy absorption threshold during tensile pulling when accompanied with appropriate step angles that are close to the pulling force.

5. Reference

[1] J.C. Weaver, G.W. Milliron, A. Miserez, K. Evans-Lutterodt, S. Herrera, I. Gallana, et al., 2012 The stomatopod dactyl club: a formidable damage-tolerant biological hammer *Science*, 336, 1275-1280.

[2] Patek S.N[1], Caldwell R.L, 2005 Extreme impact and cavitation forces of a biological hammer: strike forces of the peacock mantis shrimp Odontodactylus scyllarus, *J Exp Biol.* 208(Pt 19):3655-64

[3] S.N. Palek, W.L Korff, R.L. Caldwell, 2004 Biomechanics: Deadly strike mechanism of a mantis shrimp, *Nat Commun,* 428, 819-820

[4] Dr. Robert J. Full, Roy L. Caldwell, Shong W. Chow, 1989 Smashing Energetics: Prey Selection and Feeding Efficiency of the Stomatopod, *Gonodactylus bredini, Ethology,* 81(2), 134-147

[5] Yaraghi, Adwait A. Trikanad, David Restrepo, Caldwell, Pablo D. Zavattieri, David Kisailus et al., 2019 The Stomatopod Telson: Convergent Evolution in the Development of a Biological Shield *Adv. Funct. Mater.* 29, 1902238

[6] Douglas J. Emlen, 2008 The Evolution of Animal Weapons *Annu. Rev. Ecol. Evol. Syst* 39, 387–413

[7] P. A. Green and S. N. Patek, 2015 Contests with deadly weapons: telson sparring in mantis shrimp (Stomatopoda) *Biol Lett.* 11(9): 20150558

[8] J. R. A. Taylor and S. N., Patek Ritualized fighting and biological armor: the impact mechanics of the mantis shrimp's telson, *The Journal of Experimental Biology* 213, 3496-3504

[9] Liang Dong, Haydn Wadley, 2015, Mechanical properties of carbon fiber composite octet-truss lattice structures, *Composites Science and Technology* 119, 26-33

[10] Toshiya Kamae, Lawrence T. Drzal,2012, Carbon fiber/epoxy composite property enhancement through incorporation of carbon nanotubes at the fiber–matrix interphase – Part I: The development of carbon nanotube coated carbon fibers and the evaluation of their adhesion, *Composites: Part A* 43 1569–1577

[11] D.K.Shanmugam, S.H.Masood, 2009, An investigation on kerf characteristics in abrasive waterjet cutting of layered composites *Journal of Materials Processing Technology* 209(8), 3887-3893

[12] A. Riccio, A. De Luca, G. Di Felice, F. Caputo, 2014, Modelling the simulation of impact induced damage onset and evolution in composites, *Composites: Part B* 66, 340–347

[13] T. Apichattrabrut & K. Ravi-Chandar, 2006, Helicoidal Composites, *Mechanics of Advanced Materials and Structures*, 13:1, 61-76

[14] B.W. Kim, A.H. Mayer, 2003, Influence of fiber direction and mixed-mode ratio on delamination fracture toughness of carbon/epoxy laminates, *Compos Sci Technol,* 62(5), 695-713

[15] SHOUKAI WANG & D.D.L. Chung,1998 Interlaminar shear in carbon fiber polymer-matrix composites, studied by measuring the contact electrical resistance of the interlaminar interface during shear, *Composite Interfaces*, 6:6, 507-517

[16] J.L. Liu, H.P. Lee, V.B.C. Tan, 2018, Failure mechanisms in bioinspired helicoidal laminates, *Composites Science and Technology*, 157, 99-106

[17] Suparerk Sirivedin, Seog Young Han and Kwan Soo Lee, 2007, *Journal of Mechanical Science and* Technology, 21, 2023-2030

[18] Ijan Mohammadi, Mohammad Abbaszadeh & Alireza Keshmiri, 2018, Variational approach development in analysis of matrix cracking and induced delamination of cross-ply composite laminates subjected to in-plane shear loading, Mechanics of Advanced Materials and Structures, 25:6, 481-499

[19] Yoon-Ji Yim, Kyong-Min Bae, Soo-Jin Park, 2018, Influence of Oxyfluorination on Geometrical Pull-Out Behavior of Carbon-Fiber-Reinforced Epoxy Matrix Composites, *Macromolecular Research 26(9), 794*

Review of the Surface Treatment s of Plant Fibers as Reinforcements in Polymer Composites

Z H Zhu 1, Y Zhang1, N Zhang 1, T Wang 1 and M Y Hao 2
1 School of advanced manufacturing technology, Guangdong Mechanical & Electrical Polytechnic, Guangzhou, China;
2 The Key Laboratory of Polymer Processing Engineering, Ministry of Education, South China University of Technology, Guangzhou, China
Email: 184823174@qq.com

Abstract. The poor compatibility between plant fiber and polymer matrix leads to many defects and deficiencies in plant fiber reinforced polymer composites. Various common physical or chemical treatments of plant fibers are listed in this paper. Surface treatments can improve the interface bonding of fibers and matrix, whereas the mechanical properties of plant fibers will be greatly reduced after treatments. How to retain the properties of plant fibers to the greatest extent and improve the comprehensive properties of plant fibers reinforced polymer composites is a problem that needs attention and urgent solution when treating plant fibers by various treatments.

1. Introduction

The research of natural fiber reinforced PLA composites has been a hot spot in the field of polymer composites in recent decades. Compared with traditional composites, natural fiber reinforced PLA composites have advantages of environmental friendliness and reproducibility. [1-2] The greatest challenge in preparing natural fibers reinforced PLA composites is their large variation in properties and characteristics. [3] Various kinds of methods for pretreating of fibers had been used to improve the interface bonding of matrix and fibers in the previous research. [4-7] Though these methods improved the interfacial adhesion of fibers and matrix, they eliminated some components and destroyed the construction of fibers, which declined the fiber strength and the mechanical properties of composites couldn't be enhanced as expected. In our previous work, we compounded alkali-treated SFs (ASFs) with untreated SF (USFs) to prepare hybrid SFs (HSFs) which were then filled into PLA matrix to prepare PLA/HSF composites. It was found that HSFs could improve mechanical properties of PLA composites more significantly than ASFs and USFs. In this work, we give a further study of the dynamic mechanical properties analysis (DMA) of the PLA, PLA/ASF and PLA/HSF composite.

IOP Conf. Series: Materials Science and Engineering **744** (2020) 012019 doi:10.1088/1757-899X/744/1/012019

2. Treatments of plant fibers

The essential problem of the treatments of plant fiber is how to reduce the hydrophilicity of plant fiber and thus improve the interface property of the composites. [2] At present, the commonly used method of surface treatments of plant fibers mainly includes physical treatments and chemical treatments.

2.1 Physical treatments

Physical treatments improve the structure and surface morphology of plant fibers, which doesn't change the chemical composition of the plant fibers. After physical treatments, the surface roughness of plant fiber increases, the contact area between plant fiber and polymer matrix increases, and the physical adhesion between plant fiber and polymer matrix is effectively improved. [3] Common physical treatments mainly include heat treatment, discharge treatment, steam explosion treatment and so on.

2.1.1 Heat treatment

The presence of a large number of hydroxyl groups (-oh) makes the plant fiber have a strong water absorption and a high water content, so that the polymer matrix and the plant fiber cannot combine well. Heat treatment is the most direct method to reduce the moisture content of plant fiber. Etrissans et al. [4] studied the water absorption of wood fiber after heat treatment, and found that the contact angle of wood fiber changed after heat treatment, resulting in decreased water absorption and increased hydrophobicity, which could increase the interfacial compatibility between wood fiber and polymer matrix. Giebeler studied the influence of heat treatment on the mechanical properties and morphology of wood fiber, and found that the bending strength and tensile strength of wood fiber significantly decreased during the heat treatment of 180-200℃. Heat treatment reduced the moisture content of wood fiber, increased the insulation of wood fiber, and deepened the color of wood fiber. [5] Pétrissans et al. studied the water absorption of wood fiber after heat treatment, and found that the change of contact angle of wood fiber after heat treatment resulted in the decrease of water absorption and increase of hydrophobicity, which could increase the interfacial compatibility between wood fiber and polymer matrix. [6] Ayadi studied the resistance to ultraviolet light of wood fiber after heat treatment, and found that heat treatment could improve the resistance to ultraviolet light of wood fiber and increase the stability of wood fiber. [7]

2.1.2 Discharge technology treatment

Discharge technology treatment refers to the use of electron beam to treat plant fiber to improve the surface morphology and structure of the treatment. Among these treatments, surface corona treatment is commonly used, which is one of the effective methods to change the surface characteristics and improve the surface morphology and structure of plant fibers. Kan et al. treated the natural fiber with low-temperature plasma and studied the change of fiber surface morphology before and after treatment with scanning electron microscope. It was found that the surface roughness of natural fibers increased after low-temperature plasma treatment, and the degree of fiber orientation along the axis increased with the treatment time increasing. [8] Khan et al. studied the influence of ultraviolet irradiation on the mechanical properties of sisal fibers and found that the tensile strength and elongation at break of sisal fibers increased after ultraviolet irradiation. [9] Li et al. studied the effect of electron beam irradiation on mechanical properties of sisal fiber reinforced polycaprolactone (PCL) composites with different contents, and found that the mechanical properties of composites were the best when the SF content of sisal fiber (SF) was 45% and the irradiation dose was 150 kGy. [10]

2.1.3 Steam blasting treatment

Steam blasting treatment refers to the process of cooling plant fibers after being treated under high temperature and pressure for a certain period of time under the action of steam. After steam explosion treatment, the cell walls of plant fibers are broken, lignin and hemicellulose structures are destroyed and removed, and thus increasing the solvent accessibility of cellulose. The contact area between plant fiber and polymer matrix also increases at the same time. [11] Qu prepared the composite materials with Cotton bast fiber and LDPE with steam blasting and studied the mechanical properties of the composites. It was found that the mechanical properties of the composites were better than those of the untreated Cotton bast fiber and pure LDPE. Qu pointed out that the surface roughness of the cotton-skin fiber increased after steam explosion treatment, which increased the contact area between the fiber and LDPE matrix and thus improved the mechanical properties of the composites. [12] Kaushik et al. prepared biodegradable composites by blending Wheat straw fiber treated by steam blasting and Starch plastics, and found that the crystallinity of the fibers after steam blasting was 46.8% higher than that before treatment. [13]

2.2 Chemical treatment

Chemical treatment is the improvement the properties or structure of plant fibers, so as to improve the interfacial compatibility of plant fiber and polymer matrix, and make plant fiber distributed more evenly in the composites, eventually make plant fiber reinforced polymer composites effectively. The common chemical include alkali treatment, surface grafting, interfacial coupling and acetylation.

2.2.1 Alkali treatment

Alkali treatment is a commonly used treatment for chemical treatments of plant fibers, and also used for other pre. After alkali treatment, most of the lignin hemicellulose and pectin were effectively removed, the length-diameter ratio of the fiber was improved, and the surface roughness of the fiber was increased. The concentration, temperature, treatment time and the additives in the treatment solution or process can all affect the effect of alkali treatment. Xia et al. studied the changes in the microstructure, thermal properties and tensile properties of cotton fibers treated with alkali, and found that the tensile strength, tensile modulus and crystallinity of fibers treated with alkali improved. [14] Kabir et al. found that the tensile strength and tensile modulus of Kenaf fiber after alkali treatment were significantly improved. [15] Iannace et al. compared the changes of sisal fibers before and after alkali treatment in the mixing process, and found that the degree of fibrillation of sisal fibers with alkali treatment increased, and the length of sisal fibers with alkali treatment decreased, and the ratio of length to diameter increased. [16] Mishra et al. treated sisal fiber with a lower concentration (mass fraction 5%) NaOH solution for 1 h, and found that the surface roughness of sisal fiber increased, the tensile strength and tensile modulus decreased, while the elongation at break did not change significantly. [17]

2.2.2 Surface grafting

Surface grafting is a very effective treatments technology to change the surface properties of plant fibers, which introduces alkyl, acetyl, benzoyl, and other groups on the surface of plant fibers to improve the hydrophilic and hydrophobic surface properties of plant fibers. The surface grafting can be carried out not only before the compounding, but also during the blending process. Acrylonitrile (AN), maleic anhydride (MAH) and polylactic acid (PLA) are commonly used graft reaction monomers. Mishra et al. studied the mechanical properties of cyanoethylated sisal fiber reinforced polyethylene terephthalate (PET) composites and found that the tensile properties, bending properties and impact properties of cyanoethylated composites were improved compared with alkali treatment. [18] Chuai et al. grafted onto the surface of wood fiber with maleic anhydride (MAH) to improve the interfacial compatibility between wood fiber and polypropylene (PP) matrix. It was found that the machinability and mechanical properties of the composites were all improved. [19]

2.2.3 Interfacial conjugation

Interfacial conjugation is the process of treating plant fibers with coupling agent and thus change their surface properties and improve their interfacial compatibility with polymer matrix.

The commonly used coupling agents include silane coupling agent, titanate coupling agent, etc. Before the coupling agent treatment, alkali solution is generally used to pretreat plant fibers to increase the accessibility of cellulose. Liu et al. modified wood fiber with silane coupling agent and then mixed it with PVC to prepare composites. Study found that silane coupling agent enhanced the interaction between wood fiber and PVC matrix, and the interface strength were improved obviously. [20] Liao et al. found that titanate coupling agent could improve the interfacial bounding between wood fiber and Linear Low Density Polyethylene (LLDPE) [21]. Huda et al. prepared the PLA/KF compositeS and studied their mechanical properties. It was found that the hemp fiber treated with alkali solution and coupling agent had the best interfacial compatibility with the PLA matrix, and the corresponding composite had the best performance. [22]

2.2.4 Acetylation

Hydroxyl group (-oh) can react with acetic anhydride (ch3-co-o-co-ch3) or enone (-c =C=O)

$$R - OH + CH_3 - CO - O - CO - CH_3 \rightarrow R - O - CO - CH_3 + CH_3COOH \quad (1)$$

Chen et al. [18] pointed out that acetylation treatment can reduce the hydrophilicity of plant fibers effectively, and thus improve the interfacial compatibility of fibers and matrix. Wei et al. treated poplar fiber with Acetic acid, Benzoic acid and Propionic acid respectively and blended them with high-density polyethylene. It was found that the thermal stability and tensile strength of the composites reinforced with acetylated poplar fiber were improved, and the tensile strength of the composites treated with benzoic acid acetylated was higher than the other two composites, among which the tensile strength of the composites treated with acetic acid increased the least. [23]

Due to the complexity of plant fiber, the single physical or chemical treatment of plant fiber is often unable to achieve a better treatment effect. At present, processing technologies are often combined with various physical or chemical processing technologies to modify plant fibers to achieve ideal treatment effect. For example, the alkali treatment process is often used for the pretreatment of other processes to increase the accessibility of cellulose in plant fibers. Heat treatment is often used as the final treatment process before blending plant fibers with polymer matrix to remove moisture from plant fibers.

3. Conclusion

From the above physical or chemical treatment, it can be found that these surface can improve the surface characteristics of plant fibers and improve the interface performance between plant fibers and polymer matrix. However, the mechanical properties of plant fibers will be greatly reduced after treatments, which will be induced which will affect the comprehensive properties of the composites.

Therefore, the treatments of plant fiber becomes valuable only when the improvement of interface property is greater than the decrease of the property of plant fibers. In the following studies, how to retain the properties of plant fibers to the greatest extent and improve the comprehensive properties of plant fibers reinforced polymer composites is a problem that needs attention and urgent solution when treating plant fibers by various .

Reference

[1] Rout J, Misra M, Tripathy S S, et al. The influence of fiber surface treatments on the mechanical properties of coir-polyester composites. Polymer Composites, 2001, 22(4):468-476

[2] Rout J, Misra M, Tripathy S S, et al. The influence of fiber surface treatments on the mechanical properties of coir-polyester composites. Polymer Composites, 2001, 22(4):468-476

[3] Kim T J, Lee Y M, Im S S. The preparation and characteristics of low-density polyethylene composites containing cellulose treated with cellulase. Polymer Composites, 1997, 18(3):273-282

[4] Nair S S, Hurley D C, Wang S, et al. Nanoscale characterization of interphase properties in maleated polypropylene - treated natural fiber - reinforced polymer composites. Polymer Engineering & Science, 2013, 53(4): 888-896.

[5] Giebeler E, Bluhm B, Alscher A, et al. Process for the treatments of wood: U.S. Patent 4, 377, 040. 1983-3-22

[6] Petrissans M, Gerardin P, El Bakali I, et al. Wettability of heat-treated wood. Holzforschung, 2003, 57(3):301-307

[7] Ayadi N, Lejeune F, Charrier F, et al. Color stability of heat-treated wood during artificial weathering. Holz Als Roh-Und Werkstoff, 2003, 61(3):221-226

[8] Kan C W, Yuen C W M. Influence of Low Temperature Plasma Treatment on the Properties of Tencel and Viscose Rayon Fibers. Ieee Transactions On Plasma Science, 2009, 37(8):1615-1619

[9] Khan M A, Hassan M M, Ara J, et al. Surface treatments of sisal (Agavae sisalana) fiber by photocuring: Effect of additives. Polymer-Plastics Technology and Engineering, 2007, 46(5):447-453

[10] Li W, Qiao X, Sun K, et al. Effect of electron beam irradiation on the silk fibroin fiber/poly(ε-caprolactone) composite. Journal of Applied Polymer Science, 2009, 113(2):1063-1069

[11] Garcia-Jaldon C, Dupeyre D, Vignon M R. Fibres from semi-retted hemp bundles by steam explosion treatment. Biomass & Bioenergy, 1998, 14(3):251-260

[12] J P Qu, S X Hu, Y H Feng, et al. Effect of steam explosion pretreatment on mechanical properties of LDPE/ cotton-skin fiber composites. Plastics industry, 2009, 37(2):32-35

[13] Kaushik A, Singh M, Verma G. Green nanocomposites based on thermoplastic starch and steam exploded cellulose nanofibrils from wheat straw. Carbohydrate Polymers, 2010, 82(2):337-345

[14] Xia G, Sadanand V, Ashok B, et al. Preparation and Properties of Cellulose/Waste Leather Buff Biocomposites. International Journal of Polymer Analysis & Characterization, 2015, 20(8):693-703

[15] Kabir M M, Wang H, Lau K T, et al. Chemical on plant-based natural fibre reinforced polymer composites: An overview. Composites Part B-Engineering, 2012, 43(7SI):2883-2892

[16] Iannace S, Ali R, Nicolais L. Effect of processing conditions on dimensions of sisal fibers in thermoplastic biodegradable composites. Journal of Applied Polymer Science, 2001, 79(6):1084-1091

[17] Mishra S, Misra M, Tripathy S S, et al. The influence of chemical surface treatments on the performance of sisal-polyester biocomposites. Polymer Composites, 2002, 23(2):164-170

[18] C Z Chuai, S li, Y Q Lin. Study on compatibility and properties of polypropylene/graft wood fiber composites. Chinese Plastics, 2000(5):25-30

[19] Y Q Liu, Z M Zhao. Wood - plastic composites and their development. New Chemical Materials, 2005, 33(3):59-61

[20] B Liao, Y H Huang, M C Chen. Effect of modified wood fiber on mechanical properties of LDPE and wood fiber composites. Polymer Materials Science and Engineering, 1999, 15(3):124-126

[21] Huda M S, Drzal L T, Mohanty A K, et al. Effect of fiber surface-treatments on the properties of laminated biocomposites from poly(lactic acid) (PLA) and kenaf fibers. Composites Science and Technology, 2008, 68(2):424-432

[22] Y R Chen, L M Xia, M B Wu, et al. Advances in pretreatments of plant cellulose materials., 1999, 18(4):24-26

[23] Wei L, McDonald A G, Freitag C, et al. Effects of wood fiber esterification on properties, weatherability and biodurability of wood plastic composites. Polymer Degradation and Stability, 2013, 98(7):1348-1361

Acknowledge

IOP Conf. Series: Materials Science and Engineering **744** (2020) 012019 doi:10.1088/1757-899X/744/1/012019

This research was supported by Key Research Platforms and Projects of Universities of Guangdong Province (grant no. 2018GKTSCX111) and Funding Project for Scientific Research of High-level Talents of Guangdong Mechanical & Electrical Polytechnic (grant no. Gccrcxm-201903).

IOP Conf. Series: Materials Science and Engineering **744** (2020) 012020 doi:10.1088/1757-899X/744/1/012020

Simple fabrication of magnetically separable ZnO-based photocatalyst nanocomposites

ThianKhoon Tan[1,*], PoiSim Khiew[1], WeeSiong Chiu[2] and ChinHua Chia[3]

[1] Engineering Foundation, Faculty of Science and Engineering, University of Nottingham Malaysia, Jalan Broga, 43500 Semenyih, Selangor Darul Ehsan, Malaysia
[2] Low Dimensional Materials Research Center, Department of Physics, Faculty of Science, Universiti Malaya, 50603 Kuala Lumpur, Malaysia
[3] School of Applied Physics Studies, Faculty of Science & Technology, Universiti Kebangsaan Malaysia, 43600 Bangi, Selangor Darul Ehsan, Malaysia
[*] E-mail: TK.Tan@nottingham.edu.my

Abstract. Magnetically recyclable ZnO/Fe_3O_4 nanocomposite photocatalyst was successfully fabricated through simple mechanical mixing and low calcination temperature. The nanocomposite was characterised by XRD, FESEM, TEM and EDX analysis. The magnetic nanocomposite optimisation, such as photocatalyst loading, doping amount of Fe_3O_4 nanoparticle into the nanocomposite and sintering temperature were conducted. The optimum loading amount of the ZnO/Fe_3O_4 nanocomposite was 9.0 wt%. The most favourable amount for Fe_3O_4 doped was 10 wt% and 300 °C was the value for the sintering temperature. The ZnO/Fe_3O_4 nanocomposite was successfully extracted by applying magnetic field and able to be reused with no significant change arises in the degradation efficiency.

1. Introduction

Water contamination has brought considerable threat to public health and ecosystems. This is due to the increasing demand of the toxic organic dyes by various industries such as textile, jute, paint and pharmaceuticals where they discharged their effluent into the environment without proper processes (Mandal, Dutta et al. 2019). There was report of more than 50% of current synthetic dyes are insusceptible to natural degradation (Muhd Julkapli et al. 2014). The accumulation of wastewater from these industries will definitely lead to significant damage that brought serious threat to the environment.

Current synthesis techniques such as green chemistry processes has been a highly promising technique due to its simple, mild chemical reaction and no toxic chemicals used. Lately, nano-sized semiconductors have been monopolised as photocatalytic oxidation processes due to their non-toxic and chemical properties (Fresno, Portela et al. 2014). There were also remarkable studies on the utilisation of visible light to successfully degrade variety of organic dyes in wastewater (Anwer, Mahmood et al. 2019). The frequently studied photocatalytic semiconductors are titanium dioxide (TiO_2) and zinc oxide (ZnO). Recent research report ZnO shows better photodegradation performance for some dyes in aqueous solutions (Tudose and Suchea 2016). The performance of current photocatalysts are closely associated with low utilisation of visible light, wide energy bandgap and the duration of the electron-hole separation efficiency (Ge, Zhang et al 2019). While these major obstacles are constantly studied, to achieve high photocatalytic efficiency, another problem arises after the water was treated through photocatalytic processes. As a consequences, a simple and efficient technique to separate and recycle the photocatalyst was by slotting magnetic materials into the photocatalyst (Kim, Woo et al. 2016). Fe_3O_4 was adopted in many magnetic photocatalysts due to its good magnetic properties, low cost and environmental friendly nature (Wu, Chen et al. 2019).

There have been extensive studies on various preparation methods in obtaining a remarkable and highly effective ZnO/Fe_3O_4 nanopowder. Wang et al. (2016) synthesised ZnO/Fe_3O_4 core/shell nanoparticle by simple two-step chemical method (Wang, Yang et al. 2016) whereas Nguyen et al. (2015) prepared ZnO/Fe_3O_4/CS magnetic nanocomposite with combination of ZnO nanoparticles and chitosan by co-precipitation method (Nguyen, Nguyen et al 2015). Another group of researcher have synthesized Fe_3O_4/CuO/ZnO/NGP using a sol-gel method followed by hydrothermal method for photodegradation of MB (Tju, Taufik et al. 2016). After thorough investigation, this study aims to present a cost effective method to synthesise ZnO based magnetic nanocomposites. After scrutinising few factors affecting photocatalytic efficiency, the separable ZnO/Fe_3O_4 magnetic nanocomposite photocatalyst was effectively degrade 97.3% of MB at 12 mg·L^{-1} concentration.

2. Experimental

All the materials used in this study were of reagent grades. Zinc oxide (ZnO, R & M Chemicals), methylene blue (MB, R & M Chemicals) and iron oxide (Fe_3O_4, Systerm, ChemPur) were used directly without further purification. All solutions were prepared with Favorit Water Distiller deionised water.

In the preparation of ZnO/Fe_3O_4 magnetic nanocomposites photocatalyst materials, ZnO nanopowders was weighed followed by weighing of the Fe_3O_4 nanopowders. A different weight percentage (wt%) of Fe_3O_4 was prepared, i.e. 1, 5, 10, 12, 15 and 20 wt%. They were then placed in a porcelain bowl and were dry mixed for about 20 minutes. These premixed nanocomposites powders were then calcined for 5 hours for a set of sintering temperature, i.e. 200, 250, 300, 350, 400 and 450 ^{o}C, in the muffle furnace from Carbolite. The samples were calcined at the rate of 20 C^{o} per minute. After calcined, the process was completed by grinding all samples with the micro spatula. This method is generally used to prepare mixed oxide catalyst material (Sun, Meng et al. 2019).

The resulting magnetic nanocomposites were confirmed by X-ray diffraction (XRD, Bruker) and the field emission scanning electron microscopy (FESEM, FEI) coupled with energy dispersive X-ray analyser (EDX). The morphologies were characterised by transmission electron microscope (TEM, LEO) and the specific surface area (BET) by using Micromeritics.

Distilled water was used to prepare various loading of the magnetic nanocomposites solutions, i.e. 1.0 wt% to 10.0 wt%, whereas MB dye in the range of 2.00 ppm to 12.0 ppm. 0.20 mL of the nanocomposites were immersed into 3.00 ml of the organic dye solution. These samples were placed in the UV chamber (Uvitec, CL-508.G) for various durations, at 10 minutes interval, to observe the capability of the nanocomposites to photodegrade the dye (Figure 1, Tan, Khiew et al. 2019).

After the irradiation, each nanocomposites samples were collected from the solutions by applying a magnetic field (0.6 T). A 2 mL aliquot of all samples were taken and placed into the UV-Visible Spectrophotometer (Varian, Cary 50). The scan rate was 120 nm/min through wavelength of 200 nm to 800 nm, where their absorption maximum was analysed and studied.

The Beer-Lambert law was applied when analysing the absorbance calibration curve of all the solutions (Mantele and Denize 2017). Therefore, the decolourisation efficiency was calculated by the equation: Percentage degradation (%) = $[(C_o - C_t)/C_o] \times 100 = [(A_o - A_t)/A_o] \times 100$, where C_o and A_o are the dye concentration and the absorbance of the dye before irradiation, whereas C_t and A_t are the dye concentration and the absorbance of the dye after the irradiation of time t.

IOP Conf. Series: Materials Science and Engineering **744** (2020) 012020 doi:10.1088/1757-899X/744/1/012020

3. Results and Discussion

3.1. Samples Morphologies for Different Sintering Temperature
This experiment was conducted on the ZnO/Fe_3O_4 nanocomposites for the 20 wt% doped Fe_3O_4.

3.1.1. FESEM Analysis. The prepared samples were sintered at 200, 250, 300, 350, 400 and 450 °C, respectively, for a duration of 5 hours. The FESEM micrographs of ZnO/Fe_3O_4 nanocomposites sintered at 300 °C is shown in Figure 1(a). The micrograph shows mixture of different particle size of nanoparticles but evenly distributed across the samples. In addition, all samples display moderate porosity which will contribute significantly to the surface area to initiate the degradation process.

Overall, the small spherical shapes particles are Fe_3O_4 particles whereas the medium and large rod-like and cubic shapes are ZnO particles. The particles surfaces are quite smooth, which shows no sign of over stressed was subjected onto the powder during mixing. The average particle size is (164 ± 12) nm for 300 °C sintering temperature. The results show no significant impact on the particle size for these range of sintering temperature.

3.1.2. XRD Analysis. Figure 1(b) shows XRD patterns for nanocomposites sintered at 200, 300 and 400 °C sintering temperatures. Samples main peaks at 31.8° of (100), 34.4° for (002) and 36.3° for (101) planes for ZnO, and at 30.2° for (220), 35.6° for (311) and 43.2° for (400) planes for Fe_3O_4 crystals, coincide with the hexagonal wurtzite phase for ZnO (space group $P6_3mc$; JCPDS 36-1451) (Nandi and Das 2019), and cubic single phase structure for Fe_3O_4 (face-centered cubic; JCPDS 82-1533) (Guy, Atacan et al. 2018). All samples' peaks were quite similar with no unfamiliarity on the crystallinity of the nanocomposites. By using Scherer equation, the average crystallite size was calculated as (121 ± 13) nm for the 300 °C sintering temperature.

3.1.3. BET and EDX Analysis. The applied sintering temperature between 200 to 450 °C on the ZnO/Fe_3O_4 nanocomposites did not affecting the microstructure of the nanocomposites since the values averages for BET surface area and pore size were between 1.9 to 2.6 m^2/g. On the other hand, the EDX spectrums of ZnO/Fe_3O_4 nanocomposites for 300 and 400 °C sintering temperature acquired the elemental map which conclude Fe_3O_4 nanoparticles were deposited on the surface of the ZnO/Fe_3O_4 nanocomposites (Amornpitoksuk, Suwanboon et al 2018). Thus, the XRD results conformed with the EDX that high crystallinity single phase structure was form.

3.2. Samples Morphologies when Doped with Different wt% of Fe_3O_4
This experiment was conducted on the ZnO/Fe_3O_4 nanocomposites sintered at 300 °C.

3.2.1. FESEM Analysis. Figure 1(c) shows the micrographs for the ZnO/Fe_3O_4 nanocomposite samples with 10 weight percent (wt%) of Fe_3O_4. The image shows mixtures of obviously many different particles of various sizes but evenly scattered all around. The small particles are spherical and slightly mixed with elongated structures, while the larger particles are of cubic and rod shapes. The particles are fairly distributed without detectable agglomeration but high porosity was seen all around. The average particle size for 10 wt% doped Fe_3O_4 is (159 ± 12) nm.

3.2.2. XRD Analysis. Figure 1(d) is the XRD results for the ZnO/Fe_3O_4 nanocomposites with 20, 10 and 1 weight percent (wt%) of Fe_3O_4 nanoparticles. The diffraction main peaks at 31.8° of (100), 34.4° for (002) and 36.3° for (101) planes of wurtzite crystal structure for ZnO standard JCPDS (No. 36-1451), whereas peaks at 30.2° for (220), 35.6° for (311) and 43.2° for (400) planes of cubic single phase structure for Fe_3O_4 crystals for JCPDS (No. 82-1533). These basically explained that ZnO and Fe_3O_4 nanoparticles samples are well mixed by the dry mixing. The average crystallite size as calculated by using the Scherer equation is (137 ± 15) nm for the 10 wt% of doped Fe_3O_4 nanocomposite sample.

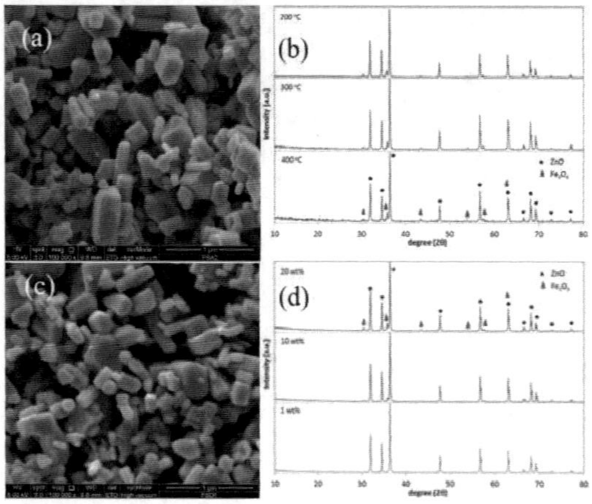

Figure 1. FESEM microstructure and XRD for ZnO/Fe_3O_4 nanocomposites.

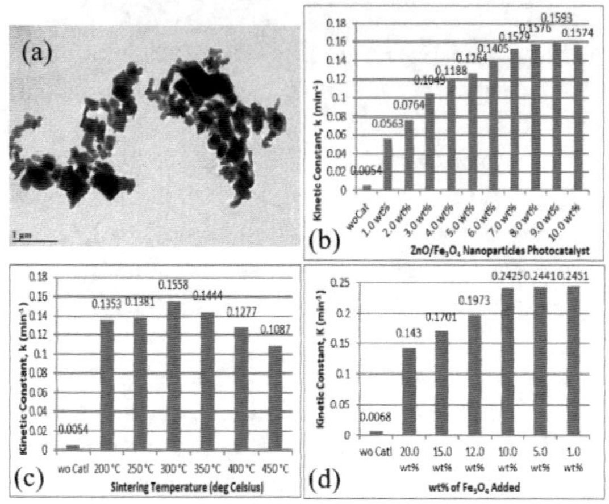

Figure 2. TEM micrograph and kinetic constant for ZnO/Fe_3O_4 nanocomposites samples.

3.2.3. BET and EDX Analysis. BET and pore size results for 10 wt% doped Fe_3O_4 is 1.676 and 22.62, respectively. No significant changes were detected in the BET result if compared to the undoped ZnO nanoparticles, except for the 10 and 1 wt% doped samples. Nonetheless, the increase of these pores would certainly contribute to the photocatalytic performance. EDX results acquire the peaks of O, Zn and Fe elements for the nanocomposites which conformed Fe_3O_4 mixed well with ZnO. These EDX and XRD analysis can definitely verified the purity of the prepared nanocomposites.

3.3. Transmission Electron Microscopy (TEM) Analysis

Figure 2(a) provides the TEM film for the 10 wt% doped Fe_3O_4 sintered at 300 °C for the ZnO/Fe_3O_4 nanocomposites. The micrograph result shows the shape of the nanocomposite was inconsistent with elongated and cubic structures but without any agglomeration. The average particle size obtained from the particle size distribution was (162 ± 12) nm.

3.4. The Study of Photocatalytic Effects

Extensive studies have been placed on the effects of photocatalyst loading for the degradation of organic dyes in wastewaters (Chanu, Singh et al. 2019). For this study, the photocatalysts loading was conducted in the range of 1.0 to 10.0 wt% by using MB dye at 12.0 ppm concentration.

3.4.1. ZnO/Fe_3O_4 Loading. Figure 2(b) depicts degradation rate increases with the catalyst loading to 0.1593 min^{-1} for 9.0 wt% but dropped to 0.1574 min^{-1} for 10.0 wt% catalyst. This was due to higher loading tends to increase the surface area which relate to the active sites and consequently improve degradation rate (Habibi-Yangjeh and Shekofteh-Gohari 2016). The rate dropped due to increase of cloudiness of the solution, which bring blockage to light absorption thus affecting the photocatalytic process (Han, Wang et al. 2012). The kinetic constant for MB follows pseudo-first-order kinetics from Langmuir-Hinshelwood (LH) model, which can be described by $\ln(C_o/C) = kt$, where C_o and C are the initial and actual concentration of MB and k is the degradation rate parameter.

3.4.2. Different Sintering Temperature. Kinetic analysis for Figure 2(c) shows all ZnO/Fe_3O_4 nanocomposite samples attributed to the LH model and coincide with the pseudo-first-order kinetics. The highest kinetic rate recorded was sample calcined at 300 °C. This result could relate to its highest pore volume and pore size as compared to other samples sintered at different temperature.

3.4.3. Different wt% of Fe_3O_4. Figure 2(d) shows highest degradation rates k, for samples with 1 wt% to 10 wt% Fe_3O_4 doped nanocomposite material irradiated in just 30 minutes. This results indicated that current amount of doped Fe_3O_4 nanoparticles was not affecting the photocatalytic performance of ZnO nanoparticles (Bhukal, Shivali et al. 2014). This clarified that the doping of Fe_3O_4 nanoparticle was within its acceptable range that can hinder the reactive surface area of the ZnO nanoparticle (Han, Liao et al. 2012).

3.5. The Magnetic and Reliability Analysis

The photocatalyst loading of 9.0 wt% ZnO/Fe_3O_4 nanocomposite with 10 wt% doped Fe_3O_4 sintered at 300 °C was conducted with the MB dye.

3.5.1. Vibrating Sample Magnetometry Analysis (VSM). Figure 3 provides the magnetization curves versus the magnetic field for ZnO nanoparticle and ZnO/Fe_3O_4 nanocomposite. Obviously ZnO nanoparticle is diamagnetic material since no magnetic properties was displayed, whereas for ZnO/Fe_3O_4 nanocomposite, apparent magnetic hysteresis loops with ferromagnetic feature was formed, which is obviously a soft ferromagnetic material. This feature was coincide with the finding by Guy et al. (2018) on various ZnO based nanocomposites photocatalysts which also found ferromagnetic feature for their ZnO/Fe_3O_4 nanocomposite (Guy, Atacan et al. 2018). The magnetic moments were simply induced by the magnetic field due to the minimal coercivity revealed by the samples. The ZnO/Fe_3O_4 nanocomposite was easily magnetised achieving 84% induced magnetization with only 1088 Oe magnetic field. The ZnO/Fe_3O_4 nanocomposite saturation magnetization was 6.124 emu/g.

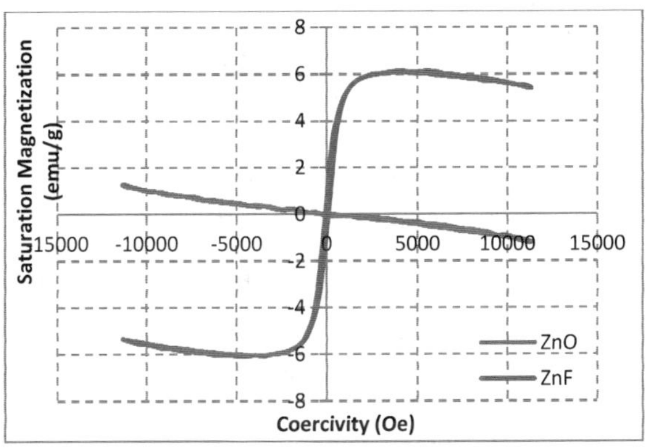

Figure 3. Hysteresis loop for ZnO (ZnO) and ZnO/Fe_3O_4 (ZnF) nanocomposite powders.

Figure 4. Separation of ZnO/Fe_3O_4 powder by applying a magnetic field.

3.5.2. Magnetic Separation Analysis. The magnetically induced separation experiment for 9.0 wt% loading of ZnO/Fe_3O_4 nanocomposites is shown in Figure 4 by applying a magnetic field strength of 0.6 T. The effect of the magnetically induced separation is clearly visible with the progressive separation of a colloidal suspension. After the magnetic separation, the colloidal ZnO/Fe_3O_4 nanocomposites remain dispersible without manifestation of irreversible agglomeration.

3.5.3. Repeatability Analysis. The repeatability analysis was performed under the similar experimental conditions. After the first run of the photodecomposition was completed, the retrieved nanocomposite was reused for the following analysis, where it was recycled for at least three runs. Based on the results, no significant changes appear in degradation efficiency with an average of 97% percentage degradation.

4. Conclusion

Novel magnetically separable ZnO/Fe_3O_4 nanocomposite photocatalyst was successfully prepared. The ZnO/Fe_3O_4 nanocomposite photocatalyst has been carefully inspected by having the optimum loading of 9.0 wt%, 10 wt% of doped Fe_3O_4 and 300 °C of sintering temperature. The synthesised nanocomposite was equally superior in the effectiveness to photodegrade organic dye as compared to its counterpart. The nanocomposite was recycled for three times and achieved the same degradation effect.

Acknowledgments
We are overwhelmed and greatly acknowledged the financial support provided by MOSTI, Government of Malaysia, with the grant No. 03-02-12-SF0019.

5. References

[1] Amornpitoksuk, P, Suwanboon, S, and Randorn, C 2018 *Mater. Sc. Semicond. Proc.* **84** 50–57
[2] Anwer, H et al 2019 *Nano Research* **12**(5) 955-972
[3] Bhukal, S, Shivali, and Singhal, S 2014 *Materials Sci. in Semiconductor Processing* **26** 467-476
[4] Chanu, L A, Singh, W J, Singh, K J, and Devi, K N 2019 *Results in Physics* **12** 1230-1237
[5] Fresno, F, Portela, R, Suarez, S, and Coronado, J M 2014 *J. Mater. Chem. A*, **2**, 2863
[6] Ge, J, Zhang, Y, Park, S J 2019 *Materials* **12**, 1916 doi:10.3390/ma12121916
[7] Guy, N, Atacan, K, Karaca, E, and Ozacar, M 2018 *Solar Energy* **166** 308-316
[8] Habibi-Yangjeh, A, and Shekofteh-Gohari, M 2016 *Separ. Purif. Technol.* **166** 63-72
[9] Han, L, Wang, D, Cui, J, Chen, L, Jiang, T, and Lin, Y 2012 *J. Materials Chem.* **22**(25) 12915
[10] Han, Z, Liao, L, Wu, Y, Pan, H, Shen, S, and Chen, J 2012 *J Hazard Mater* **217-218** 100-106
[11] Kim, S -E, Woo, J -Y, Kang, S -Y, Min, B K, Lee, J K, and Lee, S -W 2016 *Journal of Industrial and Engineering Chemistry* **43** 142–149
[12] Mandal, S K et al 2019 *Mater. Chem. Phys.* **223** 456-465
[13] Mantele, W, and Deniz, E 2017 *Spectrochim. Ac. Part A: Molec. Biomol. Spectros.* **173** 965–968
[14] Muhd Julkapli, N, Bagheri, S, Bee Abd Hamid, S 2014 *Sci. World J.* **2014** 692307
[15] Nandi, P, and Das, D 2019 *Appl. Surf. Sci.* **465** 546-556
[16] Van Cuong Nguyen et al 2015 *Adv. Nat. Sci: Nanosci. Nanotechnol.* **6** 035001
[17] Sun, Y, Meng, X, Dall'Agnese, Y, Dall'Agnese, C, Duan, S, Gao, Y, Chen, G, and Wang, X -F 2019 *Nano-Micro Lett.* **11**(79) 1-22
[18] Tan, T K, Khiew, P S, Chiu, W S, and Chia, C H 2019 *IOP Conf. Ser.: Mater. Sci. Eng.* **544** 012023
[19] Tju, H, Taufik, A, and Saleh, R 2016 *J. Phys.: Conf. Ser.* **710** 012005
[20] Tudose, I V and Suchea, M 2016 *IOP Conf. Ser.: Mater. Sci. Eng.* **133** 012040
[21] Wang, J, Yang, J, Li, X, Wang, D, Wei, B, Song, H, Li, X, Fu, S 2016 *Physica E* **75** 66-71
[22] Wu, Z, Chen, X, Liu X, Yang, X and Yang Y 2019 *Nanoscale Research Letters* **14**(147) 1-14

IOP Conf. Series: Materials Science and Engineering **744** (2020) 012021 doi:10.1088/1757-899X/744/1/012021

Magnetised photocatalyst TiO_2/Fe_3O_4 nanocomposite capable to photodegrade organic dye

ThianKhoon Tan[1,*], PoiSim Khiew[1], WeeSiong Chiu[2] and ChinHua Chia[3]

[1] Engineering Foundation, Faculty of Science and Engineering, University of Nottingham Malaysia, Jalan Broga, 43500 Semenyih, Selangor Darul Ehsan, Malaysia
[2] Low Dimensional Materials Research Center, Department of Physics, Faculty of Science, Universiti Malaya, 50603 Kuala Lumpur, Malaysia
[3] School of Applied Physics Studies, Faculty of Science & Technology, Universiti Kebangsaan Malaysia, 43600 Bangi, Selangor Darul Ehsan, Malaysia
[*] E-mail: TK.Tan@nottingham.edu.my

Abstract. In this study, methylene blue (MB) is used as a measure of the photocatalytic capabilities of the prepared TiO_2/Fe_3O_4 photocatalyst nanocomposite. The nanocomposite was suspended in the dye aqueous solutions placed in a UV chamber under UV-C (254 nm) light irradiation. The TiO_2/Fe_3O_4 nanocomposite was characterised by XRD, FESEM, TEM and EDX analysis. The TiO_2/Fe_3O_4 nanocomposite crystal structure retained TiO_2 pure anatase of tetragonal crystal structure (SG: $I4/_{a}md$) and Fe_3O_4 cubic spinel crystal structure with average crystalline size of (62.1 ± 5.7) nm. The optimum performance of the magnetic nanocomposite was monitored through few categories, such as concentration of photocatalyst, doping amount of Fe_3O_4 nanoparticle into the nanocomposite and sintering temperature. The TiO_2/Fe_3O_4 nanocomposite optimum loading was detected at 3.0 wt%. The nanocomposite performed well at 1 wt% of Fe_3O_4 doping and 350 °C of sintering temperature. The reliability of the TiO_2/Fe_3O_4 nanocomposite was also conducted where it was successfully repeated for at least 3 times with no obvious changes observed in the degradation efficiency.

1. Introduction

Many a time there are news on the increasing problem of many countries around the world due to environmental pollution, depleted water resources and energy shortages. Wastewater from industries effluents is the main cause of environmental pollution involving water resources (Lee and Park 2013). Therefore, photocatalytic processes have been the focus of attention due to its low cost and non-toxic end product (Amornpitoksuk, Suwanboon et al. 2018). In recent years, TiO_2 has been a highly active photocatalyst due to its stability, nontoxicity and economically excellence (Cheng, Zhang et al. 2016).

The TiO_2 based semiconductors has actively gone into the area of environmental purification and CO_2 photoconversion into fuels. Recently, magnetic titania has received abundance reputation due to its magnetic recovery and excellent photocatalytic properties. Among the famous magnetic iron oxide nanomaterials being incorporated with TiO_2 are Fe_3O_4 and γ-Fe_3O_2 due to their superparamagnetic properties (Tedsree, Temnuch et al. 2017).

Abundance of trials have come up through various processing methods in producing an effective and reliable magnetic titania, such as solvothermal, co-precipitation, sol-gel and chemical deposition methods (Hong, Mao et al. 2017, Tedsree, Temnuch et al. 2017). The studies have found similar stability and photocatalytic enhancement on MB and RhB dyes, as well as excellent magnetic recoverability which overcoming the limitations of separation. On the other hand, Dagher et al. (2018) was using a non-thermal synthesis methods to synthesise the magnetic nanoparticles, which improved the MB degradation rate and photoinduced charge generation and separation, besides magnetically separable (Dagher, Soliman et al. 2018).

A hybrid phototacalysts study conducted by An et al. (2018) on Fe_3O_4 magnetic particles showed promising effect on the photocatalytic activity for the MB dye under visible light irradiation. The photocatalyst was analysed and repeated for at least four successive runs by applying an external magnetic field (An, Liu et al. 2018). With a barrage of investigation on the magnetic photocatalysts, this study is trying to match TiO_2 nanoparticle with Fe_3O_4 nanoparticle in the hope of maintaining the former photocatalytic properties and utilising the later magnetic properties. Therefore, this study was aims to scrutinising the weight fractions of Fe_3O_4 nanoparticle as well as the loading and sintering temperature of the magnetic nanocomposite. The TiO_2/Fe_3O_4 magnetic nanocomposite was an effective photocatalyst and managed to degrade 87.9% of MB (12 $mg \cdot L^{-1}$) in the third run.

2. Materials and Methods

All the materials used in this study were of reagent grades. Titanium dioxide (TiO_2, R & M Chemicals), iron oxide (Fe_3O_4, Systerm, ChemPur) and methylene blue (MB, R & M Chemicals) were used without any modification. Experimental solutions were prepared using deionised water from Favorit Water Distiller.

The TiO_2/Fe_3O_4 magnetic nanocomposites photocatalyst materials were prepared by weighing TiO_2 nanopowders and added with a series of 1, 5, 10, 12, 15 and 20 weight percentage (wt%) of iron oxide (Fe_3O_4) nanopowders. All nanopowders were dry mixed for duration of 20 minutes. Each series of premixed nanocomposites was calcined in the Carbolite muffle furnace from 200 to 450 ^{o}C for 5 hours. The nanocomposite was heated at 20 C^{o} per minute. All powders were grinded by a spatula after calcination. This process is technically implemented in mixing oxides (Song and Zhang 2010).

For the characterisation, X-ray diffraction (XRD, Bruker), field emission scanning electron microscopy (FESEM, FEI) which is coupled with energy dispersive X-ray analyser (EDX), transmission electron microscope (TEM, LEO) and BET from Micromeritics were used.

The nanocomposites solutions were prepared using distilled water for concentration between 1.0 wt% to 6.0 wt%. MB dye was also premixed with distilled water for concentration between 2.00 ppm to 12.0 ppm. To initiate the experiment, 0.20 mL of premixed photocatalysts were immersed into 3.00 ml of the dye solutions. All the samples were transferred into the UV chamber (Uvitec Cambridge ultraviolet crosslinker) (Figure 1, Tan, Khiew et al. 2019). Finally, a 0.6 T magnetic field was applied on the samples to collect the nanocomposites and 2 mL solutions were tested using a UV-Visible Spectrophotometer (Varian, Cary 50).

The percentage degradation was calculated based on the Beer-Lambert law (Mantele and Denize 2017), i.e. Degradation (%) = $[(C_o - C_t)/C_o] \times 100 = [(A_o - A_t)/A_o] \times 100$, where C_o and A_o are the dye concentration and the absorbance of the dye before irradiation, whereas C_t and A_t are the dye concentration and the absorbance of the dye after the irradiation of time t.

3. Results and Discussion

3.1. Samples Morphologies for Different Sintering Temperature

The sample used is the TiO_2/Fe_3O_4 nanocomposites with 20 wt% doped Fe_3O_4.

3.1.1. FESEM Analysis. Figure 1(a) shows relatively small size and homogeneous spherical morphology. Aggregation phenomenon was not detected across all micrographs of highly porous structures. The average particle size calcined at 350 °C is (112 ± 12) nm. The FESEM results show no significant effects on the morphological properties for the current sintering conditions.

3.1.2. XRD Analysis. Figure 1(b) provides XRD results for TiO_2/Fe_3O_4 nanocomposites sintered at 350 °C. All detected peaks represented by the TiO_2 and Fe_3O_4 phases. The TiO_2 main peaks at 25.3° for (101),36.9° for (103), 37.7° for (004) and 38.6° for (112) diffraction planes of pure anatase phase of tetragonal structure (JCPDS 21-1272) (Dagher, soliman et al. 2018), and the Fe_3O_4 main peaks at 30.2° for (220), 35.6° for (311), 43.2° for (400) and 57.1° for (511) diffraction planes of cubic spinel structure (JCPDS 19-0629) (Wu, Wu et al. 2015). The XRD results for TiO_2/Fe_3O_4 nanocomposites clarify that no impurities was present throughout the preparation of the samples. In addition, an average crystallite size of (105 ± 13) nm was obtained by using the Scherer equation.

3.1.3. BET and EDX Analysis. The BET for TiO_2/Fe_3O_4 nanocomposites shows no significant difference with the values around 7.8 m^2/g. Similar with the pore size and pore volume which are around 0.05 nm and 27 cm^2/g, respectively. The EDX spectrums results show the acquired elemental map, i.e. O, Ti and Fe, which clarified Fe_3O_4 were deposited onto the TiO_2/Fe_3O_4 nanocomposites surface. This result, combined with the XRD analysis validates the purity TiO_2/Fe_3O_4 nanocomposites.

3.2. Samples Morphologies when Doped with Different wt% of Fe_3O_4
TiO_2/Fe_3O_4 nanocomposites sintered at 300 °C are used.

3.2.1. FESEM Analysis. From the micrographs in Figure 1(c), no significant different in the microstructure with all the TiO_2/Fe_3O_4 nanocomposites were generally small size and majority were spherical. No aggregations was seen from the micrographs. The average particle size was (92.6 ± 7.2) nm with observable porosity.

3.2.2. XRD Analysis. The XRD results in Figure 1(d) show highly crystalline nanocomposites without any impurities. The main peaks at 25.3° for (101),36.9° for (103), 37.7° for (004) and 38.6° for (112) diffraction planes are of pure anatase phase of TiO_2 tetragonal structure (JCPDS 21-1272), as well as the main peaks at 30.2° for (220), 35.6° for (311), 43.2° for (400) and 57.1° for (511) diffraction planes are the Fe_3O_4 crystal of cubic spinel structure (JCPDS 19-0629). This confirmed that both TiO_2 and Fe_3O_4 nanoparticles bind well through the applied mixing. Calculation from the Scherer equation yield the average value of (120 ± 17) nm. This experimental study shows the amount of Fe_3O_4 nanoparticles used does not affecting the crystallite structure of the nanocomposites.

3.2.3. BET and EDX Analysis. The amount of Fe_3O_4 incorporated shows insignificant effect on the BET and pore size of all the samples. The EDX spectrum acquired the main elements, i.e. O, Ti and Fe, which conformed Fe_3O_4 mixed well with TiO_2 in producing the TiO_2/Fe_3O_4 nanocomposites. These double confirmations of EDX and XRD analysis ascertained the purity of the crystalline structure.

3.3. Transmission Electron Microscopy (TEM) analysis
The TEM film in Figure 2(a) is the 1 wt% Fe_3O_4 incorporation calcined at 350 °C sample. It shows the mixture of small size and variation of globular structures. The average particle size is (86.2 ± 9.0) nm.

3.4. The study of photocatalytic effects
Photocatalyst loading has been the focused on the effects of photodegradation of organic dyes in wastewaters (Anwer, Mahmood et al. 2019). This study has used 12.0 ppm MB for experimenting the metal oxides photocatalysts loading range between 1.0 to 6.0 wt% with the increment of 1.0 wt%.

3.4.1. TiO_2/Fe_3O_4 Loading. Loading amount was significantly affecting the kinetic constant of the photocatalytic processes, as depicted by the 1 wt% to 3 wt% samples, which was gradually accelerating from 0.0296 to 0.0406 min^{-1}, respectively, as shown in Figure 2(b). The pronounced increase in the degradation rate was contributed by the increase of the surface area, which is the sites for the photocatalytic interactions (Salama, Mohamed et al. 2018). However, if the loading keep on increases, the degradation rate will decline as the 3 wt% showed and dropped to 0.0362 min^{-1} for the 6 wt% loading. This is due to the reducing yield of the photogenerated electrons and holes, as a consequences of reduced exposure of the photocatalyst to light (Chung, Nguyen et al. 2018).

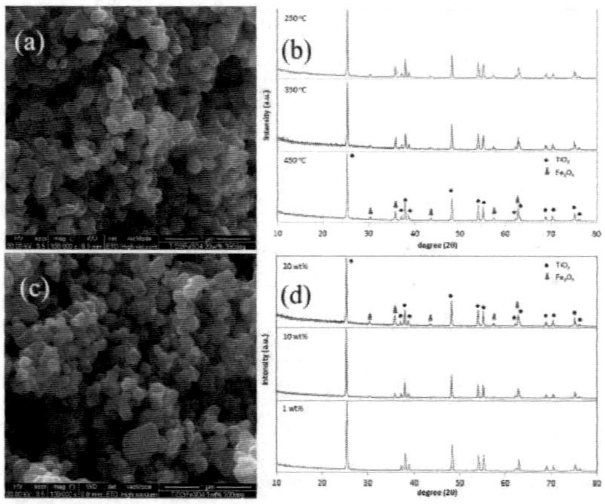

Figure 1. FESEM microstructure and XRD for TiO_2/Fe_3O_4 nanocomposites.

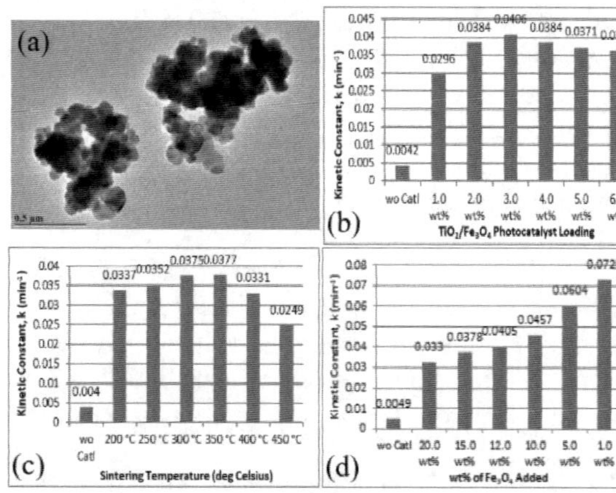

Figure 2. TEM micrograph and kinetic constant of TiO_2/Fe_3O_4 nanocomposite.

3.4.2. Different sintering temperature. Generally, kinetic rate in Figure 2(c) shows equal rate of degradation. The rate constant for the nanocomposite samples attributed to the LH model and correspond with the pseudo-first-order kinetics. This experiment infers that sample with 350 °C sintering temperature having the highest degradation rate 0.0377 min^{-1}.

3.4.3. Different wt% of Fe_3O_4. The kinetic constant was the highest when the amount of Fe_3O_4 incorporated was the lowest with the value of the kinetic constant of 0.0729 min^{-1}, as shown in Figure 2(d). However, as the incorporation values increased to 20 wt%, the value of *k* dropped more than half to 0.0330 min^{-1}. This experimental values show the incorporation of Fe_3O_4 into the nanocomposite significantly block its active sites. Similar finding was obtained by Shojaei et al. (2017) with their TiO2@Fe3O4 microsphere which show significant increase in the photocatalytic performance (Shojaei, Shams-Nateri et al. 2017). The optimum amount of Fe_3O_4 doped is therefore 1.0 wt%.

3.5. The magnetic and reliability analysis

The 3.0 wt% loading of TiO_2/Fe_3O_4 nanocomposites was used with MB dye.

3.5.1. Vibrating sample magnetometry analysis (VSM). The magnetization curves versus the magnetic field for TiO_2 nanoparticle and TiO_2/Fe_3O_4 nanocomposite were depicted in Figure 3. The TiO_2 clearly exhibits no magnetic properties, which characterized as diamagnetic material. On the other hand, TiO_2/Fe_3O_4 nanocomposite clearly displays magnetic hysteresis loops with ferromagnetic feature, which is the soft ferromagnetic material. Similar hysteresis loops were obtained by Lendzion-Bielun at al. (2020) where they deduced the significant dropped of the magnetization in Fe_3O_4–TiO_2 is attributed to the contribution of TiO_2 mass and a lower volume fraction of Fe_3O_4 (Lendzion-Bielun, Wojciechowska et al. 2020). Due to the minimal coercivity exhibited by the nanocomposite, the presence of low magnetic field could easily induced the saturation magnetisation. This is clearly indicated by the curve where 92.7% magnetization was achieved with just 1584 Oe of magnetic field. TiO_2/Fe_3O_4 nanocomposites exhibits high susceptibility because with only 4033 Oe of applied magnetic field, it able to attain magnetisation value of 5.543 emu/g.

3.5.2. Magnetic separation analysis. Figure 4 shows the induced separation experiment for 3.0 wt% loading of TiO_2/Fe_3O_4 nanocomposites with magnetic field strength of 0.6 T. The suspension of the TiO_2/Fe_3O_4 nanocomposites was clearly induced by moving toward the wall of the vial closer to the magnet and the colloidal suspension slowly becoming clear. The colloidal TiO_2/Fe_3O_4 nanocomposite remains dispersible manifestation of permanent agglomeration.

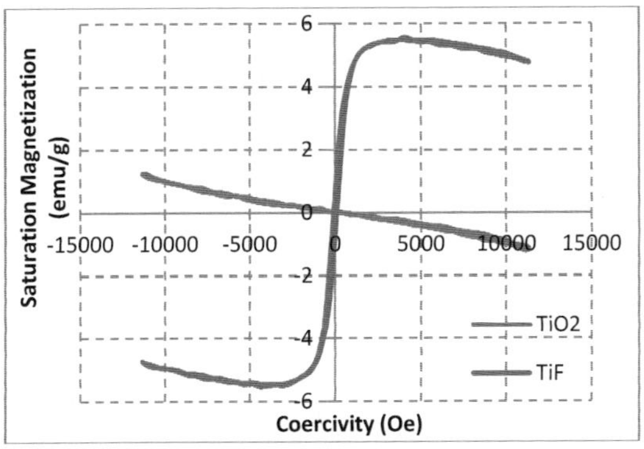

Figure 3. Hysteresis loop for TiO_2 (TiO_2) and TiO_2/Fe_3O_4 (TiF) nanocomposite powders.

Figure 4. Separation of TiO_2/Fe_3O_4 powder by applying a magnetic field.

3.5.3. Repeatability analysis. Similar experimental conditions was applied in performing the repeatability analysis for TiO_2/Fe_3O_4 nanocomposite. After each photodecomposition process was completed, the nanocomposite was retrieved and reused. The percentage degradation for all the three runs were 84.19, 87.32 and 87.92%, respectively. This experiment provides attractive results for the TiO_2/Fe_3O_4 nanocomposite which shows high photocatalytic stability, retrievable and recyclable.

4. Conclusion

A successful preparation of magnetically separable TiO_2/Fe_3O_4 nanocomposite was achieved in this study. The optimum loading for TiO_2/Fe_3O_4 nanocomposites is found to be 3.0 wt%. TiO_2/Fe_3O_4 nanocomposites for the optimisation of doped Fe_3O_4 is 1.0 wt% and sintering temperature is achieved at 350 °C. The nanocomposite was successfully recycled for up to three successive photodegradation and exhibited similar photocatalytic performance.

Acknowledgments

We are greatly acknowledged the financial support obtained from MOSTI, Government of Malaysia, with the grant No. 03-02-12-SF0019.

5. References

[1] Amornpitoksuk, P, Suwanboon, S, and Randorn, C 2018 *Mater. Sc. Semicond. Proc.* **84** 50–57
[2] Anwer, H et al 2019 *Nano Research* **12**(5) 955-972
[3] An, B, Liu, Y, Xu, C, Wang, H, and Wan J 2018 *New J. Chem.* **42** 8914–8923
[4] Cheng, L, Zhang, S, Wang, Y, Ding, G, and Jiao, Z 2016 *Mater. Res. Bull.* **73** 77–83
[5] Chung, W J, et al. 2018 *J. Environ. Management* **213** 541-548
[6] Dagher, S, Soliman, A, Ziout, A, Tit, N, Hilal-Alnaqbi, A, Khashan, S, Alnaimat, F, and Qudeiri, J A 2018 *Mater. Res. Express* **5** 065518
[7] Hong, T, Mao, J, Tao, F, and Lan M 2017 *Molecules* **22** 2044
[8] Lee, S -Y, and Park, S -J 2013 *J. Indust. Eng. Chem.* **19** 1761-1769
[9] Lendzion-Bielun, Z, Wojciechowska, A, Grzechulska-Damszel, J, Narkiewicz, U, Sniadecki, Z, and Idzikowski, B 2020 *J. Phys. Chem. Solids* **136** 109178
[10] Mantele, W, and Deniz, E 2017 *Spectrochim. Ac. Part A: Molec. Biomol. Spectros.* **173** 965–968

[11] Salama, A, Mohamed, A, Aboamera, N M, Osman, T A, and Khattab, A 2018 *Applied Nanoscience* **8** 155-161
[12] Shojaei, A F, Shams-Nateri, A, and Ghomashpasand, M 2017 *Micro & Nano Letters* **12**(3) 161-165 doi: 10.1049/mnl.2016.0620
[13] Song, L, and Zhang, S 2010 *Journal of Hazardous Materials* **174** 563-566
[14] Tan, T K, Khiew, P S, Chiu, W S, and Chia, C H 2019 *IOP Conf. Ser.: Mater. Sci. Eng.* **544** 012023
[15] Tedsree, K, Temnuch, N, Sriplai, N, and Pinitsoontorn, S 2017 *Mater. Today Proc.* **4** 6576-6584
[16] Wu, W, Wu, Z, Yu, T, Jiang, C, and Kim, W -S 2015 *Sci. Technol. Adv. Mater.* **16** 023501

IOP Conf. Series: Materials Science and Engineering **744** (2020) 012022 doi:10.1088/1757-899X/744/1/012022

Towards graphene-based new energy devices

YEFENG WU[1]
1 School of Material Science and Engineering, Hubei University, Wuhan, P. R. China
E-mail:wuyefeng2@163.com

Abstract. The seeking for "killer application" of graphene has been the bottleneck for the industrialization of graphene. From its first application in sophisticated FET devices to many newly arising environment-friendly energy efficient devices, graphene seems to be an excellent candidate material when assembled with functional materials in new energy devices, such as current collector, lithium ion battery, photovoltaic cell or solar-thermal devices. Here, we reviewed the recent reports on these field, and prove that the existence of graphene could bring about a "win-win" outcome in above mentioned representative new energy devices, providing a new design paradigm to find out the "killer application" of graphene.

1. Introduction

Current applications of graphene has been extended to multifarious fields, from electronics to energy and environmental issues. However, the killer application of graphene is still an open question for many scientists. Once supposed to be an excellent candidate material in semiconductor industry during the past years, graphene is now suffering from doubts whether it could be suitable enough for the next ordained semiconductor to substitute silicon and overcome the Moore's Law, because graphene's dark current issue originating from its zero bandgap nature has become more as the Achilles' heel than an advantage given its application to electronic devices. Besides, when used in field effect transistor (FET) devices, it always requires single crystalline high quality graphene samples, which are rare and expensive, and except for the mechanically exfoliated graphene from high quality graphite, only a few groups globally can synthesize inch-sized single crystalline graphene samples via the vastly adopted chemical vapor deposition (CVD) method[1-3]. This reality makes it difficult for graphene to find its "killer application" in electronics. However, benefitting from the synthesis of various graphene products from CVD-derived graphene to solution-derived graphene, graphene could be applied to energy and environmental areas, where such light, highly thermo-/electro-conductive, flexible film has found its application in supercapacitors, or as seawater desalination films, or as component material for improving the performance of electrodes/carrier collector/conductors in new energy devices[4-10].

Instead of searching the application of graphene in sophisticated FET devices at an expensive price, the assembly of traditional functional materials with graphene (we refer to it as "FunMat@Gr") has proved to be a win-win strategy[1c]WeiDong Fei[5] and coworkers has reported a heterostructural graphene quantum dot/MnO_2 nanosheets toward high potential window electrodes for high-performance supercapacitors. Besides its application in supercapacitors, graphene could also be found in other "FunMat@Gr" assembly to bring in a win-win effect. Such as in a series of new energy devices shown in **Figure 1**: current collector, lithium battery[12], solar cell and solar thermal devices. In the following parts, we will focus our discussion on these four aspects to reveal the magic role of graphene.

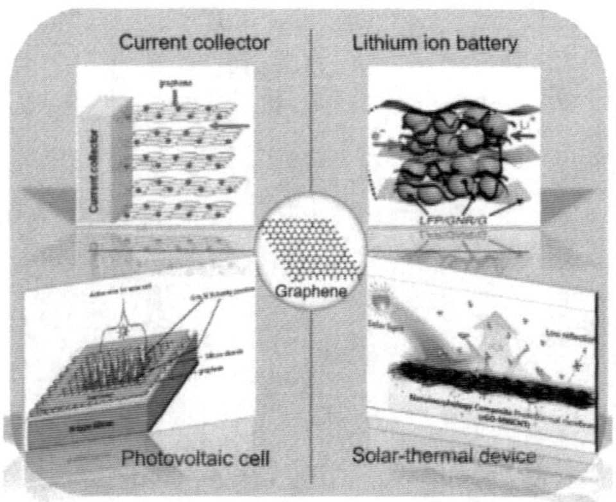

Figure 1. Four representative graphene-based new energy devices: current collector[13] (Reprinted with permission from Ref. 13. Copyright 2016 Science China Press); lithium ion battery[14] (Reprinted with permission from Ref. 14. Copyright 2019 Journal of Material Engineering); photovoltaic cell[15] (Reprinted with permission from Ref. 15. Copyright 2017 Royal Society of Chemistry) and solar thermal devices[16] (Reprinted with permission from Ref. 16. Copyright 2018 Royal Society of Chemistry).

2. Graphene-based new energy devices

Current Collector In rechargeable lithium-ion batteries (LIBs), Cu and Al are the most commonly used materials as current collectors, because of their cheap price and good electronic conductivity[17]. However, the ubiquitous corrosion to current collectors severely hinders the lifetime of LIBs during long-term cycling, where anticorrosion techniques are highly in need. For example, the long-term performances of current 4 V LIBs are deteriorated by the corrosions of Al foil in LIB, which are believed to be a bottleneck for next-generation 5 V LIBs. Many protective methods have been adopted trying to solve this problem, for example, $LiPF_6$-based electrolyte is believed to be the benchmark electrolyte for LIBs and with its generation of about 1 nm thick insoluble protective layer of AlF_3, the corrosion of Al foil has been efficiently passivated. However, this ad-layer is tenuous and because of the inhomogeneities of local current density, pit corrosion to Al foil still exists, which will be further exacerbated by the use of high-voltage positive electrode materials, such as $LiNi_{0.5}Mn_{1.5}O_4$, $LiCoPO_4$, etc.

Comparatively, the direct growth of impermeable, high conductive graphene cover layer on Al foil via plasma-enhanced CVD method (PECVD) stands out. This sheath layer is impermeable to species except proton, thus minimizing the possibility of the coordination between Al^{3+} and other species. As shown in **Fig. 2a**, the cells using graphene-armored Al foil as current collectors demonstrate enhanced electrochemical performance, including better long-term cycling performance at low rate, relieved self-discharge propensity, and enhanced rate/power performance.

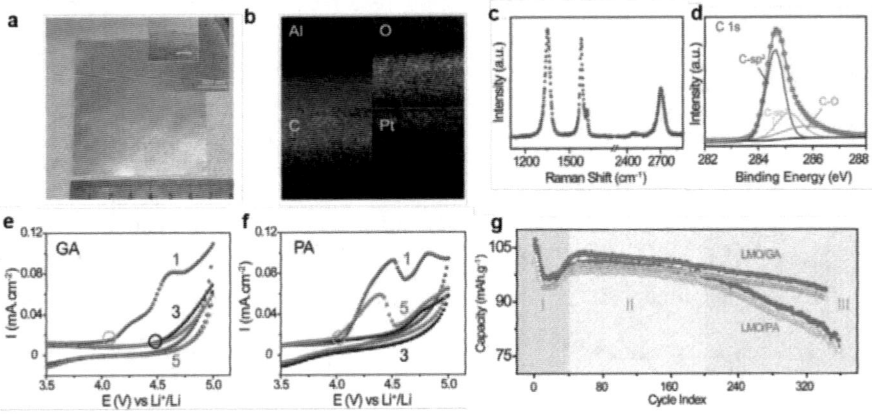

Figure 2. Synthesis and characterization of graphene-armored Al foil (GA) and comparison of performance over pristine Al foil (PA). (a) Digital camera image of GA. Inset is low and high-resolution TEM image of multilayer graphene. Scale bar: 500 nm and 5 nm, respectively. (b) Elemental mapping of GA. Platinum was thermally deposited to protect graphene from possible damage during sampling. (c) Typical Raman spectrum of GA. (d) XPS signal of C 1s of GA. (e, f) Enhanced anticorrosion performance of GA over PA: CV curves of GA (e) and PA (f) in LPF6 electrolyte. The number denotes the cycle index. (g) Enhanced electrochemical performance of GA over PA. Long-term low-rate (0.5 C, 1 C = 148 mA g^{-1}) cycling performance of LMO/PA and LMO/GA cells. Solid and circle represent the C_{cha} and C_{dis}, respectively. Reprinted with permission from Ref. 17. Copyright 2017 Wiley-VCH Verlag GmbH & Co. KGaA, Weiheim.

Lithium-ion battery Benefite from its high electronic conductivity, graphene has also been used as promising substrates in high energy and power density lithium-ion batteries[18-20]. As a key parameter to determine the reversible capacity of lithium-ion batteries, a faster and easier transfer velocity of Li ions is realized by incorporating large surface area graphene in the electrodes, and an improved final cycling performance of lithium ion battery was also achieved. For example, one work showed an enhanced Li ion and electron transfer rate in vertical graphene encapsulated SiO microparticles[23], as shown in **Fig. 3a-b**. Another representative work done by Ouyang and coworkers[21] reported an enhanced lithium-ion storage of 670 mAh g^{-1} with the capacity retention of 99% after 30 cycles at 100 mA g^{-1} based on a MoS_2 anchored free-standing three dimensional vertical graphene foam (MoS_2@3DVG), much better than the 550 mAh g^{-1} at 100 mA g^{-1} based on MoS_2 at planar graphene (MoS_2@3DG). **Fig. 3c** is the SEM image of the as-prepared MoS_2@3DVG, and from **Fig. 3d**, we can find both the charge and discharge capacity of MoS_2@3DVG are better than MoS_2@3DG after only 20 cycles. The MoS_2@3DVG sample was synthesized using the cost effective *M. alternifolia* as precursor via PECVD method. Afterwards, the binder-free MoS_2@3DVG anode to be used in the Li-ion battery was made by anchoring MoS_2 nanosheets onto the free-standing 3DVG by a hydrothermal method. The merits of such hierarchical three-dimensional vertical graphene come from its large surface area, inherent 3D network and excellent ion transport properties. The Raman spectrum indicated the existence of defects in 3DVG compared with 3DG (see **Fig. 3b**).

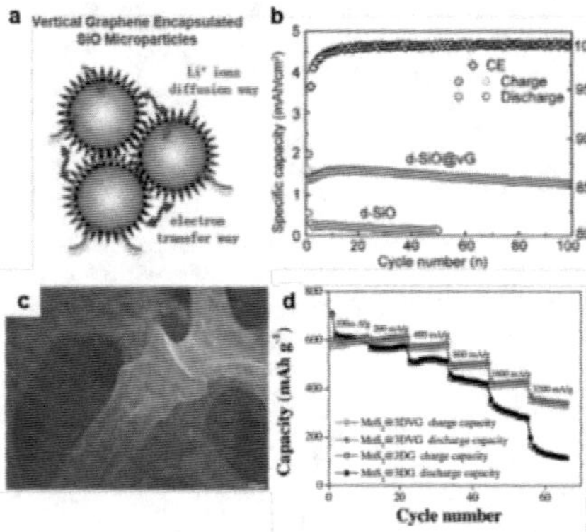

Figure 3. Typical graphene-based lithium ion batteries. (a) Schematic illustration of the coexisted pathways for Li ion diffusion and electron transfer in vertical graphene encapsulated SiO micropaticles. (b) Specific capacity and cycling efficiency of d-SiO and d-SiO@vG electrodes, and the current density is 320 mA g^{-1}. (a, b) Reprinted with permission from Ref. 23. Copyright 2017 American Chemistry Society. (c) SEM image of the as-prepared MoS_2@3DVG. (d) The rate stability of MoS_2@3DVG and MoS_2@3DG electrodes at various current densities. (c, d) Reprinted with permission from Ref. 21. Copyright 2015 Elsevier.

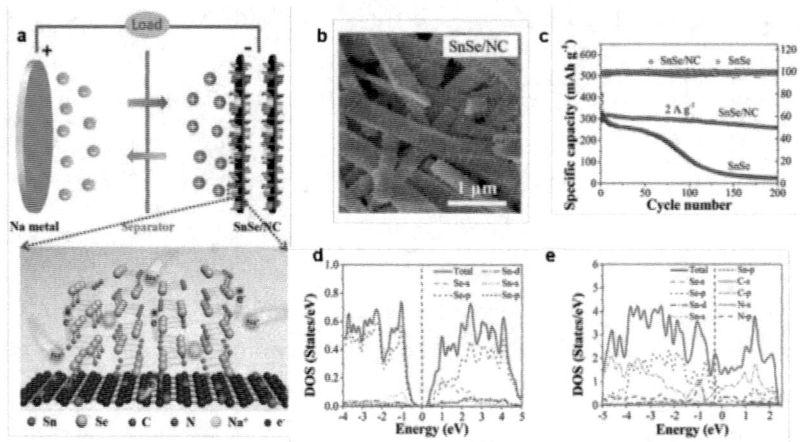

Figure 4. Schematic mechanism, morphology and performance of nitrogen doped graphene used in Na-ion battery. (a) Schematic mechanism of high rate performance of SnSe/NC. (b) SEM image of the as-prepared SnSe/NC nanobelts. (c) Comparision of cycling performance and coulombic efficiencies between SnSe/NC and SnSe. (d, e) Calculated DOSs of (d) SnSe and (e) SnSe/NC. Reprinted with permission from Ref. 24. Copyright 2018 Elsevier.

To further improve the performance, nitrogen doped graphene is becoming more and more prevalent [22]. Compared with traditional graphene, nitrogen doped graphene shows an enhanced electric conductivity and affinity to ions, providing more active site for the attachment of metal ions, as shown in **Fig. 4d-e**. Thus, compounds consisting of nitrogen doped graphene and another semiconductor have been fabricated as electrodes in Li-ion[23] or Na-ion batteries[24-26]. A typical device detection result is shown here, **Fig. 4a** shows nitrogen doped graphene could enhance the Na ion transfer because of the good conductivity of graphene and the negative electronic chemical environment coming from the n-doping effect of nitrogen. And the SEM image revealed that graphene was vertically grown covering the SnSe nanobelts (**Fig. 4b**). The enhanced cycling performance and coulombic efficiency of SnSe/NC over pristine SnSe was obvious in **Fig. 4c** and after 150 cycles, the capacity of SnSe decreased hugely, while the capacity of SnSe/NC still maintained the same level as the beginning point, due to the enhanced conductivity of SnSe/NC (**Fig. 4d-e**).

Solar energy Graphene shows its application in photovoltaic solar cells in mainly three different aspects. First, graphene may be used as electrode[27], such as counter electrode or transparent conductive electrode[28-30]. Second, graphene can be diffused in solution and then spin coated on the electrode to improve the power conversion efficiency (PCE)[31]. Third, graphene could be used in heterostructures with another semiconductor, like GaAs[32,33], Si[34,35] and so on[36].

Benefiting from its high transparency and conductivity, graphene can act as the transparent electrode. Notte[27] and cowokers tried to substitute indium tin oxide (ITO) with graphene in polymer solar cells. The graphene was fabricated via CVD method, and the multilayer graphene was synthesized by transferring complementary graphene layers onto other graphene substrate. Then the graphene was treated with oxygen plasma and $SOCl_2$ chemical doping, respectively. Afterwards, the wettability and sheet resistance of the oxygen plasma-treated graphene, $SOCl_2$ chemically doped graphene and the pristine graphene was measured for comparison. From **Fig. 5a-b**, we found that the wettability change of graphene (contact angle decrease from 90° to 58°) and the sheet resistance decreased with the increasing of graphene layer number. The five-layer $SOCl_2$ chemically doped graphene showed the lowest sheet resistance, because of the covalent attachment of Cl atoms, which served as electron acceptors in graphene, while the oxygen-plasma treatment graphene presented the highest R_s compared to the other two graphene samples, for the top layer has been etched during the plasma treatment. At last, the photovoltaic device measurement results also showed the performance of photovoltaic devices based on graphene treated by oxygen plasma is able to provide a higher photo conversion efficiency (PCE) than the graphene treated by $SOCl_2$ chemical doping. This result revealed the fact that the wettability is more crucial than the Ohmic contact in solar cell devices. Subsequently, the device based on ITO gets a highest PCE among the three devices, because the water contact angle of ITO is even lower than the oxygen plasma treated graphene.

In addition, graphene/semiconductor heterostructure are fabricated by some groups to make use of the wonderful properties of graphene. For example, Li and cowokers[28] made a photovoltaic device based on the peculiar nature of graphene that the Fermi level of graphene can be tuned by gate structure relatively independent of semiconductor substrate, the structure is shown in **Fig. 5c**. PCE of previously reported solar devices based on gated graphene/GaAs heterostructure was generally less than 10%. Herein, the author developed an improved graphene-dielectric-graphene gated structure for graphene/GaAs heterostructure. In this structure, PCE and open circuit voltage of device is able to achieve 18.5% and 0.96 V, respectively. This structure could merge the merit of anti-reflection property of the dielectric layer with high transparency and highly tunable Fermi level of graphene together. In this report, heterostructure with trilayer graphene indicates the best performance. This is mainly for the reason that sunlight absorbed by graphene cannot transfer into electricity, due to the zero band gap of graphene. Therefore, monolayer graphene would absorb 2.3% of sunlight which leads to a 2.3% decrease in the short circuit current density (J_{sc}). On the contrary, multilayer graphene with more layers in the device means lower series resistance. Finally, comprehensively considering these features affected the nature of graphene, a conclusion was drawn that photovoltaic device based on trilayer graphene could get the best performance and the highest PCE obtained by solar cells fabricated as mentioned above is 9.20%. The relation between the Graphene layer number and PCE is demonstrated in **Fig. 5d**.

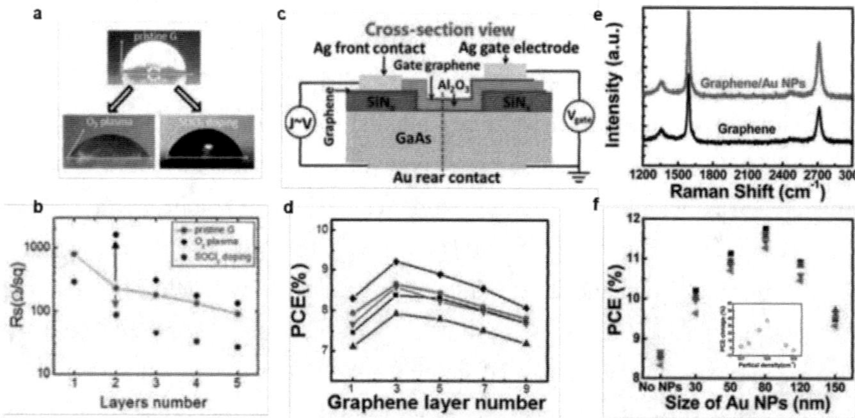

Figure 5. The structure scheme and characterization of different categories of graphene and solar cells. (a) The comparison of contact angles of pristine graphene, oxygen plasma treated graphene and $SOCl_2$ chemical doping graphene, respectively. (b)The relation between resistant and number layers of pristine graphene, oxygen plasma treated graphene and $SOCl_2$ chemical doping graphene. (a, b) Reprinted with permission from Ref. 27. Copyright 2013 Royal Society of Chemistry. (c) The scheme of the structure of the graphene/semiconductor heterostructure solar cell. (d) The relation between the graphene layer number and PCE of five sets of solar cells. (c, d) Reprinted with permission from Ref. 32. Copyright 2015 Elsevier. (e) The Raman scheme of graphene and graphene covered with Au NPs. (f) The impact of size of Au NPs on PCE of the solar cell. (e, f) Reprinted with permission from Ref. 33. Copyright 2016 Wiley-VCH Verlag GmbH & Co. KGaA, Weiheim.

Furthermore, Lin and coworkers recently fabricated a graphene/GaAs heterostructure solar cell and Au nanoparticles were deposited near the photoactive layer to enhance the PCE of the solar device by increasing the light absorption in the photoactive layer[33]. As is shown in **Fig. 5e**, the Raman spectrum of graphene coated with Au NPs and that of pristine graphene are resemble. The same position of G and 2D peaks illustrates the coating of Au NPs have no bearing on the graphene. After antireflection coating and chemical doping, the PCE is improved from 8.83% to 16.2%. In their report, Au NP with diameters of 30, 50, 80, 120, and 150 nm are spun on graphene to compare with each other's size and find out whose PCE is the highest. **Fig. 5f** shows the PCE value of solar cells with different diameters Au NP and different particle densities. Obviously, solar device with 80 nm NP could get the highest PCE which is 11.8%. Measurement of PCE of solar cells among different particle densities was also taken, and Au NPs density with 1×10^{8} cm^{-2} could best ameliorate the device performance. Eventually, after TFSA doping of graphene and ARC coating on the front surface, device performance was further improved to 16.2%.

Solar-thermal Except being used in supercapacitor, lithium-ion battery and solar cell, an aspect normally with scant attention will be introduced here. The solar-thermal property of graphene, given to the low thermal conductivity and large surface area, graphene is a promising material in this field[38,39].

Normally, the solar-thermal application serves as a "light trap", by reducing the reflection and transmittance, to convert sunlight to heat and then generate vapor to make seawater desalination. In recent reports, there are three kinds of ways to fabricate light trap, the corrugated surface of graphene glass, multi-walled carbon nanotubes of graphene oxide (GO)-based aerogels and the graphene nanoplates (GNPs) secondary structure of the hierarchical graphene foam (h-G foam). Among them, the h-G foam has the best performance that the solar thermal conversion rate of the h-G foam is 93.4%. Moreover, when the length of the GNPs grow longer, the quality of the h-G foam is better. Besides, when the corrugated surface of graphene glass grow, the length and width of it increases. Then the performance got improved. In these reports, the porous structure and the excellent wettability of the carbon material endow the application to possess a good solar-thermal property. **Fig. 6a** illustrated the porosity of h-G foam, graphene glass and GO aerogels. Furthermore, the defects of the graphene contribute to the low thermal conductivity of the application, which in turn to preserve heat. In Ren and coworkers' work, the color difference of three foams made by different kinds of material demonstrated in **Fig. 6a** shows the difference in the absorption among these three foams. In addition, h-G foam is the darkest one, which means the absorption of h-G foam is the highest among these foams. **Fig. 6b** illustrates the measurement of three samples' solar-thermal conversion, which was taken with various incident angles. Although the solar-thermal conversion performance kept stable until the incident angle was increased to 55°, the solar-thermal performance would only drop a little when the incident angle was increased to 70°, which is caused by the dramatically decreased projected area. Furthermore, the surface temperature of these three samples show rising trend when the radiation power of the solar illumination improved from 1 to 5 kW/m^2. Among these three samples the h-G foam's surface temperature increases drastically (**Fig. 6c**).Graphene based solar-thermal devices are very promising in seawater desalination applications to help tackle the environmental problems.

IOP Conf. Series: Materials Science and Engineering **744** (2020) 012022 doi:10.1088/1757-899X/744/1/012022

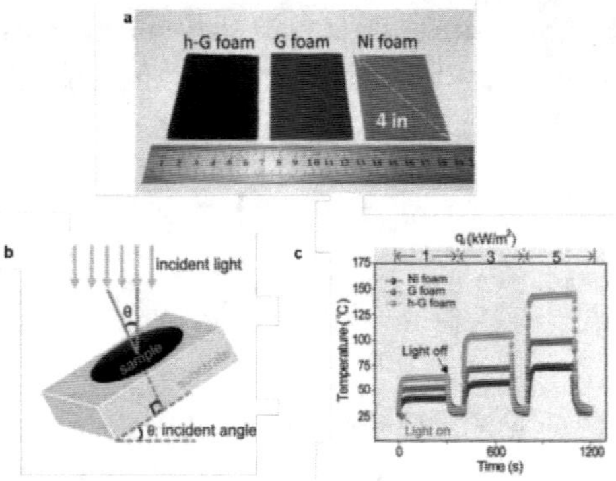

Figure 6. Characterization and measurements of the solar-thermal device. (a) The picture of foams which were made by h-G, G and Ni, respectively. (b) The scheme of the measurements of surface temperature of different samples which were taken with varied incident angles. (c) The relation between temperature and times of different samples with solar illumination whose radiation power increased from 1 to 5 kW/m^2. Reprinted with permission from Ref. 39. Copyright 2017 Wiley-VCH Verlag GmbH & Co. KGaA, Weiheim.

3. Comparison and Conclusion

Raman spectra are widely adopted to characterize the quality of graphene. Here, as shown in **Fig. 7**, the Raman spectra of graphene used in four new energy devices are compared. They share a same feature, that is, all of the Raman spectra show the appearance of a high intensity D band, corresponding to the lattice defect of graphene. This indicates that when used in new energy devices, the domain size of crystallized graphene don't have to be very large size, μm-sized domain of graphene flakes or sheets in combination with functional materials are suitable enough to guarantee both the high electronic and thermal conductivity of graphene to improve the performance than pure functional materials.

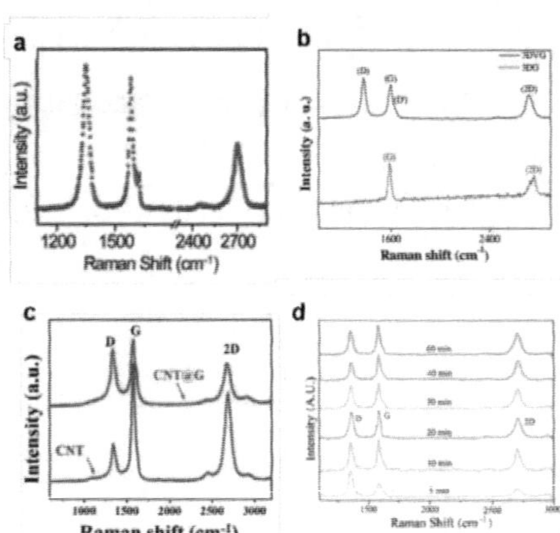

Figure 7. The Raman scheme of graphene which are applied in different kinds of devices: (a) Current collector. Reprinted with permission from ref. 17. Copyright 2017 Wiley-VCH Verlag GmbH & Co. KGaA, Weiheim. (b) Lithium ion battery. Reprinted with permission from Ref. 21. Copyright 2015 Elsevier. (c) Solar cell. Reprinted with permission from ref. 36. Copyright 2018 Wiley-VCH Verlag GmbH & Co. KGaA, Weiheim. (d) Solar thermal devices. Reprinted with permission from Ref. 39. Copyright 2017 Wiley-VCH Verlag GmbH & Co. KGaA, Weiheim.

Thus, to seek for the killer application of graphene, given with the strict synthesis and fabrication requirements on graphene used in FET device, we believe this cheaper and simpler strategy by the assembly of functional materials with less perfect graphene flakes or graphene oxides in new energy devices is providing a new design paradigm for the exploitation of graphene's "killer application".

References

[1] Lee, J. H. *et al. Science* **344**, 286-289 (2014).
[2] Chen, S. *et al. Adv. Mater.* **25**, 2062-2065 (2013).
[3] Yan, Z. *et al. ACS Nano* **6**, 9110-9117 (2012).
[4] Yoon, Y. *et al.* ACS *Nano* **8**, 4580-4590 (2014).
[5] Jia, H. *et al. Adv. Sci. (Weinh)* **5**, 1700887 (2018).
[6] Lin, J. *et al. Adv. Sci. (Weinh)* **5**, 1700687 (2018).
[7] Zheng, S. *et al. ACS Nano* **11**, 4009-4016 (2017).
[8] Jang, H. *et al. Adv. Mater.* **28**, 4184-4202 (2016).
[9] Ababtain, K. *et al. Mater. Res. Express* **5**, 014004 (2018).
[10] Bae, S. *et al. Physica Scripta* **T146**, 014024 (2012).
[11] Wei, N. *et al. J. Mater. Chem. A* **7**, 4813-4822 (2019).
[12] Ji, L. *et al. Nano Energy* **1**, 164-171 (2012).
[13] Wu, Z.-S. *et al. Scientia Sinica Chimica* **46**, 732-744 (2016).
[14] Cui, C. *et al. J. Mater. Eng.* **47**, 1-9 (2019).
[15] Bhopal, M. F. *et al. J. Mater. Chem. C* **5**, 10701-10714 (2017).
[16] Wang, Y. *et al. J. Mater. Chem. A* **6**, 963-971 (2018).
[17] Wang, M. *et al. Adv Mater* **29**, 1703882 (2017).
[18] Ghosh, M. *et al. Solid State Ionics* **296**, 31-36 (2016).
[19] Joseph, J. *et al. Adv. Mater. Technol.* **3**, 1800070 (2018).
[20] Kwon, G. D. *et al. ACS Appl. Mater. Interfaces* **10**, 29486-29495 (2018).
[21] Ouyang, B. *et al. Electrochimica Acta* **194**, 151-160 (2016).
[22] Tzeng, Y. *et al. Sci. Rep.* **4**, 4531 (2014).
[23] Shi, L. *et al. Nano Lett.* **17**, 3681-3687 (2017).
[24] Ren, X. *et al. Nano Energy* **54**, 322-330 (2018).
[25] Xu, D. *et al. Adv. Energy Mater.* **6**, 1501929 (2016).
[26] Zhao, X. *et al. Nano Energy* **47**, 224-234 (2018).
[27] La Notte, L. *et al. Nanoscale* **9**, 62-69 (2017).
[28] Garg, R. *et al. Adv. Energy Mater.* **7**, 1601393 (2017).
[29] Lin, G. *et al. Funct. Mater. Lett.* **11**, 1850009, (2018).
[30] Wang, Z. *et al. Carbon* **67**, 326-335 (2014).
[31] Lee, M. W. *et al. Carbon* **106**, 48-55 (2016).
[32] Li, X. *et al. Nano Energy* **16**, 310-319 (2015).
[33] Lin, S.-S. *et al. Adv. Energy Mater.* **6**, 1600822 (2016).
[34] Gnisci, A. *et al. physica status solidi (a)* **216**, 1800555 (2018).
[35] Rehman, M. A. *et al. Carbon* **132**, 157-164 (2018).
[36] Tsai, M. L. *et al. Nano Lett.* **16**, 309-313 (2016).
[37] Ci, H. *et al. Nano Res.* **11**, 3106-3115 (2018).
[38] Ren, H. *et al. Adv. Mater.* **29**, 1702590 (2017).

A Synchrotron Study of Molecular and Chemical Interaction at the Dental Material/Biomimetic Composite/Native Hard Dental Tissue Interface

D Goloshchapov [1], V Kashkarov[1], Yu Ippolitov [2], I Ippolitov [2], J Vongsvivut[3] and P Seredin[1]

[1]Department of Solid State Physic and Nanostructures, Voronezh State University, Voronezh, Russian Federation
[2]Dentistry Institute of Postgraduate Medical Education Department, Voronezh State Medical University
[3]Australian Synchrotron, Synchrotron Light Source Australia Pty LTD, Melbourne, Australia
E-mail: goloshchapov@phys.vsu.ru

Abstract. Based on the technology of molecular multidimensional IR- synchrotron visualization, the paper explores integration of a new generation of biomimetic composites that regenerates a mineral organic dental enamel complex with native human hard dental tissues and dental cement. The data of spectral molecular visualization was obtained from the area of the healthy hard tissue (enamel/dentine) – biomimetic transition layer – dental material/adhesive interface. The resulting data is indicative of chemical differentiation of functional groups of all the materials at the biomimetic system/natural hard tissue and prove the chosen method to be effective in analyzing integration of dental concrete and new generation biomimetic composites.

1. Introduction

Despite high levels of adhesion and strength, modern composites and bonds employed in restorative dentistry are known to have a low affinity with native human dental tissue with regards to chemical composition and morphological organization [1].

The most viable method for designing a synthetic material – enamel/dentin interface of a high-quality is to come up with a biomimetic material mimicking the composition, micro and nanostructure of hard dental tissues as part of the biomimetic approach [2,3]. Introduction of calcium hydroxyapatite (HAP) into a biocomposite is well-known to improve integration of synthetic materials [4]. Additionally, inclusion of polar amino acids of an enamel matrix [5,6] into biocomposites enables mimicking of a certain enamel or dental dentin area by means of a biomimetic material as well as improves their adhesive and strength characteristics [7].

Evaluation of bond integration with dental tissues as well as analysis of an emerging adhesive/enamel interface can be performed by means of the Fourier-transform infrared spectroscopy (FTIR) which allows a molecular composition and thin structural properties of biological objects to be identified based on analyzing molecular oscillation bands in IR-spectra that are specific to particular chemical groups [8]. While studying biological objects, large datasets can be obtained from a microarea of a sample due to a microscope included into a measurement scheme as well as a source of synchrotron radiation [9]. Using a spectrum massive obtained, IR-microspectroscopic map of an object providing a wealth of information on molecular bonds in a sample and their distribution in space can be designed.

Therefore the objective of the paper is to investigate molecular and chemical features of a dental material – biomimetic buffer layer – hard dental human tissue interface by means of multidimensional visualization of IR-microspectroscopy data.

2. Materials and methods

2.1. Samples preparation
In order to obtain biomimetic modelling materials, we employed a system containing the following: hyaluronic acid (0.01-0.05 wt. %), L-histidine (0.01-0.2 wt. %), L-lysine hydrochloride (0.05-0.4 wt. %), L-arginine hydrochloride (0.2-1.6 wt. %) as well as ethylene glycol methyl ether (30-85 wt. %), diglycidyl dimethacrylate (1-15 wt. %), urethane dimethacrylate (1-15 wt. %), ethyl alcohol (2-20 wt. %) and water (the remainder). In order to design an environment which is high-affinity to native dental tissues, a synthetic carbonate substitute calcium hydroxyapatite (CHAP) was added in the ratio of 1 ml of the mix - 0.01 g CHAP to meet a range of the characteristics of the enamel apatite and human dental dentin [5]. In order to fix the above buffer system, a universal adhesive was used for a bioactive bonding system. The CHAP resulting from the current study was added in the ratio 1 ml of the adhesive – 0.01 g CHAP into the universal adhesive for improving the affinity. The HAP components were mixed with those of the buffer system and the adhesive by means of the ultrasonic homogenizer QSonica for 30 seconds.
The integration of the buffer layers based on the biocomposites was examined using samples of teeth removed from patients aged 18 to 45 in accordance with orthodontic treatment plan. The teeth were prepared and the biomimetic buffer system was applied according to the scheme which follows.
In the first stage the enamel and dentin were prepared and the filling void was formed.
In the second stage the enamel was etched for 60 seconds using an enamel etching gel based on a 37% phosphorous acid followed by water rinsing and air flow drying. The dentin was then processed with the dentin conditioner based on a hyaluronic acid mix (0.01-0.05 wt. %), L- histidine (0.01-0.2 wt. %), L-lysine hydrochloride (0.05-0.4 wt. %), L- arginine hydrochloride (0.2-1.6 wt. %) for 20-30 seconds. After that, the void surface was dried.
In the third stage the biocomposite buffer system resulting from the current study was evenly distributed into the walls of the obtained void and 20 seconds later it was dried using an air compressor. A universal light-curing adhesive containing CHAP which was then preliminarily photopolymerized for 20 seconds was applied onto the buffer layer surface prepared in such a fashion.
Finally, in the fourth stage a minute later a commercial compomer restorative material containing the adhesive was applied onto the biomimetic buffer layer. Taking into consideration the research methodology (IR-microspectroscopy) requirements for sample geometry, parallel slices of the segments of the restored teeth samples were prepared.

2.2. Method
The molecular composition of the samples in the dental material – biomimetic buffer layer – native hard human dental tissue interface was examined using synchrotron IR-microspectroscopy as well as attenuated total reflection (ATR). The study was conducted on the Infrared Microspectroscopy (IRM) (the Australian synchrotron) containing a Hyperion 3000 IR-microscope (Bruker) and an ATR element with a germanium prism [9]. In Fig. 1a a rectangular area indicates an image of the investigated interface area used to obtain the absorption IR-spectra in the range from 3800-700 cm^{-1} (Fig. 1b).
Using the possibilities offered by the IR-microscope and the OPUS software, following molecular mapping one-dimensional IR-images (IR-maps) were designed in the sample area sized 100x100 μm (Fig. 1a) with a step 2μм based on colour coding of the intensities of the absorption bands of the IR-spectra. The maps show the distribution of the intensities of the molecular group and thus its concentration at a certain point of the investigated sample.

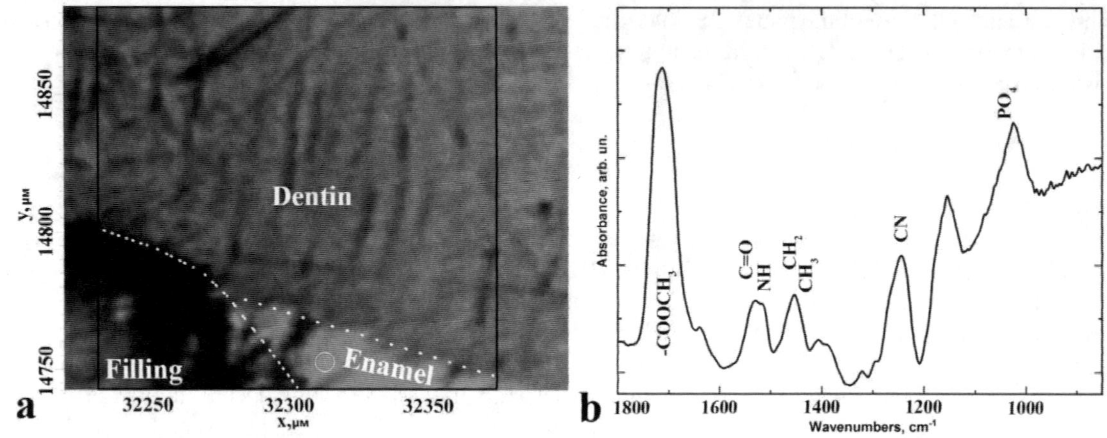

Figure 1. (a) - 100x100 mkm section containing the interphase border of a light-cured dental material / biomimetic composite / enamel from which an array of spectra was obtained. (b) - typical infrared absorption spectrum from the interface region.

3. Results

Through the course of the study of the interphase interface area (Fig. 1a) a set of the main oscillation modes was identified in the IR-spectrum (Fig. 1b) between the light-curing dental material, biomimetic composite as well as enamel and dental dentin.

Fig. 2a presents the IR distribution map of the PO_4 molecular group in the interface area (Fig. 1b). The data analysis shows that the dental material area does not contain any phosphate groups. The entire area interfacing with the enamel where a non-zero intensity of the active oscillations in the range 1163-981 cm^{-1} (PO_4 band) is sized ~30 μm and is indicated in Fig. 2a with a dotted line.

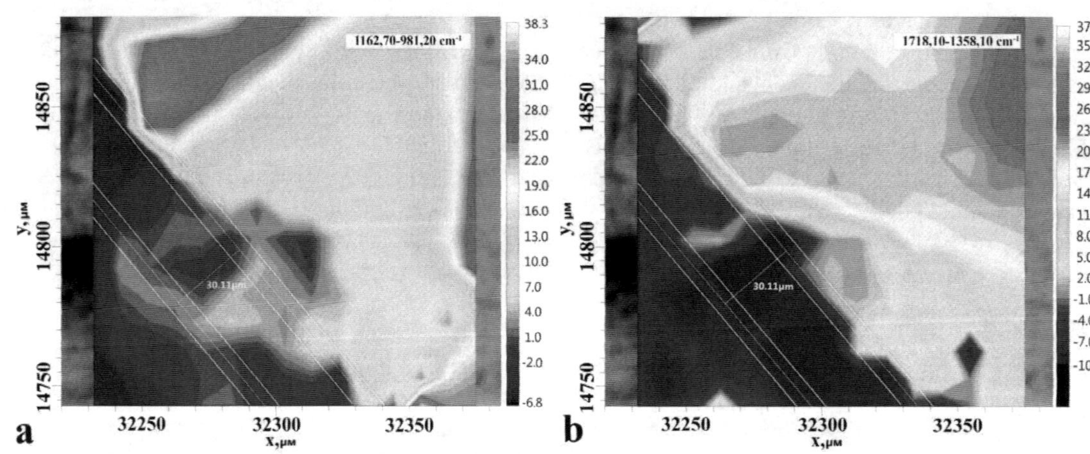

Figure. 2 The characteristic IR image obtained on the basis of color coding of the intensity of the absorption band: (a) of PO_4 1163-981 cm^{-1} and (b) organic component of 1718-1358 cm^{-1}

The IR-image presented in Fig. 2b was generated for obtaining extra information on the interphase interface area. This IR-map provides the data on the distribution of the intensities of the CN, NH, C=O, CH_2/CH_3 molecular groups in the range 1718-1358 cm^{-1} (Fig. 1b) which are part of the collagen as well as the Amide I and Amide II components that are contained in the organic component of the enamel/dentin and biomimetic buffer layer. Analytical processing of the IR-map (Fig. 2a) enables us to conclude that the distribution of the organic component in the transition buffer layer is more homogeneous compared to that of the phosphate groups. The IR-spectrum in Fig. 1b contains the absorption band in the area 1725 cm^{-1}. This oscillation is characteristic of IR-spectra of dental materials and belongs to the ether molecular group (–$COOCH_3$) [10]. Additionally, this absorption band is not overlapped by the other oscillations, which allows us to design the IR-image (Fig. 3a).

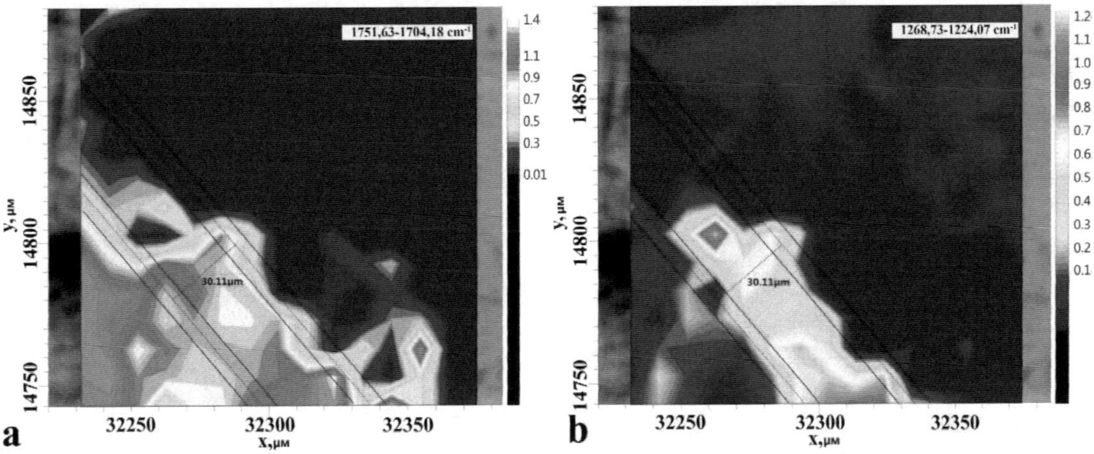

Figure. 3 The characteristic IR image obtained on the basis of color coding of the intensity of the absorption band: (a) of compomer material 1725 cm^{-1} and (b) of Amide III 1269-1224 cm^{-1}

Fig. 3a suggests that the maximum distribution of the intensities of the ether group (–$COOCH_3$) coincides with the position of the dental material observed in the optical shot (Fig. 1a).
The analysis of the resulting experimental spectrum from Fig. 1b allows an intensive absorption band Amide III to be identified in the area 1269-1224 cm^{-1} and the IR-map which only belongs to the transition buffer layer (Fig. 3b) to be designed. It should be noted that the above oscillation (Fig. 1b) is not overlapped by the absorption of the other functional groups and can act as a reference of the biomimetic composite. The comparison of the data of the optical image (Fig. 1a) and IR-map in Fig. 3b is clearly indicative of the fact that the molecular group 1269-1224 cm^{-1} is mainly distributed only in the narrow area of the sample containing the light-curing material – biomimetic buffer layer – enamel/human dental dentin interface.

4. Discussion

While analyzing IR-images of the area of the heterophase interface, it has to be remembered that the area in the IR-map of the phosphate group (Fig. 2a) with the intensity ranging from 1.0 to 7.0. is of certain interest. This area belongs to the biomimetic transition layer introduced into the sample which contains the synthesized carbonate substitute calcium hydroxyapatite which when introduced into the biomimetic layer allows the molecular and chemical affinity with the anatomical dental root to be improved [5]. Due to the inclusion of the CHAP into the biomimetic transition layer in the IR-map (Fig. 2a), the interphase interface is only clearly visible between the biomimetic transition layer and dental material where a sharp colour gradation is determined by the intensity of the oscillation mode of the PO_4 CHAP group. There is no distinct native tissue/biomimemtic composite interface, which is indicative of a high affinity of the latter with the enamel and dentin.
In addition to the IR-map of the phosphate group (Fig. 2a), the IR-image of the oscillations in the area 1718-1358 cm^{-1} which belongs to the organic component of the sample was designed (Fig. 2b). This map gives a visually clearer of the enamel and dentin interface to be identified.
When it comes to the interaction of the dental material, biomimetic buffer layer and native hard tissues, it should be mentioned that the oscillation area 1718-1358 cm^{-1} in the IR-spectrum contains a whole range of overlapping absorption bands [11], which creates certain challenges for interpreting the results and arriving at sound conclusions on the interaction in the interface area.
Analysis of the IR-maps based on the distribution of the ether molecular (–$COOCH_3$) (Fig. 3a) and Amide III (Fig. 3b) group which are not overlapped with the other oscillation bands in the IR-spectrum allows us to make the following conclusions on the character of the heterophase interface of the

biomimetic composite/natural hard tissue. Firstly, it is clear that the integration of the dental material with the enamel area where a shift in the intensity of the oscillation mode –$COOCH_3$ from maximum to minimum is observed in the area with the width ~14 μm and overlaps the area where the organic component dominates (Fig. 3a,b). Secondly, based on the analysis of the distribution of the absorption intensity of the Amide III group (Fig. 3b), which is found in the biomimetic transition layer, it can be concluded that the introduced biomimetic layer separates the native hard tissues from the dental material. The data obtained based on the analysis of all the IR-images (Fig. 1–3) are genuinely indicative of chemical differentiation of the functional groups of all the materials in the biomimetic system/natural hard tissue interface area and they also prove the chosen approach to analyizing the integration of dental cement new generation and biomimetic composites to be effective.

5. Conclusions

In this research has been shown the possibility of employing molecular multidimensional IR-visualization for analyzing integration of new generation biomimetic materials regenerating the mineral and organic enamel complex with native hard dental human tissues.

By means of IR-mapping of the intensity of a certain functional molecular group using synchrotron radiation the differences between a healthy tissue, dental material and biomimetic transition layer in the interphase areas were identified and visualized. The position and concentration of functional groups responsible for integration of the biomimetic composite and native hard human dental tissue was also found.

The biomimetic system we have developed based on the nanocrystal carbonate substitute hydroxyapatite obtained by means of the biogenic calcium source and the complex of major polar amino acids belonging to the organomineral human dental complex has been shown to form a functional bond with native human hard dental tissue.

The resulting microspectroscopy data is clearly indicative of the chemical differentiation of the materials and organomineral interaction at the biomimetic system/natural hard tissue interface.

Acknowledgments

The study was performed under support of the grant RFBR 18-29-11008 mk. The study in the part of nanocrystalline carbonate-substituted calcium hydroxyapatite synthesis was performed under support of the grant of President of Russian Federation MK-419.2019.2.

References

[1] Temel U B, Van Ende A, Van Meerbeek B and Ermis R B 2017 *Am J Dent* **30** 205–11

[2] Niu L, Zhang W, Pashley D H, Breschi L, Mao J, Chen J and Tay F R 2014 *Dental Materials* **30** 77–96

[3] Cao C, Mei M, Li Q, Lo E and Chu C 2015 *Materials* **8** 2873–86

[4] Dorozhkin S V 2017 *Hydroxyapatite and Other Calcium Orthophosphates: Bioceramics, Coatings and Dental Applications* (Nova Science Publishers, Inc New York) p 462

[5] Seredin P V, Goloshchapov D L, Prutskij T and Ippolitov Yu A 2017 *Results in Physics* **7** 1086–94

[6] Aljabo A, Abou Neel E A, Knowles J C and Young A M 2016 *Materials Science and Engineering: C* **60** 285–92

[7] Ruan Q, Zhang Y, Yang X, Nutt S and Moradian-Oldak J 2013 *Acta Biomaterialia* **9** 7289–97

[8] Baker M J, Trevisan J, Bassan P, Bhargava R, Butler H J, Dorling K M, Fielden P R, Fogarty S W, Fullwood N J, Heys K A, Hughes C, Lasch P, Martin-Hirsch P L, Obinaju B, Sockalingum G D, Sulé J, Strong R J, Walsh M J, Wood B R, Gardner P and Martin F L 2014 U *Nat. Protocols* **9** 1771–91

[9] Vongsvivut J, Pérez-Guaita D, Wood B R, Heraud P, Khambatta K, Hartnell D, Hackett M J and Tobin M J 2019 *Analyst* **144** 3226–38

[10] Hędzelek W, Marcinkowska A, Domka L and Wachowiak R 2008 *Acta Physica Polonica A* **114** 471–84

[11] Seredin P, Goloshchapov D, Ippolitov Y and Vongsvivut P 2018 *EPMA Journal* **9** 195–203

The traffic performance of the wheeled carrying liquid materials study

E P Plavelsky[1], R R Sharapov[1] and A E Plavelsky[2]

1 Moscow State University of Civil Engineering, Moscow, Russia
2 IC TsNIP SDM Co., LTD, Ivanteevka, Moscow region, Russia
E-mail: ptdm_zavkaf@mail.ru

Abstract. A feature of the wheeled vehicles capable of carrying liquid materials is the presence of the external disturbances in the form of interaction with the roadbed. This applies to peculiar transport vehicles, carrying liquid materials considered as non-Newtonian fluid. The internal disturbances in the form of the liquid material free surface interaction is the presence of the external disturbances in the form of interaction with the roadbed. Internal disturbances in the form of the wheeled concrete transport machine working vessel fluid free surface interaction should also be taken into account. The article describes the first part of the task - the interaction of the system with the roadbed. The simplest form of the system motion equations with rolling is in the accompanying coordinate system when written in the form of equations in quasi-coordinates. Based on the apparatus of nonholonomic mechanics, M.V. Keldysh built the theory of the rolling wheels with an elastic tire. Under certain assumptions, the above-mentioned theory in the article is extended to the case of curvilinear motion. These equations are represented as a system of first-order equations convenient for numerical solution.

1. Introduction

The wheeled concrete transport vehicles are specialized vehicles with variable mass geometry, among which the most widespread are the concrete mixers. Stability and controllability of the specified vehicles are a part of the main indicators of the vehicles' safety at certification tests. The accumulated experience of CNIPstroydormash (Russia) testing of the mixers in terms of the test site in accordance with CD 37.001.005-86, GOST R 52302-2004 and the standard test procedure of the mixers shows their considerable complexity and insecurity [1, 2].

The calculation of the performance characteristics of the concrete mixers' movement consists of two parts. In the first part, the forces of interaction of the mixer truck rolling wheel with the roadbed are considered. The second part deals with the characteristics of the roll, course and lateral stability of the mixer truck on a single or semi-trailer chassis. This article presents the first part of the study.

Let us consider the wheel mixer truck, rolling without slipping on a flat surface (Fig.1). The deformation of the pneumatic tire will be considered small and characterized by four parameters: the lateral displacement Y_b of the contact area center relative to the track of the middle plane of the wheel on the road surface, the angle δ_b twisting pneumatics, the angle χ slope of the plane of the wheel to the road surface and the radial deformation η pneumatics. Consideration of the parameters Y_b, δ_b, χ is determined by the theory of the elastic pneumatics rolling developed by the academician M. V. Keldysh [3]. The need to take into account the radial deformation of the truck tire mixer is due to some significant changes in the load on the wheel due to the road irregularities and the mobility of the transported material.

2. Theoretical study

The position of the mixer truck wheel is determined by the values, X^*, Y^*, χ, æ (Fig.1). Here X^*, Y^* are the coordinates of the point K of the greatest slope line intersection drawn in the plane of the wheel through its center, with the plane of the road; æ – is the angle formed by the track plane of the wheel on the road and the axis of the fixed coordinate system.

In accordance with the theory of academician M. V. Keldysh, the following conditions are imposed on the wheel rolling without slipping:

– tangent to the rolling line coincides with the axis of the contact area;

– the curvature of the rolling line is uniquely determined by the pneumatics deformation parameters y_b, δ_b, χ.

Rolling conditions without slippage can be written in the form of the following equations of nonholonomic bonds [3] superimposed on the i-th wheel of the mixer truck:

$$\dot{X}_i^* \sin(\varsigma_i + \delta_{bi}) - \dot{Y}_i^* \cos(\varsigma_i + \delta_{bi}) - \dot{Y}_{bi} = 0;$$

$$\dot{\varsigma}_i + \dot{\delta}_{bi} + (\alpha_i Y_{bi} + \beta_i \delta_{bi} + \xi_i \chi_i)[X_i^* \cos(\varsigma_i + \delta_{bi}) + Y_i^* \sin(\varsigma_i + \delta_{bi})] = 0; \qquad (1)$$

The reaction forces acting on the rolling wheel of the mixer truck are equivalent to the transverse force P_{bi}, the vertical force P_{vi} applied to the point K_i; the moment T_{bi} relative to the vertical axis and the moment $T_{\chi i}$ relative to the horizontal axis. Expressions for P_{bi} and P_{vi} forces, T_{bi} and $T_{\chi i}$ moments [3] have the form:

$$P_{bi} = a_i P_{bi} - e_i\chi_i;\ P_{vi} = -c_i\eta_i;\ T_{bi} = b_i\delta_{bi};\ T_{\chi i} = -d_i\chi_i + e_i Y_{bi}. \qquad (2)$$

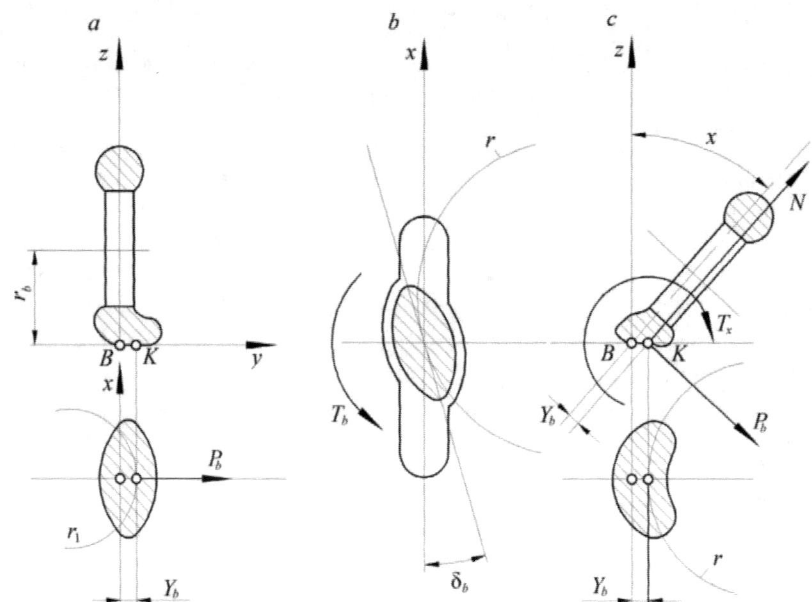

Figure 1. Interactions pneumatics with the roadbed: *a* – lateral movement; *b* – torsion; *c* – the inclination of the wheel plane.

In some cases, it is possible to simplify the equations of motion, in particular, the mixer truck, excluded from the ratio equations system (1). The order of the equations system describing the motion of the mixer truck on m wheels is reduced by 2m equations. Such a simplification is possible while neglecting transient processes in concrete mixer tires. Such neglect is acceptable if the time of the transition process in the tire τ_i is short compared to the shortest time τ of the transitional processes in the generalized coordinates describing the motion of the concrete mixer. In this case, it is possible to

use the assumption of movement with "high speed", according to which $\dot{Y}_{bi}$, $\dot{\delta}_{bi}$ are negligible compared to the wheel speed v_i. The calculation carried out by the dependences [3] shows that for the speed of the concrete mixer truck more than 10 mps $\tau_i \le 7\cdot 10^{-3}$ s. The time τ of transients at the coordinates describing the movement of the mixer truck is not less than 0.1 s [4].
The condition of the removal hypothesis applicability is in the form: (in [5]),

$$\frac{mV^2}{2} > 2k_b\Delta, \tag{3}$$

where Δ is the segment of the median line of the pneumatic tire involved in the deformation. For the concrete mixers, the limit value is $V \approx 10$ kmph. Thus, it is legitimate for the concrete mixers to use the "high speed" traffic assumption.
We write the equations of connections (1) in the coordinate system $0_i\ X_i\ Y_i\ Z_i$ associated with the center of the i-th wheel mixer truck. The axis 0_iX_i is directed along the motion of the mixer parallel to its plane of symmetry, axis 0_iY_i – to the left motion, 0_iZ_i axis – vertically upwards. The transformation is carried out in accordance with the dependencies:

$$\dot{X}_i^* = \dot{X}_i\ \cos\Omega - \dot{Y}_i\ \sin\Omega;\quad \dot{Y}_i^* = \dot{X}_i\ \sin\Omega - \dot{Y}_i\ \cos\Omega; \tag{4}$$

where Ω is the angle between the axes 0^*X^* and 0_iX_i.
Substituting the expression (4) in (1), taking into account that

$$\varsigma = \Omega + \nu_i, \tag{5}$$

where ν_i is the angle of rotation of the driven wheels.
As a result, we get the following expressions:

$$\dot{X}_i\ \sin(\delta_{bi} + \nu_i) - \dot{Y}_i\ \cos(\delta_{bi} + \nu_i) - \dot{Y}_{bi} = 0;$$

$$\dot{\varsigma}_i + \dot{\delta}_{bi} + (\alpha_i Y_{bi} + \beta_i\delta_{bi} + \xi_i\chi_i)[\dot{X}_i\ \cos(\delta_{bi} + \nu_i) + \dot{Y}_i\ \sin(\delta_{bi} + \nu_i)] = 0. \tag{6}$$

Note that $\dot{X}_i$ is equal to the speed of the i – th wheel V_i and that the deformation δ_{bi} and the rotation angles ν_i controlled wheels are small. Neglecting the small second order, we obtain:

$$\delta_{bi} + \nu_i - \frac{\dot{Y}_i}{\nu_i} - \frac{\dot{Y}_{bi}}{\nu_i} = 0;\quad \frac{\dot{\varsigma}_i}{\nu_i} - \frac{\dot{\delta}_{bi}}{\nu_i} + \alpha_i Y_{bi} + \beta_i\delta_{bi} + \xi_i\chi_i = 0. \tag{7}$$

In accordance with the theory, developed in [3], when driving at "high speed" we can neglect the speeds $\dot{Y}_{bi}$ and $\dot{\delta}_{bi}$ in comparison with V_i. So, we get the expression:

$$\delta_{bi} + \nu_i - \frac{\dot{Y}_i}{\nu_i} = 0;\quad \frac{\dot{\varsigma}_i}{\nu_i} + \alpha_i Y_{bi} + \beta_i\delta_{bi} + \xi_i\chi_i = 0. \tag{8}$$

It is possible to express the deformation characteristics Y_b, δ_{bi} from (8) in explicitly and substitute them in the expressions (2) for the forces acting on the mixer wheel. As a result, we obtain:

$$P_{bi} = -k_{1i}\left(\frac{\dot{Y}_i}{\nu_i} - \nu_i + \frac{\dot{\varsigma}_i}{\beta_i\nu_i}\right) - k_{2i}\chi_i = 0;\ T_{bi} = k_{3i}\left(\frac{\dot{Y}_i}{\nu_i} - \nu_i\right);\ T_{\chi i} = -k_{4i}\chi_i - k_{5i}\left(\frac{\dot{Y}_i}{\nu_i} - \nu_i + \frac{\dot{\varsigma}_i}{\beta_i\nu_i}\right); \tag{9}$$

$$k_{1i} = a_i\frac{\beta_i}{\alpha_i};\ k_{2i} = e_i + a_i\frac{\xi_i}{\alpha_i};\ k_{3i} = b_i;\ k_{4i} = d_i + e_i\frac{\xi_i}{\alpha_i};\ k_{5i} = e_i\frac{\beta_i}{\alpha_i}. \tag{10}$$

The conditions (9) are obtained using the Keldysh theory of rolling elastic pneumatics under the assumption of "high speed" motion. We demonstrate the relationship between these equations and the equations for the forces acting on the rolling wheel of the mixer truck, adopted in the technical theory of trucks stability.

Deviations of the motion of an auto concrete mixer from a straight line will be considered small. In this case, the coordinates X_M^*, Y_M^* of the center of the contact area of the point M are expressed by the formulas:

$$X_M^* = X^* - Y_b \sin\varsigma \cong X^*;\ Y_M^* = Y^* - Y_b \cos\varsigma \cong Y^* + Y_b, \tag{11}$$

where

$$\frac{dY_M^*}{dS} = \frac{dY_M^*}{Vdt} = \frac{1}{V}(\dot{Y}^* + \dot{Y}_b). \tag{12}$$

The curvature of the rolling line:

$$R^{-1} = \frac{d^2Y_M^*}{dS^2} = \frac{d}{dS}(\varsigma + \delta_b) = \frac{1}{V}(\dot{\varsigma} + \dot{\delta}_b). \tag{13}$$

Under the assumption of the Ω small, the speed $\dot{Y}_i^*$ can be considered equal to Y_i. Speeds $\dot{Y}_b$ and $\dot{\delta}_b$ are neglected in comparison with V_i in accordance with the assumption of movement with "high speed". Then:

$$\frac{dY_M^*}{dS} = \frac{\dot{Y}}{V};\ \frac{d^2Y_M^*}{dS^2} = \frac{\dot{\varsigma}}{V} \text{ and } \frac{\dot{Y}_i}{V_i} + \frac{\dot{\varsigma}_i}{\beta_i V_i} - \nu_i = \frac{dY_M^*}{dS} + \frac{1}{\beta_i}\frac{d^2Y_M^*}{dS^2} - \nu_i. \tag{14}$$

The coefficient β_b^{-1} in physical meaning has the dimension of length and is equal to [3]

$$\beta_i^{-1} = \frac{r_i}{2}. \tag{15}$$

Taking into account (19), the expression (14) can be written as follows:

$$\frac{\dot{Y}_i}{V_i} + \frac{\dot{\varsigma}_i}{\beta_i V_i} - \nu_i = \frac{1}{r}\left(\frac{dY_M^*}{dS} r + \frac{1}{2}\frac{d^2Y_M}{dS^2} r^2\right) - \nu_i, \tag{16}$$

where r is the length of the rolling line section equal to the rolling radius of the wheel.
The expression in brackets can be considered as the first members of the power series of the function Y_M^*. It should be noted that we can always put $Y_M^* = 0$ at $S = 0$. It is also possible to ignore the second member of the series by limiting it to the linear member only (this can be done because $r < 1$). In this case, we obtain a classical definition of the car wheel withdrawal angle [6], adopted at small angles:

$$\delta_i = \frac{\dot{Y}}{V} - \nu_i. \tag{17}$$

Substituting (21) in the expressions (9)-(11), we obtain the equations for the forces acting on the rolling wheel.

$$P_{bi} = -k_{1i}\delta_i - k_{2i}\chi_i;\ T_{bi} = k_{3i}\delta_i;\ T_{\chi i} = -k_{4i}\chi_i - k_{5i}\delta_i\ . \tag{18}$$

The formulas (18) are widely used in the vehicles' technical theory of stability and controllability [7]. The coefficients k_1, k_2, k_3, k_4, k_5 of the equations are generally nonlinear functions of the parameters χ, δ, and the wheel load G_k. They can be considered constant in the linear car model. The most significant effect on the motion mixer provides the lateral force resistance to the drift - $k_{1i}\delta_i$. In the concrete mixers, the normal load on the wheel during the movement varies in a fairly wide range [4]. The paper [7] proposes the following dependence of the drag coefficient k_1 on the normal load on the wheel G_k

$$k_1 = k_{y\ \max}\left[2.4\frac{G_k}{G_{opt}} - 1.8\left(\frac{G_k}{G_{opt}}\right)^2 + 0.4\left(\frac{G_k}{G_{opt}}\right)^{23}\right], \tag{19}$$

where $k_{y\,max}$ is the maximum value of the coefficient of resistance of the withdrawal as a function of the normal load; G_{opt} – is the value of the normal load at which $k_{y\,max}$ is achieved.
The expression (25) accurately approximates the dependence of the coefficient of resistance of the withdrawal of the normal load on the wheel in the interval $0 \le \frac{G_k}{G_{opt}} \le 2$. When $\frac{G_k}{G_{opt}} > 2$ physically unreasonable increase of the coefficient of resistance of withdrawal is observed. Therefore, the following dependence k_1 on G_k, free from this drawback, is used later:

$$k_1 = \begin{cases} k_{y\,\max}\left[2.4\frac{G_k}{G_{opt}} - 1.8\left(\frac{G_k}{G_{opt}}\right)^2 + 0.4\left(\frac{G_k}{G_{opt}}\right)^{23}\right] \\ npu \;\; 0 \le \left(\frac{G_k}{G_{opt}}\right) \le 2 \\ 0.8k_{y\,\max} \;\; npu \;\; \left(\frac{G_k}{G_{opt}}\right) > 2 \end{cases} . \tag{20}$$

The coefficients k_2, k_3, k_4, k_5 are further assumed to be constant. The normal load of G_{ki} is calculated by the formula:

$$G_{ki} = P_{vi} + G_{ki\,st}, \tag{21}$$

where $G_{ki\,st}$ is the gravity force on the i-th wheel of the mixer truck.
In the future, when determining the normal reaction of P_{vi} acting on the wheel, the force of elastic hysteresis or viscous resistance is taken into account. The expression for the normal reaction takes the form:

$$P_{vi} = -C_i(1 + j\varepsilon_i)\eta_i;\; P_{vi} = -C_i\,\eta_i - k_i\,\dot{\eta}_i, \tag{22}$$

where ε_i, k_i are the coefficients of hysteresis losses and equivalent viscosity resistance, respectively.
The expressions (18) and (22) define the forces acting on the rolling wheel of the concrete mixer truck.
The above-shown dependencies are considered from the standpoint of the established withdrawal, i.e. the curvilinear motion on a smooth surface. In real conditions of movement on the road irregularities, there are oscillations of the lateral force perceived by the wheel at a given angle of withdrawal. As a result, the lateral oscillations of the concrete mixer, caused by lateral compliance of the tires, can have a significant impact on the controlled movement of the concrete mixer. Therefore, in order to take into account, the lateral compliance of tires it is proposed [6] to introduce the concept of the contact area withdrawal angle δ_{ai}, defining it as the angle between the plane of the wheel and the velocity vector of the contact area. Then:

$$P_{bi} = -k_{1i}\,\delta_{ai} - k_{2i}\chi_j\;\; \dot{P}_{bi} = C_{1i}V_{1i}(\delta_i - \delta_{ai}), \tag{23}$$

where C_{1i} is the lateral stiffness of the pneumatics
Excluding δ_a from equations (29), we obtain:

$$-\frac{k_{1i}}{C_{1i}V_{1i}}\dot{P}_{bi} + P_{bi} = -k_{1i}\delta_i - k_{2i}\chi_i. \tag{24}$$

Thus, the expressions (18), (22) and (24) determine the forces acting on the rolling wheel of the mixer truck.
In fig. 2, according to the obtained formulas, the dependences of the drag coefficients for the lateral withdrawal of k1 tires 11.00R20 and 9.00R20 on the normal load at tire pressures of 0.84 MPa and 0.65 MPa, respectively, are calculated, proving with the results of experimental studies of MADI, NAMI. (Russia).

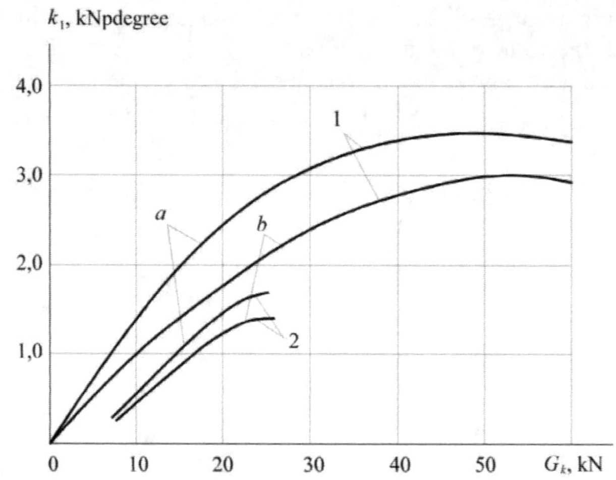

Figure 2. The dependence of the drag coefficients of lateral discharge 11.00R20 (1) and 9.00R20 (2) on the normal load at a tire pressure of 0.84 MPa and 0.65 MPa, respectively.

3. Summary

1. The wheeled concrete transport vehicles are the most vulnerable to interaction with the roadway due to additional disturbance from the side of the movable material (non-Newtonian fluid) during its transportation.
2. The use of the wheel rolling theory (elastic pneumatic), developed by the Academician Keldysh M.V., permits more precise consideration of the concrete-transporting vehicle wheels interaction with the roadbed.
3. The dependencies obtained from the developed formulas allow them to be used in the process of computer simulation of testing wheeled concrete transport machines, i.e. computational experiment, as close as possible to full-scale, which is not safe.

4. References

[1] Plavelsky E P and Sharapov R R 2017 *Int. Con. Ind. Eng. ICIE*, **206** 86-92
[2] Plavelsky E P and Sharapov R R 2017 Int. Sci. Con. En. Man. Mun. Tr. Fac. Transport EMMFT. Advances in Intelligent Systems and Computing 692
[3] Neymark Yu I and Fufaev N A 1967 Dynamics of nonholonomic systems Nauka Moscow
[4] Pavelsky E P 1987 Problems of increasing productivity and reducing the energy intensity of machines for concrete work. The issues of formation of the dynamic qualities of the wheeled concrete machines VNIIstroydormash Moscow
[5] Malinovsky E M and Gaitsgori M M 1974 Dynamics of self-propelled machines with hinged frame Mech. Eng., Moscow
[6] Khachaturov M M 1976 The dynamics of the system road–tyre – vehicle – driver, Mech. Eng. Moscow
[7] Litvinov A S 1971 Handling and stability of the vehicles Mech. Eng. Moscow numbered section of the paper

A new criterion for determination S-N curve of CFRP under tension

A G Kalinin[1], A A Lukyanchuk[1] and A V Pankov[1]

1 The Central Aerohydrodynamic Institute named after N.E. Zhukovsky (TsAGI), Zhukovsky, Russia
E-mail: aleksey.kalinin@tsagi.ru

Abstract. The article is devoted to development of fatigue failure criterion of the CFRP specimen under fatigue tension loading. The main goal of the current study is to develop a quantitative failure criterion for fatigue damage of a specimen with open hole under tension. In response to this problem, the influence of the following parameters on the stability of the criterion and on its application was determined experimentally: manufacturing technology of a material, the loading conditions during testing. The usage of the failure criterion of thickening in the hole zone made it possible to obtain S-N curve while using the classical criterion does not allow to get a S-N curve at some cases. The application of developed failure criterion of thickening made it possible to reduce the standard deviation of the logarithm of $S_{\lg N}$ durability. The dependence of the matrix damage area around the hole from the endurance at different loading levels was obtained according to the results of NDT method.

1. Introduction

The work is devoted to the second stage of experimental studies on the development of a failure criterion for fatigue tensile testing of carbon fiber reinforced polymer composites (CFRP). On the first stage [1] a method of monitoring of the specimen thickening in the hole zone during fatigue tensile tests was developed and applied. The main goal of the study is to develop a quantitative failure criterion for fatigue damage of a specimen. The influence of the following parameters on the stability of the criterion and on its application was determined experimentally: *manufacturing technology of a material, the loading conditions during testing.*

2. Background

The main documents regulating the static and fatigue testing of CFRP are currently the national standards: GOST, ASTM, EN and others. The existing list of standards is dominated by standards for static testing rather than fatigue testing of various types of specimens. For example, in the list of test standards (ASTM D4762), which contains several tens of standards, there are only three fatigue test methods – ASTM D7615, ASTM D3479 and ASTM D6115. The strength, reliability and durability of the aircraft structure is largely determined by the bearing capacity of the mechanical joints. The destruction of structures in most cases begins with the hole surface in the longitudinal and transverse joints. In such places, the material operates in a complex strain-stress state, while taking significant static and dynamic loads. To ensure the strength of the mechanical joint, it is necessary to estimate the safe allowable unit stress level in critical zones of the structure according to the strength and crack toughness. In this case, it is necessary to take into account the influence of climatic factors and material features, using existing models of CFRP failure and a definite set of fatigue characteristics of the structure material. When obtaining these fatigue characteristics of the material, the researcher is faced with the problem of the existence of a large number of types of fracture (modes). They depend both on the material and on the stacking sequence of the material, as well as on the type of loading. This work is devoted to the study of the fatigue characteristics of specimens with an open hole.

Fatigue testing of open hole or filled hole specimens is one of the most common in evaluating the fatigue characteristics of composite materials. When manufacture of specimen with a hole, it's necessary to give attention to the quality of the hole. CFRP mechanical drilling can affect the strength of the material. According to the TsAGI experience, the low quality of the hole manufacturing leads to a slight decrease in strength and, sometimes, to a catastrophic decrease in fatigue characteristics. For example, during drilling due to violation of the tool replacement time as it is worn out, breakouts and delamination of the material were produced in the hole zone, which led to a significant standard deviation of the logarithm of $S_{\lg N}$ was in the range 1.0 ÷ 2.0. The typical values of $S_{\lg N}$ for CFRP is in the range 0.15 ÷ 0.3. With such a range of characteristics, obtaining of *S-N* curve of material becomes almost impossible. The increased dispersion is associated with damage to the material on the sides of the hole. This is confirmed by research by many authors. For example, in [2], the effect of various drilling parameters on the quality of hole manufacturing is obtained. It is worth noting that due to CFRP anisotropy during drilling of the hole, formation of microcracks, fiber breakout, delamination, cracking, and temperature degradation of the material are received. To assess these damages, the parameter F_d (figure 1) was used:

$$F_d = \frac{D_{\max}}{D_0}, \tag{1}$$

where $D_{\max}$ – maximum area of the delamination zone, mm; D_0 – drill bit diameter, mm.

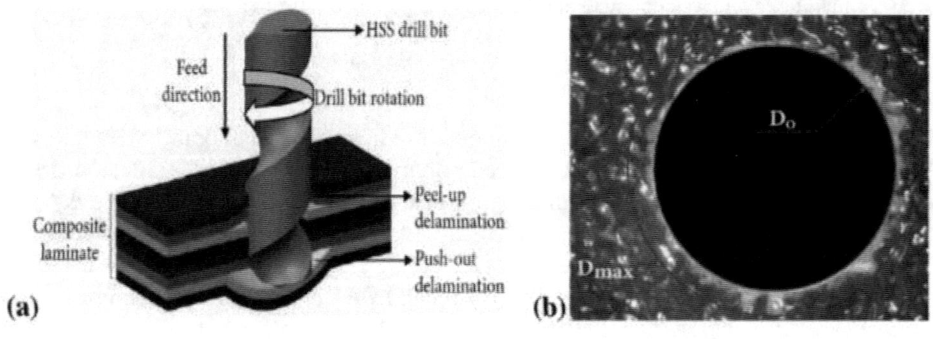

Figure 1. CFRP delamination due to drilling
(a) typical damage (b) determination of the delamination parameter of the material.

As a result, authors determinate that the values of the parameter F_d lie in the range 1.05÷1.25. Similar results were obtained in articles [3,4]. In [3] the case of manufacturing a hole using a laser machine was considered. As a result, the values of the stratification parameter F_d lie in the range 1.35÷1.58. In [4] the value of F_d lie in the range 1.12÷1.23. These initial damages affect the fatigue characteristics starting from the load level of 30% of the ultimate strength [3] and should be taken into account when obtaining the *S-N* curve of the material.

3. Experimental procedure

To define the *S-N* curve fatigue tests of CFRP specimens with a size of 210×36 mm and a 6 mm open-hole were conducted. Specimens made from two materials were tested. Fatigue tests were carried out on electrohydraulic machines by sinusoidal loading with stress ratio $R = 0$. R is the ratio of the minimum to maximum cycle stress amplitude. The maximum amplitude stress of the cycle varied in the range 0.8÷0.95 of the ultimate tensile strength. Monitoring the state of the material during the tests was carried out using an Sitescan D-20 ultrasonic flaw detector. The thickening of each specimen in the hole zone during loading was monitored using a specially designed and manufactured sensor, shown in Figure 2 (a). A typical diagram of the change in sample thickness in the hole zone from number of cycles is shown in figure 2 (b).

(a)

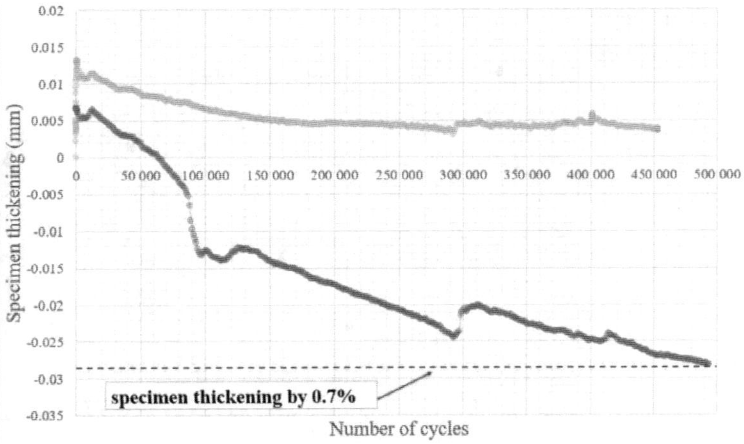

(b)

Figure 2. (a) thickening monitoring device mounted on the specimen (b) a typical diagram of the dependence of the thickness of the specimen in the hole zone on number of cycles.

4. Results and discussion

At the first stage, the dependence of the delamination area in the hole zone on the number of cycles was evaluated. Figure 3 shows the experimental data for two batches of specimens.

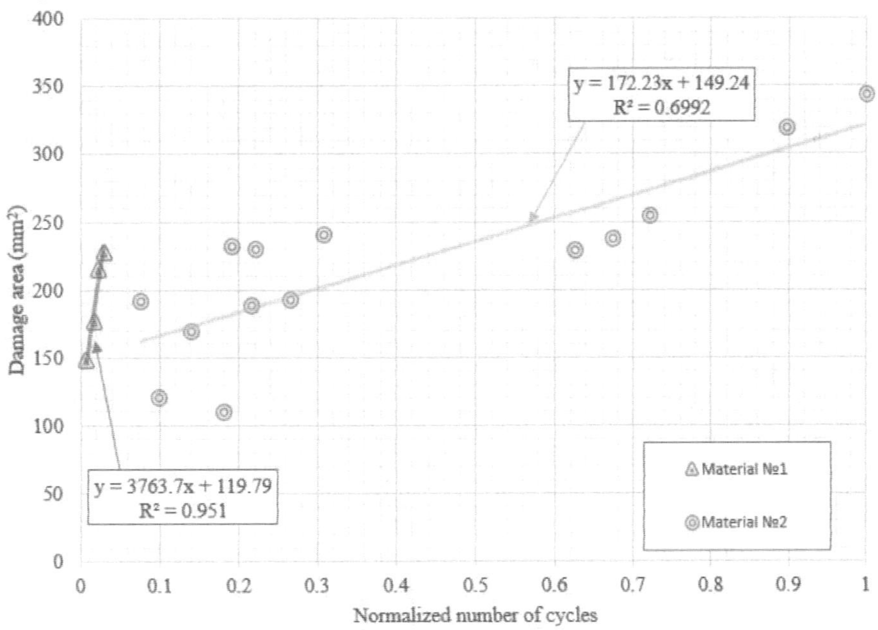

Figure 3. Dependence of the delamination area around the hole on the number of cycles.

At different number of cycles, each sample was stopped and ultrasonic testing was carried out, the results of which determined the shape and area of delamination around the hole. Figure 4 shows photos of the typical delaminations by ultrasonic testing in the process of fatigue loading. It is worth noting that the data approximation for both materials shows good convergence and the dependence of the delamination area on the number of cycles in the studied load range can be considered linear. The main reason for the appearance of delamination is the relatively low strength of the polymer matrix compared to the carbon

IOP Conf. Series: Materials Science and Engineering **744** (2020) 012025 doi:10.1088/1757-899X/744/1/012025

fiber. Over the past half century, the strength of carbon fiber has grown significantly, and the properties of the matrix have changed slightly. Therefore, if earlier delaminations appears mainly due to an improperly selected technical process, which led to increased porosity, unadhesiveness, low adhesive strength of the fiber-matrix interface, etc., now they are often the main mode of fracture of high-quality CFRP even in the absence of loads from the plane, which does not allow to use the maximum strength properties of the fiber.

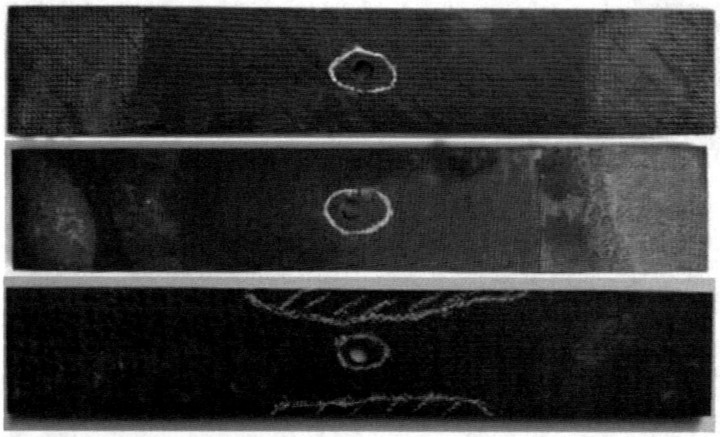

Figure 4. Specimens delamination areas, estimated by NDT method.

The problem of delamination is compounded by the lack of reliable methods for detecting them, which is related to the nature of the appearance of delamination: microcracks appear in the beginning, which practically do not reduce the strength characteristics of the material, the union of which into a macrocrack has an explosive, brittle nature. In addition, the exponent of the *S-N* curve for fractures by the delamination mechanism has significantly lower values compared to fracture forms along the fiber. The fracture of specimens with a free hole according to the delamination mode when determining the tensile *S-N* curve leads to a number of problems: it is impossibility to produce a *S-N* curve for the destruction of the specimen — a sample with extensive delamination does not reduce the residual strength; the need for continuous monitoring of the state of the sample, which in the absence of an automated procedure leads to increased dispersion of fatigue properties. It becomes the most critical, since according to the existing procedure [5] the value of $S_{\lg N}$ is definitive in the designation of safety factors in determining the life characteristics of an aircraft structure. The duration of the durability structure tests required to certify the design depends on this procedure.

At the second stage of the work, the application of different values of the developed criterion (Δ_{del}) was evaluated. The specimen was considered destroyed when the thickening of the specimen in the hole area reached a predetermined value $\Delta_f = \Delta_{del}$, where Δ_f – is the current value of the thickening, %. The fracture criterion Δ_{del} is measured in % of the initial thickness of the specimen.

The obtained experimental results are presented in table 1. $S_{\lg N}$ – is the dispersion of the *S-N* curve fatigue characteristics using the corresponding criterion, *CV* – is the coefficient of variation of the specimen delamination area determined using the corresponding criterion. Using the value of the criterion below 0.5% is impractical due to the fact that at such values it is not possible to define the delamination in the hole area by NDT method. Values higher than 0.75% cannot be used due to the fact that the thickness of the sample does not increase by such a value even with number of cycles of 10^7 cycles. Of the remaining options, the most stable results were shown by the fracture criterion equals to the specimen thickening by 0.7% of the initial thickness in the hole zone.

Table 1. Experimental results depending on the value of the fracture criteria and the scattering of durability and delamination area

Δ_{del}, %	$S_{\lg N}$	CV, %
0.50	0.32	–
0.63	0.20	39
0.70	0.17	26
0.75	–	25

5. Conclusions
The second stage of the study was carried out, as a result a criterion for fracture of a specimen under fatigue tensile loading was developed and tested. The use of the value of the fracture criterion Δ_f=0.7% made it possible to obtain the tensile *S-N* curve and reduce the standard deviation of the logarithm of the $S_{\lg N}$ durability to a level of 0.17 ÷ 0.32. A linear dependence of the delamination area around the hole on the number of cycles was obtained for different loading levels according to the results of ultrasound studies.

6. Acknowledgements
The research was financed by Ministry of Education and Science of the Russian Federation, under agreement No.RFMEFI62518X0044 (FENIKS). The authors wish to greatly acknowledge the project partner of FENIKS, Irkut Corporation.

References

[1] Lukyanchuk A A, Kalinin A G, Pankov A V and Svirskiy Y A 2018 Fracture criterion for tested CFRP specimens under tension *Procedia Structural Integrity* vol 13 (Elsevier B.V.) pp 1285–90
[2] Ismail S O, Dhakal H N, Popov I and Beaugrand J 2016 Comprehensive study on machinability of sustainable and conventional fibre reinforced polymer composites *Eng. Sci. Technol. an Int. J.* **19** 2043–52
[3] Kalyanasundaram D, Gururaja S, Prabhune P and Singh D 2018 Open hole fatigue testing of laser machined MD-CFRPs *Compos. Part A Appl. Sci. Manuf.* **111** 33–41
[4] Panchagnula K K and Palaniyandi K 2018 Drilling on fiber reinforced polymer/nanopolymer composite laminates: a review *J. Mater. Res. Technol.* **7** 180–9
[5] Pankov A V. and Konovalov V V. 2018 On equivalence of polymer composite materials structures loading programs under full-scaled service cycle testing *31st Congress of the International Council of the Aeronautical Sciences, ICAS 2018* (International Council of the Aeronautical Sciences)

Study on Preparation of Ceramsite from Shale Slag and Its Application

Changwei Xu[1] and Xinci Deng[2]
1 Professor, Shenyang JianZhu University, Shenyang, China
2 Postgraduate, Shenyang JianZhu University, Shenyang, China
E-mail: 1301663317@qq.com, 517499113@qq.com

Abstract. This paper studies the proportion of raw materials and process system for firing shale slag ceramsite. The raw materials and raw material ratio of the shale slag ceramsite were determined by the composition analysis of the shale slag; Through basic experiments, it is concluded that when the proportion of raw materials is shale slag: coal gangue: fly ash: expansion agent = 65 : 16 : 9 : 10 . Through optimization analysis, when the tube pressure strength is used as the evaluation index, the optimal plan A2B1C2 , the bulk density as the evaluation index, the optimal plan A1B3C2 , and the water absorption rate as the evaluation index, the optimal plan A3B1C3 . Conclusion: The best sintering system for shale slag granule firing is preheating temperature 50°C.heating time 20min , holding time 15min , heating to 1100°C, heating for 40min , and finally holding for 20min . Through the optimization analysis of the analysis and analysis of variance results, it is known that the firing temperature has the most significant effect on the ceramsite performance.

1. Introduction

With the deepening of people's understanding of ceramsite, ceramsite has become more and more widely used as artificial aggregate. Shale slag is a waste residue from shale retorting or combustion. In China, especially in Liaoning, a large amount of shale slag is discharged every year, which occupies land and pollutes the environment. The traditional artificial light aggregates mostly use natural resources such as clay as the main raw materials, and a large amount of mining will destroy the ecological environment. In this test, the shale slag is ground, a certain auxiliary material is added, and the spherical material is prepared according to the proportion of the designed raw materials, and then fired into ceramsite to study the influence of the sintering system on the performance of the ceramsite.

2. experiment

2.1. Experimental raw materials

The shale slag is from shale slag from the industrial waste shale slag (semi-coke) of Beipiao City, Liaoning Province. The chemical composition is shown in Table 1 ; the fly ash is used in Shenhai Thermal Power Plant, and the chemical composition is shown in Table 1 .

Table 1 Chemical composition analysis / %

Raw material	SiO_2	Al_2O_3	Fe_2O_3	CaO	MgO	K_2O	Na_2O
Coal gangue	51.17	30.73	9.45	1.12	1.60	1.10	0.70
Shale slag	54.54	14.48	6.27	8.95	3.80	3.39	1.00
Fly ash	53.47	30.48	4.73	1.12	1.40	2.52	0.90
Quartz sand	96.89	0.61	0.87	0.19	0.07	0.08	0.04

2.2. Experimental methods

The shale slag and coal gangue are ground by a ball mill in the laboratory, and the grinding time is about two hours. Then the ground raw materials are taken out, and the shale residue, coal gangue and fly ash powder are passed through a 200 mesh sieve. Screening, and then selecting the auxiliary materials to be mixed according to the established ratio. During the mixing process, a small amount of water is added (the amount of water is 35%~38% of the mixture) until the mixture becomes sticky and the ball is used. The disc is made into a spherical object of 10 mm to 15 mm . After granulation is completed, it is naturally dried in the curing room for 8 hours, then placed in a blast drying oven and dried at 105°Cfor 6 hours. Then, it is taken out from the blast drying oven, placed in a muffle furnace, and calcined according to a predetermined sintering system. After the sintering is completed, it is naturally cooled to room temperature in a muffle furnace, and the sintered ceramsite is taken out. The effects of raw material ratio, preheating temperature, holding time and heating rate on the properties of ceramsite were studied.

2.3. Test plan

The proportion of various raw materials in the experiment is shown in Table 2.

Table 2 Experimental ratio scheme/%

	Shale slag	Coal gangue	Quartz sand	Fly ash	Toner	Pulverized coal	manganese dioxide	Sodium carbonate	Sodium bicarbonate	Glass powder
1	57	38	-	-	5	-	-	-	-	-

2	56	8	20	13	-	3	-	-	-	-
3	56	8	20	13	3	-	-	-	-	-
4	65	16	-	9	10	-	-	-	-	-
5	65	16	-	9	-	10	-	-	-	-
6	65	16	-	9	-	-	10	-	-	-
7	65	16	-	9	-	-	-	10	-	-
8	65	16	-	9	-	-	-	-	10	-
9	65	16	-	9	-	-	-	-	-	10

3 results and discussion

3.1.Optimization scheme of sintering process system

Orthogonal experimental protocol:In addition to the influence of the type of foaming agent, the performance of shale slag ceramsite is also affected by factors such as the amount of blowing agent, firing temperature and sintering time. Therefore, this orthogonal test adopts three-factor and three-level orthogonal test. The foaming agent dosage is selected at three levels of 3% , 5% and 7% . The firing temperature is selected from three levels of 1050°C, 1075°C and 1100°C. The orthogonal experiment was carried out at three levels of 15 min , 20 min and 25min . The orthogonal table is shown in Table 4. The influence factors of fired shale slag ceramsite were studied by range analysis and analysis of variance.

(1) Evaluation index: 24h water absorption, bulk density, compressive strength.

(2) Influencing factors: A : firing temperature; B : holding time; C : foaming agent dosage

(3) Because this experiment uses the method of blending in the foaming agent, that is, the foaming agent is calculated into the mass ratio, so the method of adding the glass powder can be used to change the blending amount of the foaming agent while ensuring the quality of the remaining raw materials. The proportion remains unchanged.

Table 3 Orthogonal experimental scheme

Level	Burning temperature / °C	Holding time / min	Foaming agent dosage / %
	A	B	C
1	1050	15	3
2	1075	20	5
3	1100	25	7

Table 4 Orthogonal experimental data sheet

Test number	factor			Evaluation index
	A	B	C	24h water

				absorption / %
1	1	1	1	24.69
2	1	2	2	19.49
3	1	3	3	17.87
4	2	1	2	11.45
5	2	2	3	11.94
6	2	3	1	17.35
7	3	1	3	0.23
8	3	2	1	17.79
9	3	3	2	7.3
K1	20.68	12.12	19.94	
K2	13.58	16.41	12.75	128.11
K3	8.44	14.17	10.01	
R	12.24	4.29	9.93	

Range analysis: This table is based on a water absorption of ceramic as an evaluation index can be seen from Table Factors A very poor R maximum, is 12.24, indicating that the effect of firing temperature on the water absorption ratio of the maximum ceramisite. Under the influence of factor A , the average values of water absorption at three levels were 20.68 %, 13.58 %, and 8.44 %, respectively. Since the water absorption of ceramsite is as small as possible, the third level, ie, the firing temperature is 1100°C. Time is optimal. Whereas the comparative holding time B and a foaming agent dosage C, both poor R were 4.29 and 9.93 , so the dosage of the foaming agent C for the water absorption of the ceramic is greater than the influence of holding time B. In this two factors selected minimum level of water absorption, wherein the dosage of the foaming agent C corresponds to the first three levels of best, while the holding time B corresponds to the first is an optimal level.

Table 5 Analysis of variance of orthogonal test

Source of error	SS	f	S	F ratio	Significant
A	226.776	2	113.388	12.920	* (significant)
B	27.537	2	13.769	1.569	O (not significant)
C	157.868	2	78.934	8.994	O (not significant)
error	17.553	2	8.776	12.920>F0.1 (2 , 2) =9.00	

Analysis of variance: It can be seen from the data table of variance analysis that the effect of firing temperature on water absorption is significant in the three test factors, and the amount of foaming agent and holding time are not significant. From the comparison of F ratio, it can be concluded Effect of water absorption of the order of three factors: a firing temperature A > dosage of the foaming agent

C > holding time B . Corresponds to the results of the range analysis.In summary, the optimal solution for ceramsite water absorption is A3B1C3

Table 6 Variance experimental data sheet

Test number	factor			Evaluation index
	A	B	C	Bulk density / kg / m^3
1	1	1	1	740
2	1	2	2	790
3	1	3	3	630
4	2	1	2	810
5	2	2	3	900
6	2	3	1	700
7	3	1	3	600
8	3	2	1	700
9	3	3	2	170
K1	720	716.67	713.33	
K2	803.33	796.67	590	6040
K3	490	500	710	
R	313.33	296.67	123.33	

This table is based on the bulk density of the ceramic as an evaluation index can be seen from Table Factors A very poor R maximum, is 313.33 , indicating the effect of firing temperature on the ratio of the maximum bulk density of ceramic. Under the influence of factor A , the average values of the three levels of bulk density are 720 kg/m^3, 803.33 kg/m^3, 490 kg/m^3, respectively, because the bulk density of ceramsite is as small as possible, so the third The level, that is, the firing temperature is optimal at 1100°C. Whereas the comparative holding time B and a foaming agent dosage C , both poor R are 296.67 and 123.33 , the holding time B effect is greater than the bulk density of the foaming agent dosage ceramic C . Among the two factors, the lowest level of bulk density is selected, wherein the third level corresponding to the holding time B is optimal, and the blowing agent amount C corresponds to the first2 level is optimal. In summary, the order of influence on the bulk density of the three factors is: firing temperature A > holding time B > blowing agent dosage C ; the optimal solution is A3B3C2 .

Table 7 Variance Analysis Data Sheet

Source of error	SS	f	S	F ratio	Significant
A	158022.222	2	79011.111	7.493	O (not significant)
B	141355.556	2	70677.778	6.703	O (not significant)
C	29,622.222	2	14811.111	1.405	O (not significant)

error	21088.889	2	10,544.444

Analysis of variance: It can be seen from the data table of variance analysis that the influence of the three test factors on the bulk density is not significant, but from the comparison of the F ratio, the order of influence on the bulk density of the three factors can be obtained: the firing temperature A > holding time B > Volume foaming C .

In summary, the optimal solution for ceramsite bulk density is A3B3C2 .

Table 8 Orthogonal experimental data sheet

Test number	factor			Evaluation index
	A	B	C	Compressive strength / MPa
1	1	1	1	5.23
2	1	2	2	6.65
3	1	3	3	3.75
4	2	1	2	10.79
5	2	2	3	20.77
6	2	3	1	9.44
7	3	1	3	7.04
8	3	2	1	10.47
9	3	3	2	9.22
K1	5.21	7.69	8.38	83.36
K2	13.67	12.63	8.89	
K3	8.91	7.47	10.52	
R	8.46	5.16	2.14	

This table is a single grain of ceramic compressive strength as an evaluation index can be seen from Table Factors A very poor R maximum, is 8.46 , indicating that the effect of firing temperature on compressive strength ratio of the maximum Ceramisite . Under the influence of factor A , the average values of the three levels of compressive strength are 5.21 MPa , 13.67 MPa , 8.91 MPa , respectively. The higher the compressive strength of ceramsite, the better the ceramsite quality, so the second level is burnt. It is optimal when the temperature is 1100°C. Whereas the comparative holding time B and a foaming agent dosage C , both poor R were 5.16 and 2.14 , so the holding time B impact compressive strength greater than a foaming ceramic Volume C . Among these two factors, the highest level of compressive strength is selected, wherein the second level corresponding to the holding time B is optimal, and the blowing agent amount C corresponds to the third level. The level is optimal. In summary, the order of influence on the bulk density of the three factors is: firing temperature A > holding time B > blowing agent dosage C ; the optimal solution is A2B2C3 .

Table 9 analysis of variance data

Source of error	SS	f	S	F ratio	Significant
A	107.831	2	53.916	3.822	O (not significant)
B	51.109	2	25.555	1.811	O (not significant)
C	7.504	2	3.752	0.266	O (not significant)
error	28.216	2	14.108		

Analysis of variance: It can be seen from the data table of variance analysis that the effects of the three test factors on the compressive strength of single grain are not significant, but the influence of the three factors on the compressive strength of single grain can be obtained from the comparison of F ratio. order: firing temperature A > holding time B > Volume foaming C .In summary, the optimal scheme for the single-grain compressive strength of ceramsite is A2B2C3 , but it can be seen from the analysis of variance that the data error of this group is too large and cannot be used as the main reference data.

4 conclusions

In this stage, six groups of orthogonal experiments were carried out to investigate the effects of firing temperature, holding time and foaming agent dosage on the performance of ceramsite. The water absorption and bulk density of a total of 18 groups of ceramsite were studied. The test of compressive strength finally analyzed the orthogonal experimental data and obtained the following results:

(1) The experiment finally chose sodium carbonate as a blowing agent.

(2)It is concluded by orthogonal experiment that among the three factors of firing temperature, holding time and foaming agent dosage, the most important influence on the performance of ceramsite is the firing temperature.

(3) The optimum firing temperature is 1100°C, the optimum holding time is 15 min , and the blowing agent is added to7%. At this time, the water absorption rate was 0.23%, the bulk density was 600 kg/m3 , and the compressive strength was 7.04 MPa .

references

[1] Deng Jiaping , Zhang Minghua , Li Zelin , et al . Research and process design of ultralight ceramsite fired from oil shale slag [J].

[2] Wang Shujuan, Jin Yujie et al . Study on physicochemical properties of oil shale semi-coke from different producing areas . Journal of Jilin Jianzhu University .

[3] Li Yunfei , Liu Ji , Deng Jiaping . Comprehensive utilization of oil shale ash in building materials . "Brick and Tile" No. 7 of 2018 .

[4] Tan Liquan, Hu Xianghong, Yu Mei, Zhou Jianmin et al . Progress in the application of oil shale ash slag . Chemical engineer .

[5] Li Zhaofu, Zhao Xin et al . Feasibility study on smelting silicon-aluminum-iron alloy in Fushun shale waste residue . Recycling .

[6] Zhou Jianmin , Niu Xianchun et al . Source and comprehensive utilization technology of oil shale

ash . Journal of Guangdong Institute of Petrochemical Technology .
[7] Yang Shi . Discussion on laboratory test and small test operation of ceramsite raw material sample . Brick World .

IOP Conf. Series: Materials Science and Engineering **744** (2020) 012027 doi:10.1088/1757-899X/744/1/012027

Design of Joint Test Device for Fuel Delivery Shut-off Valve

Guangming Kong[1,*], Lonxiang Zhu[1], Gang Zou[1], Zhanyong Wang[1], Ding Zhang[1], Dehui Yu[1]

1 Qingdao Branch of Naval Aeronautical Institute, Qingdao, China
E-mail: kguangming@126.com

Abstract. In view of the fact that the new aircraft units were used, it was common to lack the fuel accessory inspection equipment, a set of combined test equipment for fueling and fuel delivery shut-off valve was designed. The basic work of the fuel delivery shut-off valve was introduced, the composition and principle of fueling test system and electric control system were described, and the realization methods of the hardware and software were given. The practical application shows that the test equipment can meet the test requirements of this type of fuel accessories, fully meeting the design goals.

1. Introduction

The fuel delivery shut-off valve (referred to as the shut-off valve) is the core control accessory of the baby fuel delivery system of certain types of aircrafts. The main functions of shut-off valves include: (1) when the baby is fully fueled normally or in emergency during pressure fueling, the shut-off valve cuts off fuel fueling path; (2) when the fuel in the baby of the machine is exhausted, the shut-off valve cuts off fuel delivery path automatically to prevent the dangerous fuel spills caused by failing to cut off the fuel circuit in time.

The monitoring of the baby pressure on the surface of the aircraft is automatically executed by a shut-off valve located on the fuel line, the performance of which has significant impact on the security and reliability of the aircraft fueling and transportation system. Thus, special performance testing is required on a regular basis in the daily maintenance of the aircraft. However, the lack of land performance testing system for the shut-off valve obstacles the repair work of the aircraft. In that case, we design and develop an fuel delivery shut-off valve testing system for target aircrafts, which demonstrates the capacity for all kinds of performance testing and fully meets the design requirements

2. Working principle of shut-off valve

The basic structure of the shut-off valve is shown in Figure 1. It mainly consists of a case, a fueling control component, an oil delivery control component, a valve component and an emergency annunciator.

The case is installed with a fueling joint A and a delivery joint D; the fueling control component consists of a fuel electromagnetic switch 1 that control the fueling status, and a check valve; the oil delivery control component is composed by a delivery electromagnetic switch 2 that controls the delivery status, and a check valve 5; the valve component mainly include a large valve of rubber diaphragm 4, a check valve 5, a spring, etc., which controls the switch of main channel; the emergency annunciator mainly consists of a shutter, a spring and a micro switch, controlling the emergent cut-off of fueling status. The basic working principle is described in the follows text.

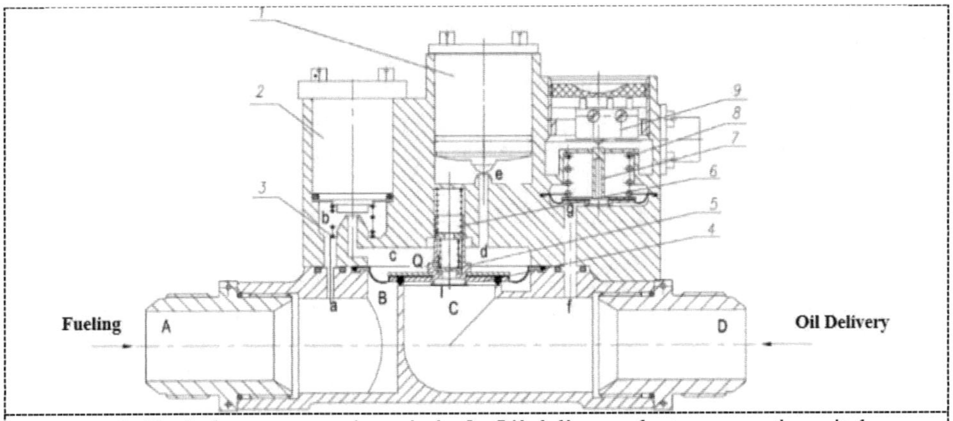

1- Fuel electromagnetic switch; 2- Oil delivery electromagnetic switch;
3- Spring; 4- Large valve of rubber diaphragm; 5- Check valve; 6- Spring;
7- Small valve of rubber diaphragm valve; 8- Spring; 9- Emergency annunciator;
Figure 1. The basic structure of the shut-off valve.

2.1. Pressure fueling state control

When the aircraft fueling normal pressure fueling to the baby through the shut-off valve, it is in a high flow rate. The delivery electromagnetic switch 2 keeps open and the fueling electromagnetic switch 1 is energized to open. Then the fuel of the aircraft fueling control path flows downstream through the path a-b-c-d-e-f, so that the pressure of Q chamber is lower than the pressure of B chamber. The large valve of rubber diaphragm 4 opens under pressure, leading fuel flow to the baby of the aircraft through A-B-C-D to complete pressure fueling. When the baby is full, the full fuel annunciator in the baby sends a power off signal to control the shut-off valve of the fueling electromagnetic switch 1 to be powered off, closing port e. And the fuel is guided into the Q chamber through flow a-b-c, balancing the pressure between the Q chamber and the fueling inlet chamber B. Under the action of the spring 6, the large valve of rubber diaphragm 4 is closed to stop the fueling.

If the full fuel annunciator in the baby fails to send control signal, the fueling process may continue. Then when the pressure of g port reaches 0.08 MPa, the shutter compression spring in the emergency annunciator 9 will opens the micro-switch, so that the fuel-operated electromagnetic switch 1 is turned off and the fueling is stopped.

2.2. Fueling status control

It is in a small flow working state when the baby shut-off valve supplies fuel to the aircraft. In this state, delivery electromagnetic switch 2 keeps open while the fueling electromagnetic switch 1 keeps close at gate e. The fuel is guided to flow to engine through h-c-b-a, which makes the pressure in chamber Q is lower than that in chamber C. Under the action of the pressure, the rubber valve 4 is opened, and the fuel flows to the main tank of the aircraft through D-C-B-A to achieve fuel delivery. When the baby is exhausted, the full fuel signal transmitter in the main tank sends a power-on signal to control the delivery electromagnetic switch 2 to be energized, which makes b port closed and the check valve 5 open, controlling air flows into chamber Q through port h. So that the pressure between chamber Q and transportation entrance port C equals. Then under the elastic force of the spring 6, the rubber shutter 4 is closed to prevent a large amount of air from flowing into the tank.

3. The composition and function of the fuel delivery shut-off valve test device

There are three technical standards for dynamic testing of shut-off valves: delivery flow resistance and closing time, fuel flow resistance and closing time, and fueling emergency shutdown time test. Following the technical standards, our integrated test device is mainly composed of a fueling test system and an automatic control system.

3.1. Fueling Test System

According to the technical standards of the shut-off valve, the composition and working principle of the shut-off valve test system is shown in Figure 2 Since the flow of fueling is about 9 times of transportation flow, they cannot share the same pump source. So the fueling system has three separate components: fueling, delivery and flow resistance test. When performing fueling and delivery performance tests, the flow resistance test section must be connected to pressure point X and pressure point Y.

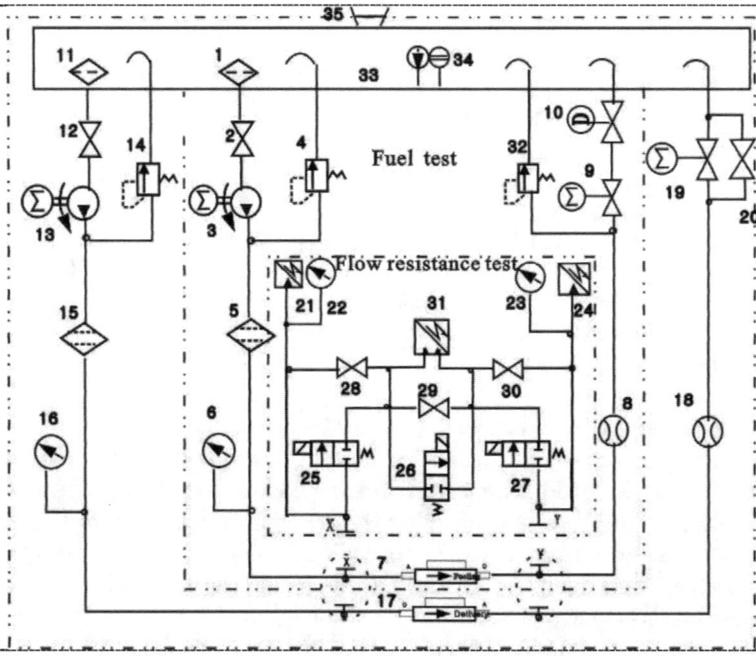

1、5、11、15-filter; 2、12-fuel valve;3、13-oil pump; 4、14- pump outlet relief valve ;16-pressure meter; 7、17-Fuel delivery shut-off valve; 8、18-turbine flow transmitter; 9、19-Sero regulating valve; 10-electric flow-off valve; 20-bypass ball valve; 21、24- differential pressure transmitter of accessory import/ outlet; 25、26、27-differential pressure transmitter control solenoid valve; 28、29、30- differential pressure transducer manual control switch; 31- differential pressure transmitter;

Figure 2. The composition and working principle of the shut-off valve test system.

By setting the large and small flow pump, the performance of the shut-off valve can be detected, which is fueling and delivery flow, pressure and pressure loss when passing shutting off valve. Therefore, there are quantitative requirements for the flow capacity and pressure of the system, and it's required to be able to adjust the flow capacity and pressure. We use variable frequency motor and intelligent servo regulating valve which can work coordinately to achieve fueling system flow and pressure adjustment. The closing time test of the valve is strictly executed in accordance with the definition of the navigational aid, which is real time collected by the sensor and measured by the PLC.

(1) Fueling test section

The main function of fueling test section is to provide a pressurized fuel source with a rated fueling flow to the fueling joint A of the shut-off valve. For the pressure fueling test, the fueling joint A shall be imported and connected to the large pipeline of the fueling test section. Mainly composed of oil pump 3, relief valve 4, filter 5, turbine flow transmitter 8, servo regulating valve 9, electric flow-off valve 10 and accessory outlet relief valve 32. During this part of operation, the fueling electromagnetic switch is opened and the shut-off valve is energized to open, then the oil pump 3 is driven by the variable frequency motor to suck fuel from the tank 33, and the fuel is pressurized by the fuel pump and sent to the fueling port A of the shut-off valve 7. When the fuel passes through the accessory, the inlet pressure, outlet pressure and flow resistance are detected by the flow resistance test section, and then flow back to the fuel tank through the turbine flow transmitter, servo regulating valve and electric flow-off valve.

(2) Delivery test section

The main function of delivery test section is to provide the pressurized fuel source with the rated fuel flow rate to the fuel delivery port D of the shut-off valve. During the fuel delivery test, the fuel connection connector D should be used as the inlet and connected to the small pipeline of the fuel delivery test section. So the fuel flows in reverse direction with that when fueling. The system composition is similar to that of the fueling part, except that the oil flow is small, the electric outlet valve and the safety valve are not provided at the outlet, and only one auxiliary ball valve is added to adjust with the servo regulating valve. When this part is working, the fuel supply solenoid valve of the shut-off valve does not need to be energized, and it is normally open.

The pump 13 is composed of a three-phase motor and a pipeline centrifugal pump (IHGB25-160) with continuously variable transmission. The power is 1.5KW; the rated flow is 4m3/h.

(3) Flow resistance test section

The flow resistance test section mainly detects the inlet and outlet pressure and the pressure difference, in manual or automatic modes.

In the manual test mode, it first confirms that the left oil inlet switch 28, the balance switch 29, and the right oil inlet switch 30 of the differential pressure transmitter 31 are all in the close position. When the test conditions are met, the front and rear pressures of the tested accessories are observed through the accessory inlet pressure gauge 22 and the accessory outlet pressure gauge 23. When the current pressure difference does not exceed the range of the differential pressure transmitter 31, three switches are opened according to a preset sequence: the balance switch 29 of the differential pressure transmitter, the left oil inlet switch 28 and the right oil inlet switch 30. After that the differential pressure transmitter balance switch 29 is closed. The signal of the differential pressure transmitter 31 is to PLC or the digital display to display the differential pressure of the shut-off valve value. When the flow resistance test is completed, the differential pressure transmitter balance switch 29 is opened first, and then the differential pressure transmitter left oil inlet switch 28 and the differential pressure transmitter right oil inlet switch 30 are closed.

When the PLC is working automatically, those three switches (28/29/30) of the manual function are in the close position. Three solenoid valves (25/26/27) are used to replace the work of the three switches. When the fuel flow entering the shut-off valve inlet reaches the test condition, the flow resistance value can be detected. The inlet /outlet pressure is collected by the accessory inlet pressure transmitter 21 and the accessory outlet pressure transmitter 24, respectively. And the pipeline inlet and outlet pressure difference is collected by the differential pressure transmitter 31. The three solenoid valves are controlled by the PLC according to the preset working procedure, and the inlet and outlet pressures and differential pressure values are collected in real time.

3.2. Automatic Control System

The automatic control system is based on tablet PC, PLC, intelligent regulating valve and frequency converter. It monitors and adjusts the working process of the shut-off valve, fully meeting the requirements of automatic testing, such as two-way flow, the inlet and outlet pressure difference of the accessory, the shutter opening and closing characteristics and the emergency closing characteristics.

(1) Hardware composition

The structure block diagram of hardware is shown in Figure 3. The system hardware is mainly composed of flat panel controller, PLC, servo regulating valve, electric flow-off valve, frequency converter, flow transmitter, pressure transmitter, input and output module, solenoid valve, switch button and printer. The upper computer uses an industrial tablet. The lower computer PLC and the industrial tablet calculator are applied to the electronic control system as well. Advanced sensor and actuator are also used to build the hardware foundation of the automatic control system. The logic control of the working process can be programmed and implemented by programmable controllers.

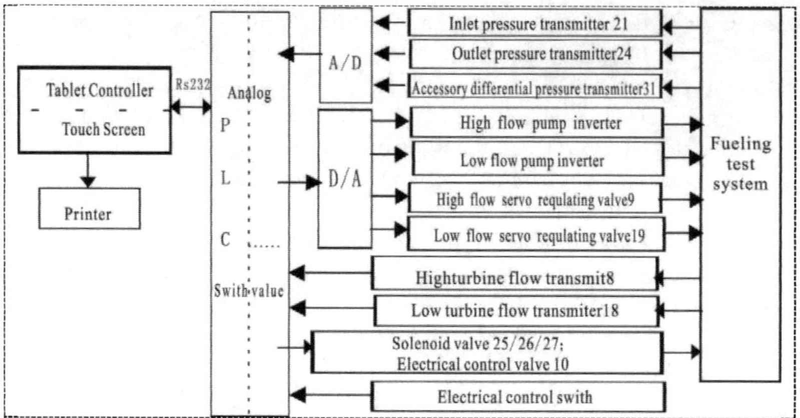

Figure 3. The structure block diagram of hardware.

(2) Software part

The software is the core of the automatic control system monitoring hardware work. The tablet controller program is compiled by Delphi2010. It communicates serially with the PLC and other devices through the RS232 interface, and displays all the status information in the field test process in real time. The PLC program is shown in Figure 4.

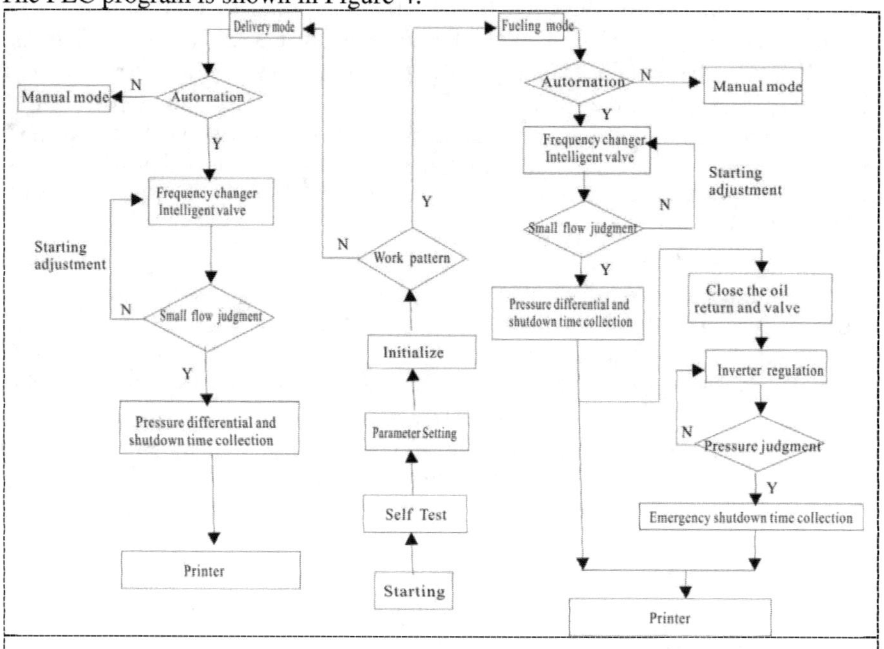

Figure 4. The PLC program.

4. Technical characteristics

4.1. Dual pump source settings

The fueling flow rate of the shut-off valve is relatively much larger than the transportation flow rate, the difference of which can be nearly 10 times. Therefore the test system is designed with two pump sources and a double-line test section for large and small flow. Taking into account the pressure and flow requirements of the tested components, as well as the noise, the pump source consists of two three-phase frequency conversion motors and centrifugal pump.

Connect the fuel delivery port D of the shut-off valve to the inlet, and connect it to the pipeline of the fuel test part. The pipeline status is the fuel delivery state (see Figure 5(a)). The test procedure is designed as follows: ① energize the product fuel delivery solenoid valve, start the fuel delivery pump 13, adjust the opening speed of the fuel pump and adjust the opening degree of the fuel return valve 19 to achieve the fuel delivery rated test flow. PLC collects the fueling pressure loss (pressure difference)of the shut-off valve in real time; ② power off the fuel delivery solenoid valve. The valve is closed after receiving the electric signal. PLC collects the flow rate through the turbine flow meter and calculates the valve fueling closing time in real time. The closing time is calculated from time of the electromagnet power-off signal, and ended when the flow rate from the turbine flow sensor is less than 15% of the rated flow rate.

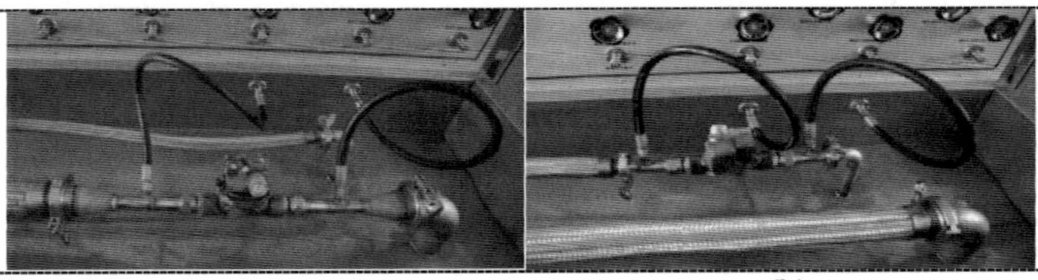

(a) (b)

Figure 5. Fuel delivery test process.

4.2. Detection of fueling emergency shutdown time

The function of shutting off the valve and fueling: When the aircraft ground pressure is fueling, if the normal full oil level control fails, the fuel oil surface continues to rise, and after the baby is full of oil, when the pressure in the baby increases to 0.08 MPa, the diaphragm in the valve is cut off. Pressing the micro-actuated electric door to close the fueling solenoid valve, the oil-removing valve is closed under the pressure of the membrane chamber, and the fueling circuit is cut off.

In order to simulate the actual working condition of the shut-off valve, the design of the fueling and returning oil pipeline is to adjust the rotation speed of the fuel pump 3 through the frequency converter to close the outlet pressure of the oil pump and the shut-off valve after the electric oil return valve 10 is used to close the fuel returning oil circuit. When the outlet pressure reaches 0.08 MPa, the valve should be able to be shut down in an emergency, and the PLC collects the flow value in real time through the turbine flow meter and calculates the emergency shutdown time.

5. Conclusion

We have designed and implemented a fuel delivery shut-off valve test system and tested it by an aviation repair factory. The evaluation demonstrates that the fuel delivery shut-off valve test system is sophisticatedly and reasonably designed. The operating of the system is reliable and stable, and the electronic control system based on tablet PC, PLC and inverter technology is powerful and efficient. The control and test automation of the shut-off valve is also supported. In conclusion, the shut-off valve test system is not only fully functional, comprehensive, compact and safe, but also has the advantages of flexible control and convenient operation. It can complete all the performance testing of a certain type of aircraft fueling cut-off valve, and fully meets the design requirements.

6. References

[1] Wang Zhanyong. The Design of Flow Proportional Control Valve Test Equipment[J]. Hydraulics Pneumatics & Seals, 2017 (8): 173-176.

[2] Bai Jie. Design of Aircraft Pneumatic Accessories Test Stand's Testing System [J]. Journal of civil aviation university of china, 2010(2): 29-31.

[3] Honeywell Engines. Systems & Services, Component Maintenance Manual with Illustrated Parts List (CMM36-10-99) [G]. Arizona, 2002.

[4] Fang Changlin. Hydraulic pneumatic transmission and control. China machine press, 2000.

[5] ZHAO Weiyi, Wang Zhanyong. The design of pneumatic components integrative testing system for an airplane [J]. Machinery Design & Manufacture, 2012, 2(2): 236-238.

IOP Conf. Series: Materials Science and Engineering **744** (2020) 012028 doi:10.1088/1757-899X/744/1/012028

Study on Activation of Waste Clay Brick Powder

Xu Changwei1,a and Yang Xutong1,b

1 Professor, Shenyang JianZhu University, Shenyang, China
2 Postgraduate, Shenyang JianZhu University, Shenyang, China
E-mail: a1301663317@qq.com, b574904865@qq.com

Abstract. Recycling and recycling of construction waste is a hot research topic nowadays. In this paper, clay bricks in building waste are activated by physical method (grinding) and chemical method (alkali activation). The effects of fineness after grinding time, types and dosage of activator on the activity of waste brick powder are studied. The results show that with the increase of grinding time, the specific surface area of waste brick powder increases, and the activity coefficient of waste brick powder increases. When grinding time is 90 minutes, the activity index reaches 67%. When the specific surface area of waste brick powder is 467 m2/kg, the activity index of waste brick powder increases first and then decreases with the increase of calcium hydroxide content. When calcium sulfate dihydrate is added alone, the activity index of waste brick powder increases first and then decreases with the increase of calcium sulfate dihydrate content. When calcium hydroxide and calcium sulfate dihydrate are mixed together, the increase of activity index is better than that of calcium hydroxide and calcium sulfate dihydrate alone. The 4% calcium hydroxide + 3% calcium sulfate dihydrate composite system has the best activation effect on brick powder.

1.Introduction

Abandoned clay brick produces a huge amount, accounting for more than 50% of the total construction waste, but it has been in a low utilization rate situation [1-2]. The main components of waste clay brick are SiO_2 and Al_2O_3, which have potential activity. In this paper, the possibility of using waste clay brick as admixture after grinding into fine powder is discussed. The waste clay brick is recycled [3-4]. After crushing, grinding for a certain time, grinding into powder. By means of physical grinding and chemical activation, the waste brick powder can be activated. By compounding with ordinary Portland cement with different dosage, the influence of dosage on the properties of composite cementitious materials is studied, and the optimum fineness and dosage of waste brick powder are finally determined. It has an important guiding effect for the application of waste brick powder as admixture in concrete, mortar and

IOP Conf. Series: Materials Science and Engineering **744** (2020) 012028 doi:10.1088/1757-899X/744/1/012028

other aspects.

2.Experimental scheme

2.1 Raw materials

Cement: in this experiment, 42.5 common portland cement was produced in the east of Hebei Province. Its chemical composition is shown in Table 1, and its physical properties are shown in Table 2[5-6].

Table 1 Chemical composition of cement (%)

SiO_2	Al_2O_3	Fe_2O_3	CaO	MgO	SO_3	R_2O	Loss
21.72	5.81	4.33	62.41	1.73	2.56	0.54	1.47

Table 2 Physical properties of cement

Strength grade	Fineness /(m^2/kg)	Stability	Setting time		Compressive strength /MPa		Flexural strength /MPa	
			Initial	Final	3d	28d	3d	28d
42.5	300	Qualified	45min	600min	17.0	42.5	3.6	6.5

Waste clay brick and brick powder: the abandoned sintered bricks used in experiments were taken from Wang Jia Tun, Shenyang. The buildings being demolished were served for 50 years, and the intensity was 5.05MPa. The waste clay brick is crushed and grinded for a certain time to obtain brick powder. The specific surface area of waste clay brick powder was determined by nitrogen adsorption analyzer. The chemical composition of waste clay brick powder is shown in Table 3.

Table 3 Chemical Constituents of Waste Clay Bricks (%)

Component	SiO_2	Al_2O_3	Fe_2O_3	K_2O	MgO	Na_2O	CaO	TiO_2	Other
Content	62.01	21.11	7.00	3.37	2.00	1.60	1.45	0.99	0.43

Activator: in this experiment, calcium hydroxide and calcium sulfate two hydrate were used as activators in the production of industrial grade calcium hydroxide and Chemical Reagent Co., Ltd. produced by Liaoning Yi Dan calcium Co., Ltd.

Water reducing agent: Polycarboxylic acid superplasticizer, solid content (>20%), water reducing rate (>30%).

Water: tap water

2.2 Design of Experimental Scheme

In order to obtain better properties of composite cementitious materials, physical grinding, chemical activation were used and we ran a comparison with different amounts of waste clay brick powder [7], among these three kinds of waste clay brick powder with grinding time of 70 min, 80 min and 90 min, brick powder with large specific surface area and high activity index after grinding is selected as cementitious material; calcium hydroxide stimulation, calcium sulfate dihydrate reagent stimulation and their combination are used. Alkali excitation, select the two groups with the highest activity index; in the composite cementitious materials with different brick powder content, choose the one with the highest strength; synthesize above three conditions, configure the samples to verify whether the performance indicators meet the national standards of cement.

3. Activation of waste clay brick powder

3.1 Activation of Physical Grinding Activity

After crushing the waste clay brick, the same size of the debris is used for ball milling. During the grinding process, the debris of the waste clay brick gradually decreases until it becomes powder, the specific surface area increases, the surface energy increases, and the active points on the surface of the waste clay brick powder increase [8-9], so that the activity of the waste clay brick powder increases. The specific surface area and activity index of brick powder change with the increase of grinding time during physical grinding as shown in Table 4.

Table 4 Activity Coefficient of Grinded Brick Powder

	Specific surface area / m^2/kg	Compressive strength /MPa		Activity coefficient /%	
		3d	28d	3d	28d
Cement	352	18.9	40.8	100	100
70min	408	6.9	23.4	36.5	57.4
80min	432	7.2	25.3	38.1	62.0
90min	467	7.7	27.4	40.7	67.2

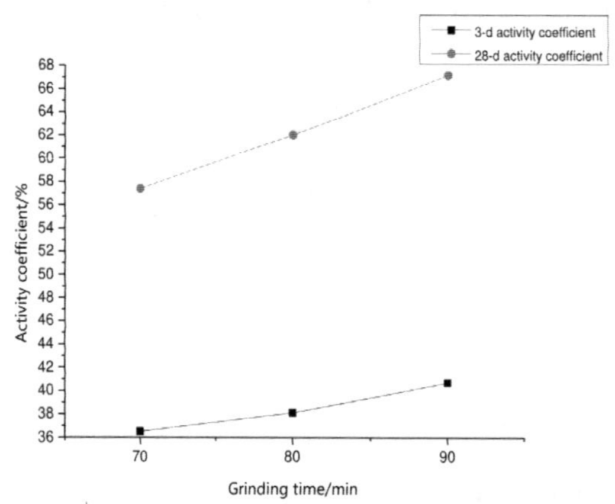

Fig. 1 Effect of grinding time on activity of brick powder

As it can be seen from Fig. 1, the activation coefficient of brick powder increases with the increase of grinding time. Because the grinding time is prolonged, the fineness and specific surface area of waste clay brick powder increases gradually, which improves the activity of waste clay brick powder[10]. However, with the prolongation of grinding time, the increasing rate of activity coefficient decreases gradually. Considering the economic cost and other aspects, 90 minutes of grinding brick powder is selected for the test.

3.2 Activation of Brick Powder by Calcium Hydroxide

Calcium hydroxide was used to stimulate the activity of waste clay brick powder. The activity coefficients of waste clay brick powder were tested with different dosages of calcium hydroxide (0%, 2.5%, 3%, 3.5%, 4%). Fig. 2 shows the effect of calcium hydroxide dosage on the activity coefficients.

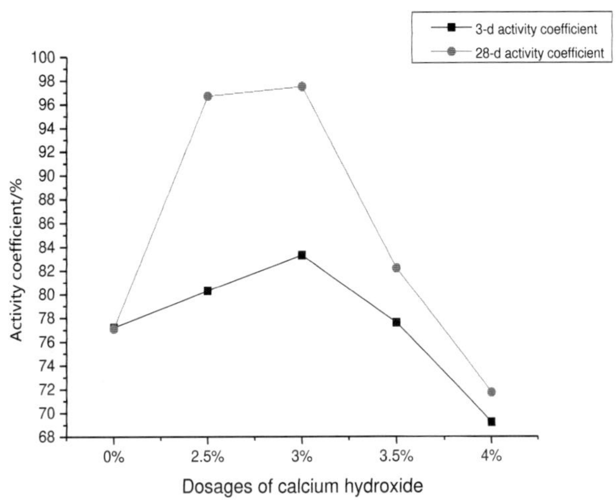

Fig. 2 Effect of calcium hydroxide content on activity

As it can be seen from Figure 2, with the increase of calcium hydroxide content, the activity coefficient of alkali-activated waste clay brick powder first increases and then decreases. When the content of calcium hydroxide is 3%, the activity coefficient is the highest, the activity coefficient is 97.5%.

3.3 Activation of Brick Powder by Calcium Sulfate Dihydrate

Calcium sulfate dihydrate was used to stimulate the activity of waste clay brick powder. Calcium sulfate dihydrate with different dosages (0%, 2.5%, 3%, 3.5%, 4%) were used to stimulate the activity coefficient. The influences of different calcium hydroxide dosages on the activity coefficient were tested. Figure 3 shows the effect of calcium hydroxide dosage on the activity coefficient.

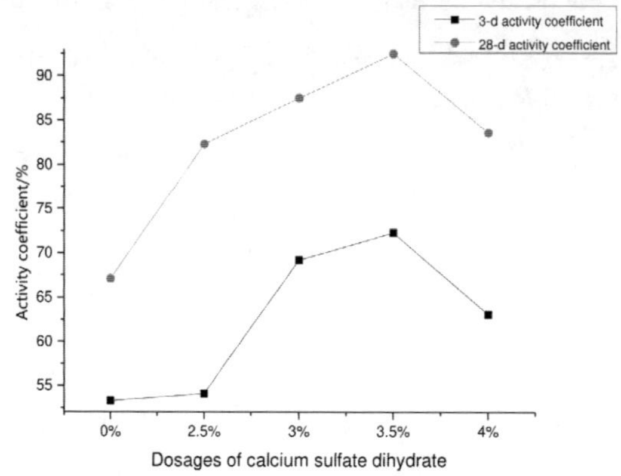

Fig. 3 The effect of calcium sulfate dihydrate on the activity of brick powder

As it can be seen from Fig. 3, with the increase of calcium sulfate dihydrate content, the activity coefficient of alkali-activated waste clay brick powder first increases and then decreases. When the content of calcium sulfate dihydrate is 3.5%, the activity coefficient is the highest, and the activity coefficient is 92.5%.

3.4 Complex Activation of Brick Powder by Calcium Hydroxide-Calcium Sulfate Dihydrate

The mixed system of calcium hydroxide and calcium sulfate dihydrate was used to activate the activity of waste clay brick powder. The activity coefficient was tested. Fig. 4 shows the effect of calcium hydroxide-calcium sulfate dihydrate complex system on the activity coefficient.

Among them, group 1:3% calcium hydroxide + 3.5% calcium sulfate dihydrate; group 2:4% calcium hydroxide + 3% calcium sulfate dihydrate; group 3:2% calcium hydroxide + 4% calcium sulfate dihydrate.

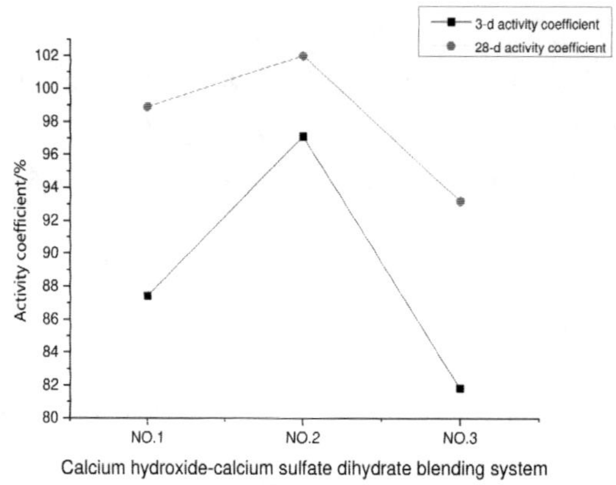

Fig. 4 Effect of calcium hydroxide-calcium sulfate dihydrate blending system on activity coefficient

As it can be seen from Fig. 4, the activity index of compound doping 2 is the highest among the three groups. Although the compound group 1 is the best content of single doping, the compound effect is not the best.

3.5 Dosage Test of Waste Clay Brick Powder

In order to determine the maximum proportion of brick powder, the ratio of 0 to 60% substitution was selected to explore the influence of substitution amount on the strength of mortar. Fig. 5 and 6 respectively show the effect of brick powder substitution on the flexural strength and compressive strength of cement mortar specimens.

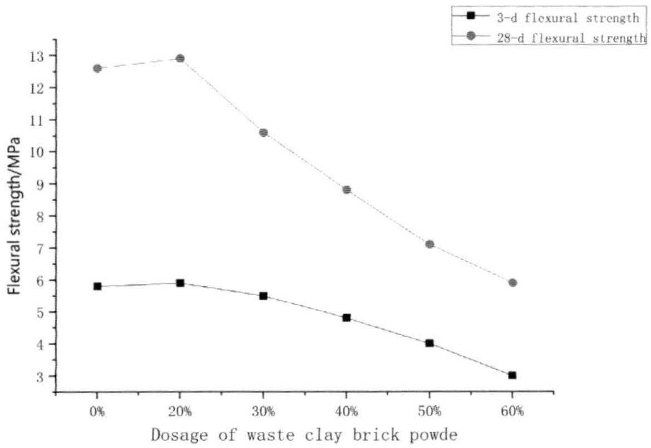

Fig. 5 Effect of brick powder substitution on flexural strength

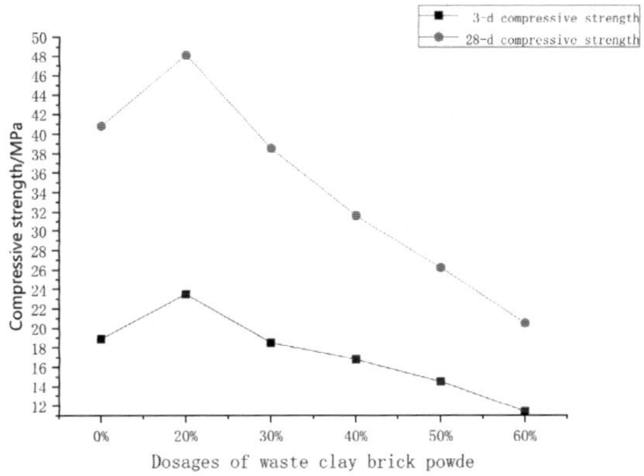

Fig. 6 The influence of brick powder substitution on compressive strength

From Fig. 5 and Fig. 6, it can be seen that the flexural strength and compressive strength of mortar specimens increase first and then decrease with the increase of the replacement amount of waste clay brick powder. When the replacement amount of brick powder is 20%, the 3-day flexural strength is 5.9 MPa; the 3-day compressive strength is 23.5 MPa; the 28-day flexural strength is 12.9 MPa; and the 28-day compressive strength is 48.1 MPa. When the replacement amount of brick powder is more than 20%, the flexural strength and compressive strength of mortar specimens have been declining, and the mechanical properties have declined significantly.

4 Conclusion

(1) Activated by physical grinding, the specific surface area and activity coefficient of waste clay brick powder increase with the increase of grinding time. When grinding time is 90 minutes, the activity coefficient reaches 67%.

(2) When the waste clay brick powder activated by chemical activation is activated by calcium hydroxide alone, the activity coefficient of the waste clay brick powder first increases and then decreases with the increase of calcium hydroxide content. When the content of calcium hydroxide is 3%, the activity coefficient is the largest, and the activity coefficient is 97.5%. When calcium sulfate dihydrate is taken as the only activator, the activity coefficient increases with the increase of calcium sulfate dihydrate content. The activity coefficient of waste clay brick powder increases first and then decreases. When the dosage of calcium hydroxide is 3.5%, the activity coefficient is the highest, 92.5%. When the dosage of calcium hydroxide-calcium sulfate dihydrate is 4% calcium hydroxide + 3% calcium sulfate dihydrate, the activity coefficient is the highest, and the activity coefficient is 102%.
(3) In the experiment of brick powder content, the flexural strength and compressive strength of mortar specimens increased first and then decreased with the increase of replacement amount of waste clay brick powder. When the replacement amount of brick powder is 20%, the flexural strength and compressive strength are the highest, the 28-day flexural strength is 12.9 MPa, and the 28-day compressive strength is 48.1MPa.

Acknowledgement
This project is funded by the National Key R&D Program of China（2018YFD1101001）.

References:
[1] Cheng Haili, Tian Haize. Potential Activity of Waste Clay Brick Powder [J]. Renewable Resources and Circular Economy, 2014, 7 (12): 25-27.
[2] Han Tao, Jin Xiuzhi. Evaluation and Application of Recycled Cementitious Material for Waste Clay Bricks [J]. Building Material Technology and Application, 2015 (03): 10-12.
[3] Lu Jing, road preparation, Cao Sai, Li Peixin, Wang Huiting, Shen Hong Li. Application of construction waste brick powder in preparation of cement admixture. [J]. comprehensive utilization of fly ash, 2012 (02): 37-39+42.
[4] Abdul Basit,M Aftab Khan,Ibrar Ahmed,M Nasir Ayaz Khan,M Umar. Effect of brick dust and lime powder on the performance of plain cement concrete[J]. IOP Conference Series: Materials Science and Engineering,2018,414(1).
[5] GB 175-2007 General Portland Cement
[6] Reading guide for the thirteenth five year plan outline for national economic and social development [J]. new bibliography of the country, 2016 (05)
[7] Cao Cao Lei, Wang Yinsheng, Jia Meixia. Waste brick powder was modified to prepare cement mixture [J]. China cement, 2011 (12): 48-50.
[8] Li Xinjian, Han Tao, Jin Xiuzhi, Han Tiebao, Gongpeng. Effect of waste clay brick granules on mortar strength [J]. Building materials technology and application, 2011 (03): 1-3.
[9] Liu Rongtao, Zhu Jianhui, Zhu Weijie, Hou, Wu Wenhan. Summary of comprehensive utilization of building waste clay brick resources [J]. silicate bulletin, 2016,35 (10): 3191-3195.
[10] Yan Fang. Effect of waste clay brick on firing and properties of recycled cement clinker [D]. 2013.

IOP Conf. Series: Materials Science and Engineering **744** (2020) 012029 doi:10.1088/1757-899X/744/1/012029

Ultrasonic welding aluminium alloy 7075 to titanium alloy Ti-6Al-4V: parameter optimization and mechanical property

LIU Shibo[1], Bi Guijun[1], Man Yijun[1], Sun Xianchong[1], Lee Bing Yang[1], Pan Dayou[1], and Cheng Chek Kweng[1]

1 Singapore Institute of Manufacturing Technology, 73 Nanyang Drive, 637662, Singapore
E-mail: liu_shibo@simtech.a-star.edu.sg

Abstract. The demand for lightweight materials has increased for automobile and airplane industry. Traditional fusion welding techniques always generate brittle intermetallic compound (IMC). Ultrasonic spot welding (USW) is a promising technique to joint dissimilar metals and suppress IMC formation. This study aims to investigate the role of process parameters as well as surface roughness on the joint shear strength of titanium alloy Ti-6Al-4V and aluminium alloy 7075. Process parameters namely penetration depth, welding time and pressure were varied and the joint shear strength obtained was investigated. Results obtained were indicative that penetration depth and pressure should be used with highest values and surface roughness with lowest. In summary, optimal parameters are USW penetration depth 0.7mm, welding time 1.8s, pressure 0.7 MPa and surface roughness of AA7075 0.1457 Ra μm and Ti-6Al-4V 0.0651 Ra μm (both ground by P1200 SiC paper) in this study.

1. Introduction

Lightweight materials, such as titanium alloy and aluminium alloy, have high ratio of strength to weight, which can reduce weights of automobiles and airplanes with sufficient mechanical properties and thus can improve fuel economy, reduce CO_2 emissions and save overall cost. Welding titanium alloy and aluminium alloy is necessary but difficult due to their inherent different properties. For example, the melting points of Al and Ti are 660 ℃ and 1668 ℃, respectively. The difference brings problems to concurrently melting of base materials. The density of Al and Ti are 2.70 g/cm2 and 4.51 g/cm2, respectively. At high temperature, Al liquid tends to float upon Ti liquid. Thermal conductivity of Al and Ti are 247 W/(m·K) and 11 W/(m·K), respectively. The difference of thermal conductivity leads to melting volume difference and cooling rate difference during welding.

Besides the above problems, the often observed intermetallic (IMC) which is hard and brittle also increase the dissimilar welding difficulty. As fusion welding, laser welding, laser roll welding, and arc welding always generate brittle intermetallic compound (IMC), such as $TiAl_3$, $TiAl_2$, TiAl and Ti_3Al_5, at the joint interface due to high reaction rate of liquid phase. Casalino et al. [1] suggest IMC mostly made of $TiAl_3$ is formed in laser butt joining of T40 and 5754 Al alloy. Ozaki et al. [2] observed Ti_3Al_5 and TiAl during laser roll welding of commercial pure titanium H4600 to A5052 aluminium alloy. As a subdivision of both liquid-phase welding and solid-state welding, diffusion bonding produces IMC at interface of Al/Ti alloy dissimilar joining. Sohn et al. [3] identified $Al_5Si_{12}Ti_7$ and $Al_{12}Si_3Ti_5$ in diffusion bonding of commercial pure Ti to 1050 Al using an Al-10.0wt.%Si-1.0wt.%Mg filler at 620℃. Friction stir welding (FSW) is a solid state welding, whose process temperature is lower than that of the fusion welding. However, IMCs still occasionally form in FSW of Al/Ti and their alloys. Choi et al. [4] investigated FSW of pure Ti and pure Al. They concluded that the formation order of the Ti and Al intermetallic compound layer is $TiAl_3$, TiAl and Ti_3Al. The formation of $TiAl_3$ is attributed to the lowest free energy of formation of $TiAl_3$, and the formations of TiAl and Ti_3Al are attributed to the Al diffusion from the $TiAl_3$. Because IMC can deteriorate the quality of welded joint, the formation of IMC should be prevented or, at least, its thickness should be kept to the lowest possible value. Some works suggest that acceptable joint strength can be achieved in the condition that IMC layer thickness is less than 10 μm [5].

Ultrasonic spot welding is a promising technique in the dissimilar joining of Al/Ti and their alloys. In ultrasonic welding process, processing parameters and welded materials play an important role in controlling the formation of IMC of Al/Ti and their alloys. Zhang et al. [6] investigated aluminium alloy AA6111–TiAl6V4 dissimilar joint using high power ultrasonic spot welding (USW). No visible intermetallic compounds were found, even using transmission electron microscopy. However, they also claimed that by prolonged isothermal heat treatment at high temperature, IMC can be made to form in Al–Ti USW. Wang et al. [7] investigated USW of AA 5754 and Ti64. Using pure Al interlayer, $TiAl_3$ was identified by XRD. However, using same USW parameters without Al interlayer, no IMC was found by XRD. Sridharan et al. [8] claimed that bond formation occurred without any intermetallic formation. Zhang et al. [9] observed no obvious intermetallic reaction layer in the AA2139–TiAl6V4 welds by USW, even using transmission electron microscopy. According to Zhang et al. [6], no visible IMC formation is due to kinetics of the process rather than solubility limit effects alone. The inter-diffusion rate of Al-Ti system and energy barrier of IMC nucleation are both important to IMC nucleation speed. However, inter-diffusion in the Al–Ti system has a relatively high activation energy, the ultrasonic welding energy input is low, and the welding time is short, all of the factors limited the growth of IMC layer. The objective of this paper is to investigate effects of process parameters on mechanical property of shear strength of USW joint.

2. Experimental procedure

2.1. Materials

The materials used in this research were Ti-6Al-4V in dimension of 1mm×10mm×48mm and Aluminium Alloy 7075 (AA7075) - T6 in dimension of 1.02mm×10mm×48mm. The true stress-strain curve of the received Ti-6Al-4V sheet is in figure 1(a). The yield strength, ultimate tensile strength and maximum strain are about 1000 MPa, 1100 MPa, and 0.1 for Ti-6Al-4V alloy, respectively. AA7075 is solution heat treated and artificially aged (T6) with chemical composition shown in table 2. The true stress-strain curve of the received AA7075-T6 is in figure 1(b).

Table 1. Chemical composition of base material AA7075-T6.

Al	Si	Fe	Cu	Mn	Mg	Cr	Zn	Ti
Bal.	0.07	0.29	1.6	0.03	2.5	0.19	5.70	0.02

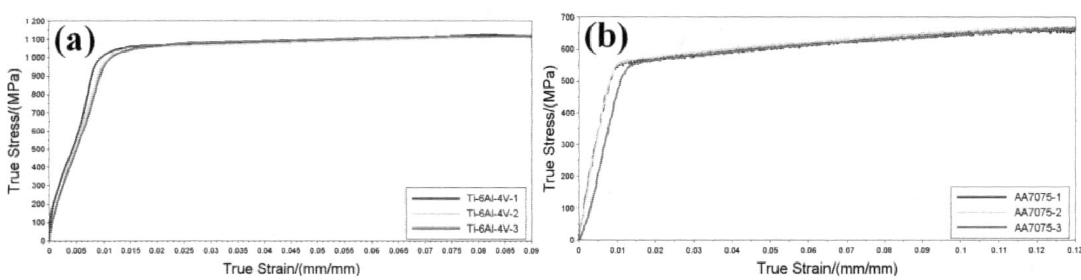

Figure 1. True stress-true strain curve obtained at strain rate $10^{-3}s^{-1}$: (a) T-6Al-4V, (b) AA7075.

2.2. Ultrasonic spot welding (USW) process

AA7075-T6 with dimensions of 1mm×10mm×48mm was welded on top of the Ti-6Al-4V with dimensions of 1.02mm×10mm×48mm. The overlap length was about 15 mm. The schematic diagram is shown in figure 2. Because of the existence of oxide films of AA7075 and Ti-6Al-4V, such as Al_2O_3 and TiO_2, faying surfaces were gradually sanded to grit number of P1200 by SiC paper to remove the oxide film and reduce surface roughness. Faying surface were washed by ethanol before USW. USW process was performed on ultrasonic metal welding SPM 4200 (serial no. NP-M-20-4200) manufactured by Guangzhou SUNTOP Ultrasonic Co Ltd. The welding power of the machine is 2.2kW and the vibration frequency is 20kHz.

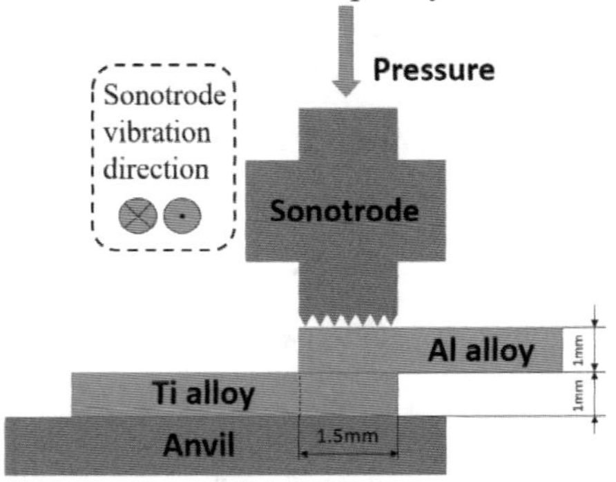

Figure 2. A schematic diagram of ultrasonic spot welding (USW) of AA7075 and Ti-6Al-4V.

After preliminary experiments, a DOE screening, using 13 different group of process parameters and repeating 3 times for each group, was randomly conducted to study the USW factors. DOE screening parameters are shown in table 2. The USW factors are pressure, welding time and penetration depth. The factor pressure is the pressure of compressed air, which is used by the USW machine to drive the sonotrode downward. The factor welding time is the vibration time during welding process. The factor penetration depth is the maximum depth that sonotrode tip can penetrate into the soften aluminium. To analyse effect of welding time, 15 samples were welded by USW process. Holding pressure at 0.7MPa and penetration at 0.7mm, 5 levels of welding time from 1.4s to 2.2s with 0.2s increment were used to join AA7075 plate to Ti-6Al-4V plate.

Effect of surface roughness on weld shear strength in USW process were analysed. AA7075 plate and Ti-6Al-4V plate were gradually ground to different roughness with grit size P400, P800 and P1200 SiC papers. Then plates were welded by USW process at pressure 0.7MPa, penetration depth 0.6mm, and welding time 1.6s.

2.3. Tensile shear test

Tensile shear test was conducted on INSTRON 5982 machine with a constant speed of 1mm/min. A typical test sample after welding and a typical test sample in tensile shear test are shown in figure 3. AA7075 plate and Ti64 plate were attached to each sample to ensure the applying force was parallel to ultrasonic welded surface. The red rectangular area was taken as weld area and was measured for every sample. Shear strength was calculated as shear force divided by the red rectangular area.

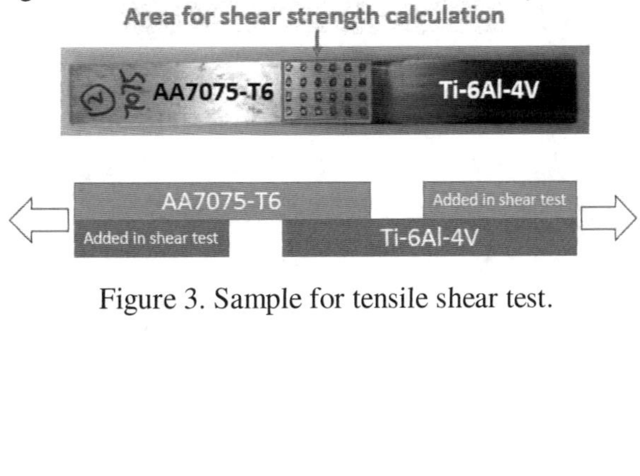

Figure 3. Sample for tensile shear test.

3. Results and discussion

3.1. Mechanical shear tests

Tensile shear strengths of USW samples are presented in table 2. For each USW parameter, 3 samples were welded. Their tensile shear strengths were marked as shear strength 01, strength 02 and strength 03. There were 13 different groups of USW parameters, thus 39 weld joints in total were shear tested.

Table 2. DOE screening of USW parameters on joining Ti-6Al-4V and AA7075-T6.

Pressure, MPa	Welding time, s	Penetration depth, mm	Shear strength 01, MPa	Shear strength 02, MPa	Shear strength 03, MPa
0.5	0.8	0.6	10.21	4.04	0
0.5	0.8	0.7	0.03	8.98	7.13
0.5	1.1	0.5	0.03	0	0
0.5	1.4	0.5	13.06	0	0
0.5	1.4	0.7	0	8.96	7.27
0.6	0.8	0.5	2.7	0	0
0.6	1.1	0.6	14.25	6.19	4.75
0.6	1.4	0.7	7.75	8.91	15.04
0.7	0.8	0.5	0	0	0
0.7	0.8	0.7	0	9.34	5.66
0.7	1.1	0.7	11.43	12.81	12.99
0.7	1.4	0.5	8.8	0	0
0.7	1.4	0.6	10.94	6.31	6.07

USW pressure, welding time and penetration depth all have positive effects on weld shear strength within the studied range. Based on statistical results of shear strengths in DOE screening, effects of USW parameter on shear strength are analysed by MINITAB and results are shown in figure 4.

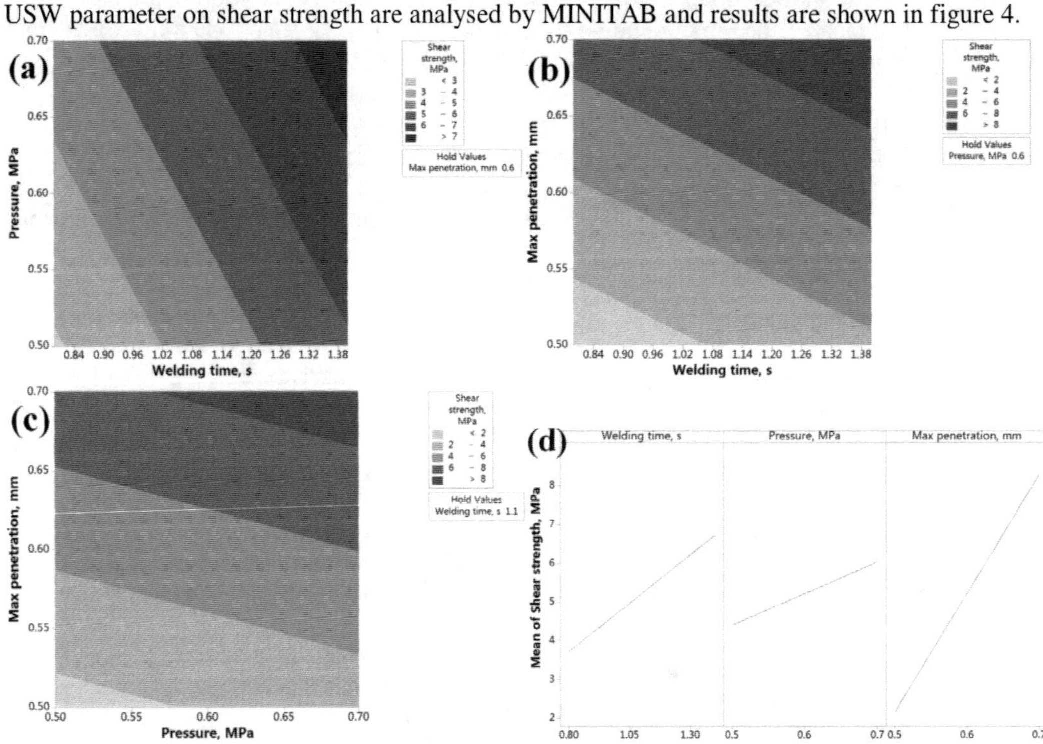

Figure 4. Processing parameters effects on shear strength of USW joint: (a) Shear strength vs. pressure and welding time, (b) Shear strength vs. penetration and welding time, (c) Shear strength vs. penetration and pressure. (d) Main effects plot for shear strength by fitted means.

3.2. Effect of pressure

In the studied pressure range, from 0.5MPa to 0.7MPa, shear strength increase as pressure increase while holding penetration depth and welding time constant. During USW process, ultrasonic vibration frequency and vibration distance is determined by USW machine and should no vary. If pressure is increased, the friction force at the interface of AA7075 and Ti64 will be increased, leading to the increase of input vibration energy at the surface. With little energy dispersed into atmosphere, the increased vibration energy transforms into higher material temperature. The strengths of interface materials are reduced due to temperature increase. With higher pressure and reduced material strength, plastic deformations of interface materials happen more easily to form intimate contact between metallic surfaces which leads to a solid state bond and joint strength increase.

Because 0.7MPa is the maximum value that can be achieved at present lab condition, the analysis of shear tests of DOE screening samples leads to the conclusion that the optimal welding pressure is 0.7 MPa for present ultrasonic welding.

However, it should be noted that too high pressure would lead friction force too large, causing relative friction motion in the specimen, or even reduces the effective amplitude. The bonding area is reduced and weakens the contact interface surface for the material being crushed. [10]

3.3. Effect of penetration depth

In the studied range of penetration depth from 0.5mm to 0.7mm, shear strength increase as penetration depth increase while holding pressure and welding time constant. During USW process, sonotrode sinks into softened AA7075 plate. Penetration depth prevent sonotrode go too deeply into welded plate by limiting sonotrode downward length. Sonotrode teeth may not firmly hold AA7075 plate if penetration depth is small. When penetration depth is reached in middle of USW process, the sonotrode stops moving downward. The pressure on AA7075 plate would be released gradually due to softening effect, causing an effect similar to gradually decreased pressure, and thus reduced joint shear strength. In this point of view, a penetration depth that is large enough is crucial to a good USW bonding. However, AA7075 cross-section area would be decreased with a large penetration depth, causing the decrease of load bear ability of welded AA7075 plate. AA7075 plate may even be broken into pieces with too large penetration depth.

At the penetration depth of 0.5mm, samples generally yield low shear strengths due to not sufficient penetration depth. Despite of two samples that yield 0.03MPa and 2.7MPa (relatively low comparing with shear strength of other samples), 11 samples out of 15 test samples are not welded with penetration depth of 0.5mm. According to analysis of figure 4, shear strength increase as penetration depth increase within the studied range. Preliminary experiments showed that AA7075 would be broken into pieces when penetration depth is higher than 0.7mm. Thus, the penetration depth of 0.7mm is the applicable highest value. The optimal value of penetration depth is 0.7mm for present USW process.

3.4. Effect of welding time

In the studied range of welding time from 0.8s to 1.4s, shear strength increase as welding time increase while holding penetration depth and pressure constant. Enough welding time should be used in USW process, to break surface oxides and bring Ti and Al alloys surfaces into intimate contact to form metallurgical bond. Sufficient welding time also promote elements diffusion across interface of dissimilar Ti and Al alloys.

Since shear strength increase as welding time increase, and 1.4s has no reach the machine limit, a further analysis of variance (ANOVA) is conducted focusing on welding time, while holding pressure at 0.7MPa and penetration depth at 0.7mm. The ANOVA experiments and their results are shown in table 3 and in figure 5. The data of welding time of 0.8s and 1.1s are from previous DOE screening analysis.

Table 3. ANOVA analysis of USW[a] welding time on joining Ti-6Al-4V and AA7075-T6.

Welding time, s	Strength 01, MPa	Strength 02, MPa	Strength 03, MPa
1.4	18.06	14.76	10.21
1.6	14.58	12.34	12.05
1.8	20.72	12.40	13.08
2.0	5.83	14.9	14.19
2.2	16.26	3.65	20.69

[a] other process parameter were 0.7MPa pressure and 0.7mm penetration depth.

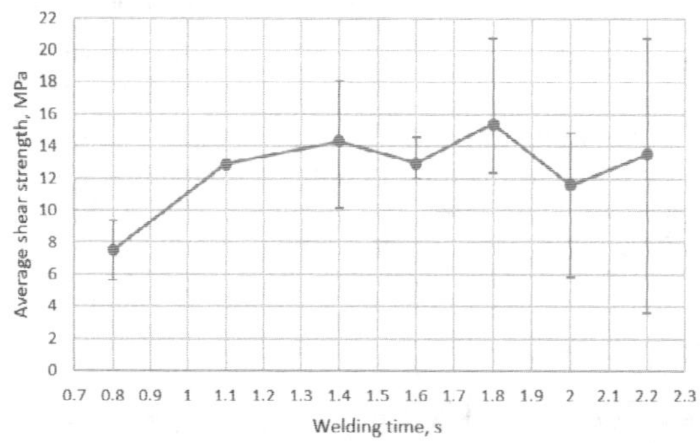

Figure 5. Average shear strength vs. welding time.

In figure 5, beyond the DOE screening studied range from 0.8s to 1.4s, in higher welding time range from 1.4s to 2.2s, average shear strength gradually increase as welding time increase and approximately reach a stable value as welding time is higher than 1.4s. The highest shear strength and highest average shear strength are taken from samples with welding time of 1.8s. At higher welding time, such as 2.2s, shear strength of weld is also high. However, at such high welding time, USW welds may be re-broken by the ultrasonic vibration, leading to a decrease in shear strength of weld. Based on the ANOVA analysis, the optimal welding time is 1.8s for present ultrasonic welding.

3.5. *Effect of surface roughness*

The surface roughnesses of AA7075 and Ti-6Al-4V ground by P400, P800 and P1200 grit size SiC papers are present in figure 6. As increase of grit number of SiC paper that is used for AA7075 or Ti-6Al-4V, the surface roughness decrease. Using SiC paper with the same grit number to grind AA7075 and Ti-6Al-4V, the obtained roughnesses of AA7075 and Ti-6Al-4V are different. With higher hardness and lower "spring back" effect during grinding, Ti-6Al-4V is ground to a lower roughness than that of AA7075 under the same grinding condition. The studied roughness range of Ti-6Al-4V is from 0.0651 Ra µm to 0.2264 Ra µm, while the roughness range of AA7075 is from 0.1457 Ra µm to 0.5211 Ra µm.

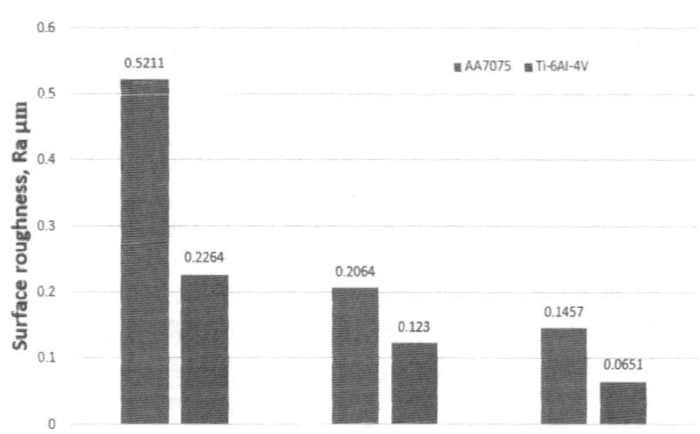

Figure 6. Surface roughnesses of AA7075 and Ti-6Al-4V plates which are finally ground by P400, P800 and P1200 grit size SiC papers.

IOP Conf. Series: Materials Science and Engineering **744** (2020) 012029 doi:10.1088/1757-899X/744/1/012029

The grit number of SiC paper that are used on AA7075 or Ti-6Al-4V, the corresponding roughness obtained, and the shear strength of USW welds manufactured by AA7075 and Ti-6Al-4V plates with corresponding surface roughness are shown in table 4. For un-welded samples, their strength is noted as 0MPa.

Table 4. Surface roughnesses of AA7075 and Ti-6Al-4V plates and their corresponding shear strength.

Grit number, AA7075[a]	AA7075 roughness, Ra μm	Grit number, Ti-6Al-4V	Ti-6Al-4V roughness, Ra μm	Shear strength, MPa
1200	0.14565	400	0.2264	0
1200	0.14565	400	0.2264	9.7
800	0.20635	1200	0.0651	29.93
400	0.52105	1200	0.0651	25.76
400	0.52105	800	0.10295	0
400	0.52105	400	0.2264	0
1200	0.14565	1200	0.0651	26.65
800	0.20635	400	0.2264	0
400	0.52105	400	0.2264	0
400	0.52105	1200	0.0651	0
1200	0.14565	1200	0.0651	26.13
1200	0.14565	800	0.10295	0
800	0.20635	800	0.10295	0

[a] The grit number of SiC paper that are applied on AA7075 plate surface.

The contour plot of shear strength with respect to surface roughnesses of AA7075 and Ti-6Al-4V are shown in figure 7. It can be concluded from figure 7 that shear strength of USW joint increases as the roughness of AA7075 or Ti-6Al-4V decreases. Within the studied roughness range, smallest roughnesses of AA7075 and Ti-6Al-4V should be used to get the highest shear strength of USW joint. It is supported by the surface roughness analyses that faying surfaces with low roughness is good for robustness of USW joint. The reason may be that low roughness reduce the amount of material deformation needed to form metal to metal contact and adhesion, which is a key factor for atomic diffusion during USW process.

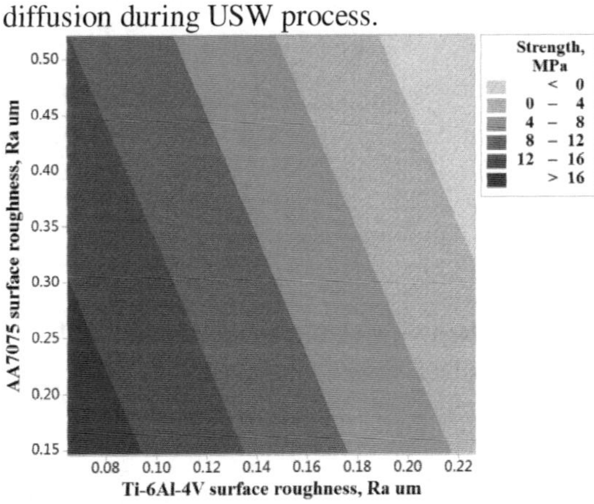

Figure 7. Effects of surface roughness of AA7075 and Ti-6Al-4V on shear strength of USW weld.

4. Conclusion

A DOE screening utilizing 13 different groups of process parameters on USW AA7075 and Ti-6Al-4V for 3 process parameters (pressure, penetration depth, and welding time), a further ANOVA analysis on welding time, and a DOE screening of surface roughness of AA7075 and Ti-6Al-4V were carried out. 3 times repetition was used for the analyses. Tensile shear tests were conducted. There was a big variability of shear strength of samples in tensile shear tests. The greatest shear strength observed is about 20.7 MPa. The tensile shear strengths are all much less than the ultimate tensile strength of both parent materials (Ti-6Al-4V around 1100 MPa and AA7075 around 650 MPa). The analyses of DOE screenings and ANOVA indicate that the following combination of USW process parameters should yield the highest strengths: USW penetration depth 0.7mm, welding time 1.8s, pressure 0.7 MPa and surface roughness of AA7075 0.1457 Ra μm and Ti-6Al-4V 0.0651 Ra μm (both ground by P1200 SiC paper).

5. References

[1] G. Casalino, M. Mortello, and P. Peyre, "Yb–YAG laser offset welding of AA5754 and T40 butt joint," *Journal of Materials Processing Technology*, vol. 223, pp. 139–149, Sep. 2015.
[2] H. Ozaki, S. Hayashi, and M. Kutsuna, "Laser roll welding of dissimilar metal joint of titanium to aluminium alloy," *Welding International*, vol. 23, no. 7, pp. 501–509, Jul. 2009.
[3] W. H. Sohn, H. H. Bong, and S. H. Hong, "Microstructure and bonding mechanism of Al/Ti bonded joint using Al–10Si–1Mg filler metal," *Materials Science and Engineering: A*, vol. 355, no. 1–2, pp. 231–240, Aug. 2003.
[4] J.-W. Choi, H. Liu, and H. Fujii, "Dissimilar friction stir welding of pure Ti and pure Al," *Materials Science and Engineering: A*, vol. 730, pp. 168–176, Jul. 2018.
[5] M. Kreimeyer, F. Wagner, and F. Vollertsen, "Laser processing of aluminum–titanium-tailored blanks," *Optics and Lasers in Engineering*, vol. 43, no. 9, pp. 1021–1035, Sep. 2005.
[6] C. Q. Zhang, J. D. Robson, O. Ciuca, and P. B. Prangnell, "Microstructural characterization and mechanical properties of high power ultrasonic spot welded aluminum alloy AA6111–TiAl6V4 dissimilar joints," *Materials Characterization*, vol. 97, pp. 83–91, Nov. 2014.
[7] S. Q. Wang, V. K. Patel, S. D. Bhole, G. D. Wen, and D. L. Chen, "Microstructure and mechanical properties of ultrasonic spot welded Al/Ti alloy joints," *Materials & Design*, vol. 78, pp. 33–41, Aug. 2015.
[8] N. Sridharan, P. Wolcott, M. Dapino, and S. S. Babu, "Microstructure and texture evolution in aluminum and commercially pure titanium dissimilar welds fabricated using ultrasonic additive manufacturing," *Scripta Materialia*, vol. 117, pp. 1–5, May 2016.
[9] C. Q. Zhang, J. D. Robson, and P. B. Prangnell, "Dissimilar ultrasonic spot welding of aerospace aluminum alloy AA2139 to titanium alloy TiAl6V4," *Journal of Materials Processing Technology*, vol. 231, pp. 382–388, May 2016.
[10] Z. Zhu, K. Y. Lee, and X. Wang, "Ultrasonic welding of dissimilar metals, AA6061 and Ti6Al4V," *The International Journal of Advanced Manufacturing Technology*, vol. 59, no. 5–8, pp. 569–574, Mar. 2012.

IOP Conf. Series: Materials Science and Engineering **744** (2020) 012030 doi:10.1088/1757-899X/744/1/012030

A Comparative Investigation of Magnetoelectric Characteristics of Nano Bi-Layer Bar and Plate Structures

T Saengow and R Silapunt

Department of Electronic and Telecommunication Engineering,
King Mongkut's University of Technology Thonburi Bangkok, Thailand
Email: treetep.marksaengow@mail.kmutt.ac.th

Abstract. In this paper, the magnetoelectric (ME) characteristics of bar and plate structures of the nano bi-layer ME materials are investigated and compared. The mathematical models of ME coefficients of the nano bi-layer bar and plate structures operated in the longitudinal-transverse (L-T) mode are developed using the constitutive equations and Newton's second law. The nano bi-layer of interest consists of Terfenol-D and Lead Zirconate Titanate (PZT) as ferromagnetic (FM) and ferroelectric (FE) phases, respectively. In the low frequency regime, the optimal thickness ratios of bar and plate structures that yield maximum ME coefficients are around 0.43 and 0.33, respectively. The optimal thickness ratios are applied in the high frequency regime to obtain the conditions where the maximum ME coefficients can be achieved. It is found that the resonant frequencies of the bar and plate structures decrease exponentially with the bi-layer length and thickness, respectively. The results also indicate that at the same resonant frequency, the dimension of the bar structure is significantly smaller than the plate structure, it thus becomes more preferable for nano scale sensing applications.

1. Introduction

Magnetoelectric (ME) material is one of mulitiferroic (MF) composites that combine ferromagnetism and ferroelectricity phenomena and exhibits the ME effect, the reversible coupling between the magnetic and electric properties of a material [1-2]. The magnetoelectric multiferroics have been extensively researched and proved that they have tremendous potential for electric and magnetic sensing applications [2-3]. The sensing capability is typically measured by the ME coefficient, which is the ratio of the electric (magnetic) field output to the electric (magnetic) field input. In order to achieve the optimal ME coefficient, types of sensing materials and their properties, operating frequency, and sensor geometries are all important factors. Most of the bi-layer, tri-layer and multilayer ME materials are studied in the bar structure, which defines the strain deformation only in the longitudinal direction [4-7]. Until [8], the first report of the bi-layer plate structure where the strain deformation in the thickness direction was published. However, the comparison between the two structures has never been reported.

This paper aims to investigate and compare the ME characteristics of the nano bi-layer bar and plate structures. Note that, the nano bi-layer ME materials are proposed here as they have become increasingly attractive for nano scale sensing applications. It is also important to point out that a desirable film sensor such as ones depicted in Figures 1-2, is commonly designed by following the geometric rule that is $l \gg t, w$, where l is length, t is thickness, and w is width [4]. The mathematical model of ME coefficient

IOP Conf. Series: Materials Science and Engineering **744** (2020) 012030 doi:10.1088/1757-899X/744/1/012030

for the longitudinal-transverse (L-T) mode bar structure will first be developed based on the constitutive equations and Newton's second law. The mathematical model of ME coefficient for the L-T mode plate structure is obtained from the previous report [8]. Then, in the demonstration of ME effects, Terfenol-D and PZT are selected as the nano bi-layer ME material, owing to their good magnetostrictive and piezoelectric properties. The ME coefficient profiles as a function of thickness ratios will be determined in the low frequency regime for both structures. The optimal thickness ratios will then be used as a baseline to determine the conditions to achieve maximum ME coefficients in a typical high frequency operation. Finally, the relation between resonant frequency and bi-layer geometry will be evaluated.

2. ME Coefficients Derivation

The magnetostrictive constitutive equations are used for the FM layer and piezoelectric constitutive equations are used for the FE layer [7-8]. Both constitutive equations are linked by the Newton's second law via mechanical coupling. The L-T mode bar and plate structures are shown in Figures 1-2, respectively.

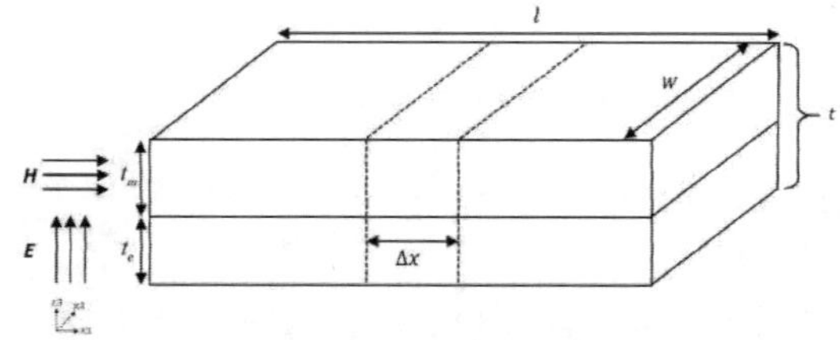

Figure 1. The L-T mode bar structure.

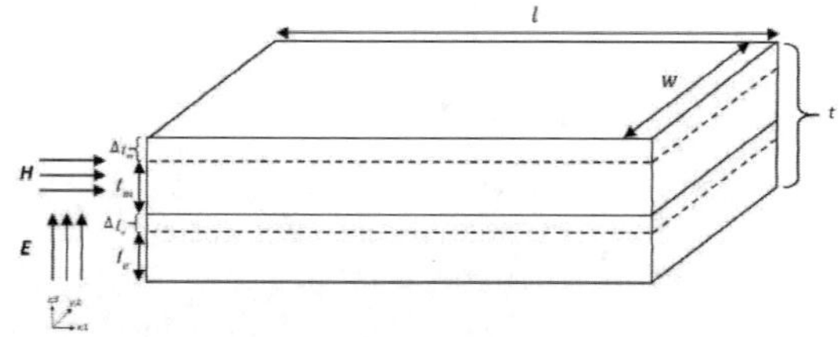

Figure 2. The L-T mode plate structure.

2.1. Coupling between the FM and FE phases using the Newton's second law

In order to combine the FM phase and FE phase, the Newton's second law with the elastic displacement, ξ_j is applied. For bar and plate structures, the displacements occur in the longitudinal direction, ξ_1, and thickness direction, ξ_3, respectively. The results are shown in equations (1) and (2) for bar and plate structures, respectively.

$$\eta \frac{d\sigma_1^e}{dx} + (1-\eta)\frac{d\sigma_1^m}{dx} = \left[\eta\rho_e + (1-\eta)\rho_m\right]\frac{\partial^2 \xi_1}{\partial t^2} \tag{1}$$

$$\frac{d\sigma_3^e}{dz} + \frac{d\sigma_3^m}{dz} = \left[\eta\rho_e + (1-\eta)\rho_m\right]\frac{\partial^2 \xi_3}{\partial t^2}, \tag{2}$$

where σ is strain, ρ is the material density, and η is the thickness ratio and is defined by $\eta = \frac{t_e}{t_e + t_m}$, t_e and t_m are the thickness of FE layer and FM layer respectively. Note that, the superscripts 'm' and 'e' are referred to FM layer and FE layer, respectively. The subscripts '1' is the longitudinal axis, and the subscript '3' is the thickness direction.

Equations (1) and (2) can be rewritten in the time domain as

$$\frac{\partial^2 \xi_1}{\partial x^2} + k^2 \xi_1 = 0 \tag{3}$$

$$\frac{\partial^2 \xi_3}{\partial z^2} + k^2 \xi_3 = 0, \tag{4}$$

where $k^2 = \omega^2 \frac{[\eta\rho_e + (1-\eta)\rho_m]s_{11}^e s_{11}^m}{s_{11}^e(1-\eta) + s_{11}^m \eta}$ and $k^2 = \omega^2 \frac{[\eta\rho_e + (1-\eta)\rho_m]s_{33}^e s_{33}^m}{s_{33}^e(1-\eta) + s_{33}^m \eta}$ for equations (3) and (4), respectively, $\xi_1 = A\cos(kx) + B\sin(kx)$, $\xi_3 = A\cos(kz) + B\sin(kz)$, and ω is the angular frequency.

2.2. ME coefficient expressions

2.2.1. Bar structure
From the constitutive equations , the stresses can be determined as shown in equations (5) and (6).

$$\sigma_1^m = \frac{\left[Ak\cos(kx) - Bk\sin(kx)\right]}{s_{11}^m} - \frac{d_{11}^m H_1}{s_{11}^m} \qquad (5)$$

$$\sigma_1^e = \frac{\left[Ak\cos(kx) - Bk\sin(kx)\right]}{s_{11}^e} - \frac{d_{31}^e E_3}{s_{11}^e} \qquad (6)$$

Using the boundary condition at $x = \frac{l}{2}$ and $x = -\frac{l}{2}$, with no external force, the coefficients A and B can be solved and obtained as $A = \frac{(1-\eta)s_{11}^e d_{11}^m H_1 + \eta s_{11}^m d_{31}^e E_3}{k\left[(1-\eta)s_{11}^e + \eta s_{11}^m\right]\cos(kl)}$ and $B = 0$

Equation (6) is then applied to the constitutive equations to calculate the current I flow through the sensor from $I = \frac{dQ}{dt}$ and $D = \frac{dQ}{dA}$, where dA is the differential area element through which the current flows.

In order to find the ME coefficient, the relation of electric field across the FE layer and magnetic field along the FM layer are solved by applying the open-circuit condition ($I = 0$) . The ME coefficient of the bar structure is shown in equation (7).

$$\alpha_{31}^{H^{bar}} = \frac{(1-\eta)s_{11}^e d_{31}^e d_{11}^m}{d_{31}^{e^2}\left[(1-\eta)s_{11}^e + \eta s_{11}^m\left[1 - \frac{\tan(kl/2)}{kl/2}\right]\right] - \varepsilon_{33}s_{11}^e\left[(1-\eta)s_{11}^e + \eta s_{11}^m\right]} \frac{\tan(kl/2)}{kl/2} \qquad (7)$$

It can be seen that the ME coefficient of the bar structure is varied by η and kl that are a function of thickness, sensor length, and operating frequency.

2.2.2. Plate structure
Referred to the previous report [8], the ME coefficient of the plate structure is shown in equation (8)

$$\alpha_{31}^{H^{plate}} = \frac{\eta s_{33}^e d_{33}^e d_{31}^m \sin(k\eta t)}{\left\{d_{33}^{e^2}\sin(k\eta t - kt) - \left[d_{33}^{e^2} - s_{33}^e\varepsilon_{33}\right]\sin(k\eta t)\right\}\left[\eta s_{33}^m - (1-\eta)s_{33}^e\right] + \eta s_{33}^m d_{33}^{e^2}\sin(kt)} \qquad (8)$$

It can be seen that the ME coefficient of the plate structure is varied by η and kt that are a function of thickness and operating frequency.

3. Results and Discussion

3.1. Optimal thick ness ratio

In the low frequency regime, the operating frequency is set close to zero or $k \to 0$. Applying equations (7) and (8) yields the ME coefficients for difference thickness ratios. Figure 3 shows the ME coefficients of the bar structure and Figure 4 shows the ME coefficients of the plate structure.

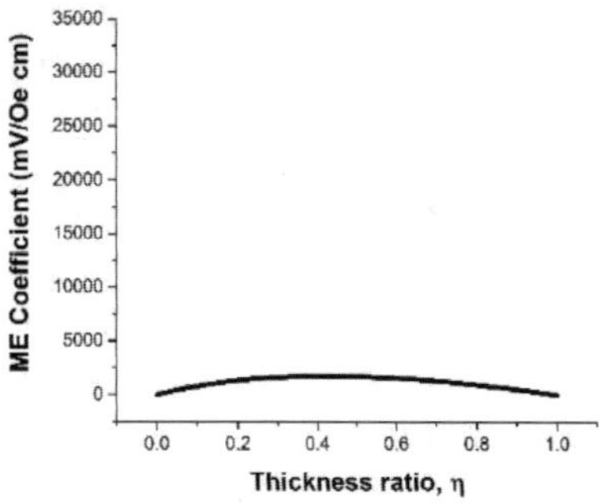

Figure 3. ME coefficient vs. thickness ratio for the bar structure.

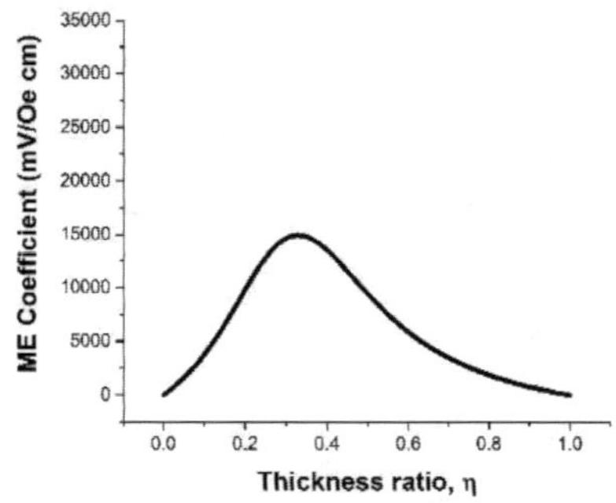

Figure 4. ME coefficient vs thickness ratio for the plate structure.

The optimal thickness ratios for both bar and plate structures are equal to 0.43 and 0.33, respectively. However, the overall ME coefficient of plate structure is significantly higher than that of the bar structure since the denominator of equation (8), the plate structure model, converges to zero while that of equation (7), the bar structure model, remains constant.

3.2. Geometry dependent frequency characteristics

By applying the optimal thickness ratios to equations (7) and (8), the conditions where the maximum ME coefficients can be achieved that are $kl = \pi$ and $kt = 0.854747\pi$ for bar and plate structures, respectively. It is clear that the maximum ME coefficients occur at the resonant frequencies for both bar and plate structures and they are influenced by length l for the bar structure and thickness t for the plate structure. The resonant frequency profiles of the bar and plate structures are plotted as a function of bi-layer geometry as shown in Figures 5 and 6, respectively. It can be seen that the resonant frequencies decrease exponentially with length for the bar structure and with thickness for the plate structure. Although the overall ME coefficient of the plate structure is higher than the bar structure, it is not suitable for current nano sensing applications due to dimension constraint. The results confirm that at the same resonant frequency, the dimension of the bar structure is always smaller than the plate structure [4].

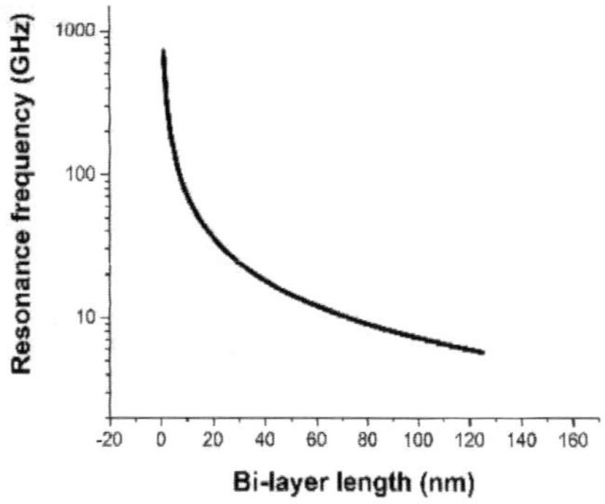

Figure 5. Resonant frequency vs bi-layer length for the bar structure

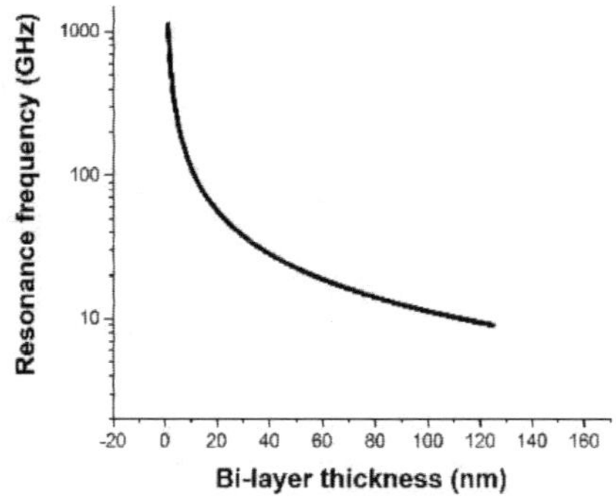

Figure 6. Resonant frequency vs bi-layer length for the plate structure

4. Conclusions

The ME coefficient mathematical models are derived in order to investigate and compare the ME characteristics of the nano bi-layer bar and plate structures. The nano bi-layer of interest consists of Terfenol-D and PZT. The calculation in the low frequency regime yields the optimal thickness ratios of 0.43 and 0.33 for bar and plate structures, respectively. Applying the optimal thickness ratios yields the conditions for maximum ME coefficients in high frequency operation that are $kl = \pi$ and $kt = 0.854747\pi$ for bar and plate structures, respectively. The relations of resonant frequency as a function of bi-layer lengths and thicknesses of the bar and plate structures respectively, are determined. At the same resonant frequency, the dimension of the bar structure is always smaller, suggesting it as preferable choice for nano sensing applications. Future work includes the study of the nano bi-layer bar structure as the magnetic read head sensing element and its performance when integrated with the additional bias layer and other complementary layers.

5. References

[1] Wang Y, Hu J, Lin Y and Nan Ce W 2010 Multiferroic magnetoelectric composite nanostructures *NPG Asia Mater.* **2**(2) 61-68

[2] Palneedi H, Annapureddy V, Priya S and Ryu J 2016 Status and Perspectives of Multiferroic Magnetoelectric Composite Materials and Applications *Actuators* **2016** 1-31

[3] Kambale R, Jeong D and Ryu J 2012 Current Status ofMagnetoelectric Composite Thin/Thick Films *Advances in Condensed Matter Physics* **2012** 15pages

[4] Vopsaroiu M, Blackburn J, Muniz-Piniella A and Cain Markys G 2008 Multiferroic magnetic recording read head technology for 1 Tbit/ in^2 and beyond *J. Appl. Phys.* **103** 07F506 1-3

[5] Choowitsakunlert S, Satitchantrakul T and Silapunt R 2014 A 1D Analysis of Nano Multiferroic Composites for the Novel Read Head Technology *Advanced Materials Research* **1052** 149-54

[6] Vopsaroiu M, Blackburn J and Cain Markys G 2007 A new magnetic recording read head technology based on the magneto-electric effect *J. Phys. D: Appl. Phys.* **40** 5027-33

[7] Dong S, Li Jie F and Viehland D Theory Analysis on Magnetoelectric Voltage Coefficients of the Terfoneol-D/PZT Composite Transducer Materials Science & Engineering, Virginia Tech, Blacksburg, VA24061

[8] Saengow T, Satitchantrakul T and Silapunt R 2018 Investigation of Magnetoelectric Effect in the Bi-Layer Plate Structure *Solid State Phenomena* **280** 9-14

6. Acknowledgments

This work has been supported by Thailand Research Fund under Research and Researchers for Industries-RRI program with the contract number PHD/58I0004. The authors would like to thank King Mongkut's University of Technology Thonburi, the University of Portsmouth, and Seagate Technology, Thailand for their equipment, software and any valuable advices.

Repair feasibility of SS416 stainless steel via laser aided additive manufacturing with SS410/Inconel625 powders

Fei Weng[1], Yongfeng Liu*[2], Youxiang Chew[1], Leilei Wang[1], Bing Yang Lee[1], Guijun Bi *[1]

1 Singapore Institute of Manufacturing Technology, 73 Nanyang Drive, 637662, Singapore
2 Beijing Key Laboratory of Performance Guarantee on Urban Rail Transit Vehicles, School of Mechanical-Electronic and Automobile Engineering, Beijing University of Civil Engineering and Architecture, 100044 Beijing, China,
E-mail: liuyongfeng@bucea.edu.cn; gjbi@simtech.a-star.edu.sg

Abstract. Laser based additive manufacturing technique has been found beneficial in fabricating complex components efficiently. In this paper, laser aided additive manufacturing with powder feeding (LAAM) was used to deposit SS410 or Inconel625 on SS416 substrate. Interfacial microstructure and mechanical properties of the deposited samples are investigated accordingly. Results indicated that deposit without obvious defects can be obtained. The interface samples exhibit comparative or slight lower tensile strength compared with the SS416 substrate. Fracture tends to occur at the SS416 side, either far from the fusion line for the SS410/SS416 sample or near the heat affected zone (HAZ) for the Inconel625/SS416 sample. Therefore, the LAAMed section are reliable enough. SS410 or Inconel625 both can be used for repair of SS416 parts.

1. Introduction

Nowadays, laser additive manufacturing has been a research hotspot, either in fabricating complex components or developing novel materials [1]. Among which, powder-feeding based laser aided additive manufacturing (LAAM) [2,3], also known as laser based direct energy deposition (DMD) [4], is one of the most popular techniques. Compared with the powder-bed based AM techniques (e.g. SLM [5] or SLS [6]), LAAM can be used for repair of certain worn components [7,8]. Moreover, the repair can even be carried out in spot without dismounting of the components from the machine.

As one kind of free-cutting stainless steel, SS416 stainless steel has been widely used in fabrication of components such as pump shaf and valve parts, which tend to be worn or damaged regionally [9,10]. It is costly to replace the components by a new one. In this case, LAAM can be used to carried out repair work, by utilizing proper additive materials under suitable laser process parameters. After repair and post machining, the worn parts are expected to be put into service again. It is meaningful to carried out relevant study on interfacial characteristic between deposit and substrate. A metallurgical bonding of deposit/substrate and superior or comparative mechanical property is of great help.

In this paper, LAAMed SS410 and Inconel625 alloys are prepared successfully on SS416 substrate. Microstructure and mechanical properties of the prepared samples are investigated accordingly. The repair feasibility of SS416 with LAAMed SS410 or Inconel625 is discussed.

2. Experimental

SS416 alloy plates were used as the substrate material. Gas-atomized SS410 and Inconel625 alloy powders (20~45 μm) were used as the additive materials. Nominal compositions of the materials are

listed in **Table.1**. Compared with SS416 substrate, SS410 has similar composition but less sulphur (S), making it easier to deposit without pores and cracks. While, Inconel625 is a traditional Ni-based alloy used for laser cladding [11]. LAAM was carried out in a self-developed 6-axis robot based LAAM system [3]. Ar was used as the shielding gas and powder carrier gas. After preliminary parameter selection, the laser power, powder feed rate, and scanning velocity were set at 800 W, 10.5 g/min, 600 mm/min, respectively. Laser beam size was kept at ~1.5 mm and layer thickness was around 1.0 mm.
LAAMed samples were sectioned and prepared via traditional grinding and polishing procedure. Microstructure was observed by optical microscope (MX51, *Olympus*). Micro-hardness was measured by micro-hardness tester (MMT-X3, *Matsuzawa*). 3 points were tested and averaged at the same depth. Instron 4505 universal testing machine was used to evaluate the tensile properties by reference to ASTM E8 standard. Crosshead displacement speed was kept at 1.0 mm/min. A schematic diagram of tensile coupons extraction is shown in Figure 1. Gauge dimensions for the tensile coupons are 25 mm*6 mm*3 mm. After tensile test, fracture position was identified by referring to the interface. A typical fractured samples were cut off, mounted, and polished. Furthermore, the interface characteristic was observed.

Table 1. Nominal composition of the substrate and additive materials (wt.%)

Elements	Fe	Ni	C	Si	Cr	Mn	Co	Mo	Al	Ti	NbTa	P	S
SS416	Bal.	≤0.6	≤0.15	≤1.0	12-14	≤1.25		≤0.6				≤0.06	≥0.15
SS410	Bal.	≤0.6	≤0.15	≤1.0	11.5-13.0	≤1.0						≤0.04	≤0.03
Inconel 625	≤5.0	Bal.	≤0.1	≤0.5	20.0-23.0	≤0.5	≤1.0	8.0-10.0	≤0.4	≤0.4	3.15-4.15	≤0.015	≤0.015

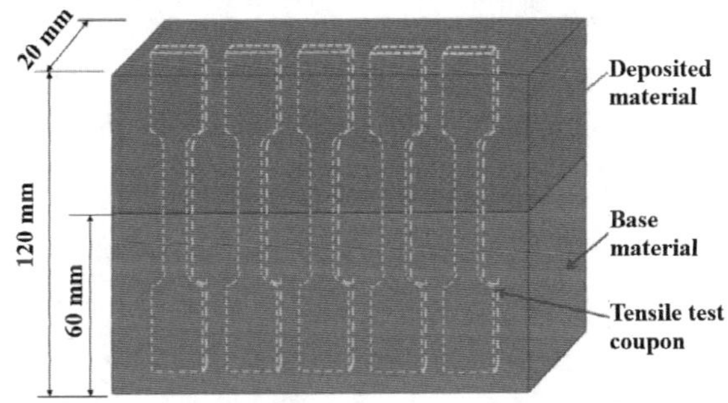

Figure 1. Schematic diagram for fabrication of tensile test coupons

3. Results and discussion

3.1. Microstructure

Figure 2 shows the overall cross-section morphologies of the single layer deposited samples. No obvious crack or pores appear at the interface area, showing that the deposits combine well with the substrate.

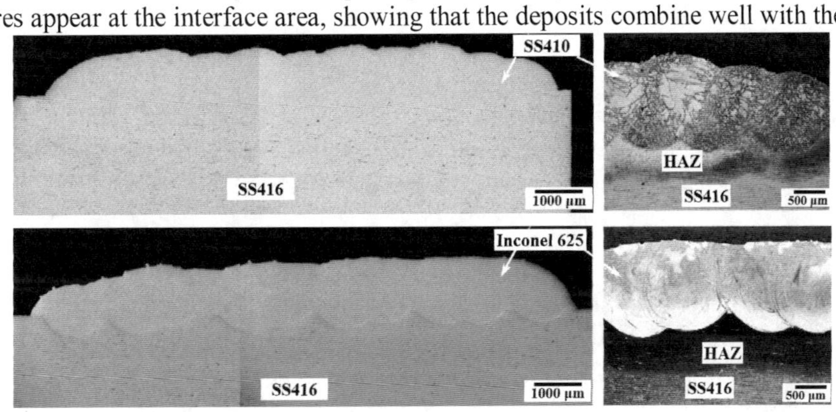

Figure 2. Cross-section of single layer deposited samples before (left) and after (right) etching

From the image after chemical etching, a heat affected zone (HAZ) appear obviously, due to the regional heating effect generated by high energy density laser beam [12].
Figure 3 shows the zoom-in microstructure of the deposited Inconel625, SS410, HAZ, and SS416 substrate, respectively. Similar with our previous work [13], the Inconel625 deposit is mainly comprised of fine cellular and cellular/dendritic microstructure, which is derived from the heat conduction condition during LAAM. While, SS410 deposit is dominated by martensite (grey area) and ferrite (light area). The formation of martensite is mainly due to the rapid cooling after solidification. The formation of ferrite is probably due to martensite decomposition during re-heating of subsequent laser scanning. Compared with printing thin wall structures [14], the thermal history in the present study is more complicated and microstructure evolution during LAAM process deserves further study in future work. HAZ and SS416 substrate are mainly comprised of martensite. Actually, SS416 is a kind of martensitic stainless steel [15]. In comparison, HAZ shows finer martensitic microstructure compared with the substrate. This is attributed to the fast heating and cooling cycle in the HAZ during the deposition process.

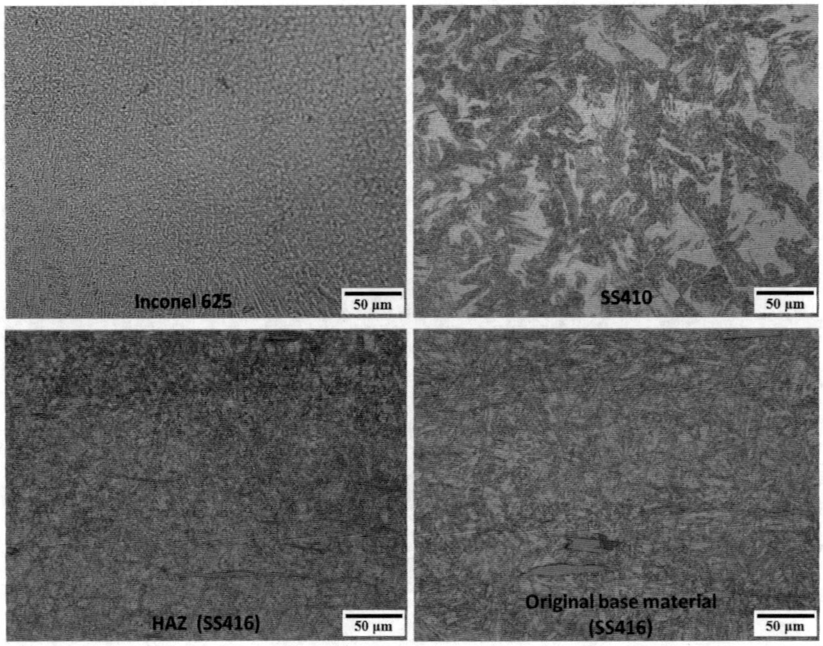

Figure 3. Microstructure of deposited Inconel625, SS410, HAZ and original substrate (SS416)

3.2. Micro-hardness
Micro-hardness distributions of the SS410 and Inconel625 deposits are shown in Figure 4. SS410 deposit shows higher micro-hardness than Inconel625 deposit and also slightly higher than the SS416 substrate. Higher hardness in HAZ can be derived from the refined martensite structure (shown in Figure 3) and partially due to the high carbon martensite formed in HAZ [15].

3.3. Mechanical properties
Figure 5 shows the tensile test results for SS416 substrate, SS410/SS416, and Inconel625/SS416 interface samples. SS410/SS416 sample exhibits comparative ultimate tensile strength (UTS) and yield strength, compared with the SS416 substrate. Inconel625/SS416 shows both lower UTS and yield strength. As demonstrated in Figure 5, all tensile coupons tend to fracture in the substrate material. For Inconel625/SS416, fracture occurs near the HAZ, while it is far from the HAZ for the SS410/SS416 case. As previously reported [16], the LAAMed Inconel625 showed higher UTS (~840 MPa), which is higher than that of the SS416 substrate used in this paper. Hence it is reasonable that the fracture occurs at the substrate side in the Inconel625/SS416 case. The result is also similar with the report in welding

of Inconel718 and SS416 [15]. For the SS410/SS416 case, LAAMed SS410 shows slightly higher hardness, implying higher strength.

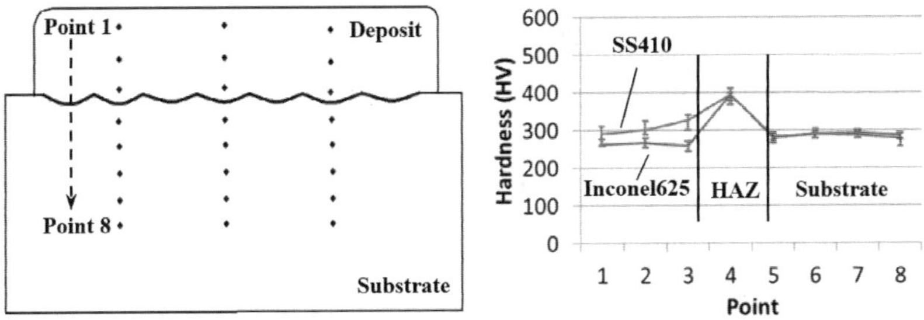

Figure 4. Micro-hardness distributions along the cross-section of the deposited SS410 and Inconel625 materials (distance along depth ~0.5 mm)

However, it is noteworthy that the deformation during tensile is nonuniform for the SS410/SS416 and Inconel625/SS416 interface samples. Hence, the fracture position is not only affected by the strength difference of the two materials, but also the internal deformation mechanism. In this study, as a sulphur-rich easy-cutting stainless steel, the sulphur inclusions in the SS416 substrate would act as weak point in the composite gauge area and subject to premature fracture during tensile test. This is probably one of the reasons for fracturing at the substrate side. In future, more work would be carried out in the deformation mechanism of the interfacial samples.

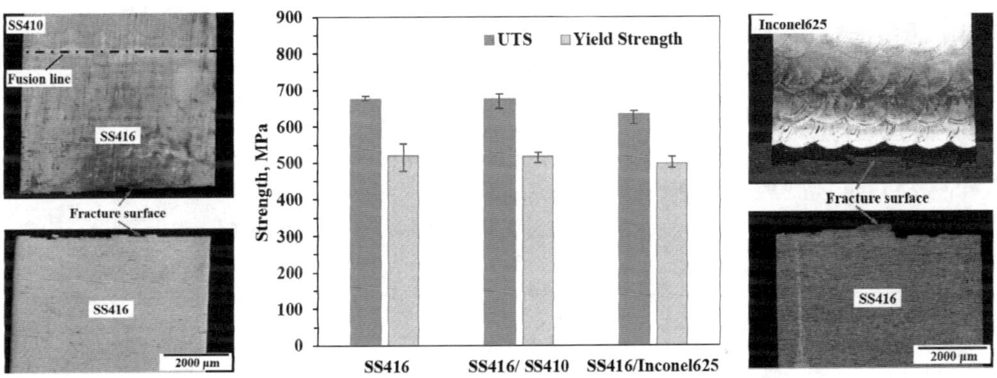

Figure 5. Tensile strength of SS416 substrate, SS410/SS416, and Inconel625/SS416 samples, and typical fracture cross sections

4. Conclusion

In this paper, SS410 and Inconel625 were successfully deposited on SS416 substrate by LAAM. Microstructure and mechanical properties demonstrated that the deposits are defect-free and show good integrity throughout the interface. LAAMed SS410 is mainly comprised of martensite and ferrite, while Inconel625 shows obvious cellular and cellular/dendritic microstructure. Fracture both occur at the substrate side for the two samples, either near or far from the interface. With comparable strength with the substrate, SS410 can be applied in repair of SS416 parts. Though the Inconel625/SS416 interface sample shows lower UTS and yield strength, Inconel 625 can also be used for repair SS416, considering the good interfacial integrity.

5. References

[1] DebRoy T, Wei H L, Zuback J S, Mukherjee T, Elmer J W, Milewski J O, Beese A M, Wilson-Heid A, De A and Zhang W 2018 Additive manufacturing of metallic components – Process, structure and properties *Prog. Mater. Sci.* **92** 112–224

[2] Zhang B, Bi G, Chew Y, Wang P, Ma G, Liu Y and Moon S K 2019 Comparison of carbon-based reinforcement on laser aided additive manufacturing Inconel 625 composites *Appl. Surf. Sci.* **490** 522–34

[3] Chew Y, Bi G J, Zhu Z G, Ng F L, Weng F, Liu S B, Nai S M L and Lee B Y 2019 Microstructure and enhanced strength of laser aided additive manufactured CoCrFeNiMn high entropy alloy *Mater. Sci. Eng. A* **744** 137–44

[4] Weng F, Gao S, Jiang J, Wang J and Guo P 2019 A novel strategy to fabricate thin 316L stainless steel rods by continuous directed energy deposition in Z direction *Addit. Manuf.* **27** 474–81

[5] Song J, Chew Y, Jiao L, Yao X, Moon S K and Bi G 2018 Numerical study of temperature and cooling rate in selective laser melting with functionally graded support structures *Addit. Manuf.* **24** 543–51

[6] Gu D, Shen Y and Lu Z 2009 Microstructural characteristics and formation mechanism of direct laser-sintered Cu-based alloys reinforced with Ni particles *Mater. Des.* **30** 2099–107

[7] Saboori A, Aversa A, Marchese G, Biamino S, Lombardi M and Fino P 2019 Application of Directed Energy Deposition-Based Additive Manufacturing in Repair *Appl. Sci.* **9** 3316

[8] Zhang X, Li W and Liou F 2018 Damage detection and reconstruction algorithm in repairing compressor blade by direct metal deposition *Int. J. Adv. Manuf. Technol.* **95** 2393–404

[9] Harsha A P, Limaye P K, Tyagi R and Gupta A 2016 Effect of Temperature on Galling Behavior of SS 316, 316 L and 416 Under Self-Mated Condition *J. Mater. Eng. Perform.* **25** 4980–7

[10] Nandhakumar S and Prakash R S 2017 Parametric Optimization in CNC Turning of Martensitic Stainless Steel 416 using Taguchi Method *J. Chem. Pharm. Sci. ISSN* **974** 2115

[11] Hong C, Gu D, Dai D, Alkhayat M, Urban W, Yuan P, Cao S, Gasser A, Weisheit A, Kelbassa I, Zhong M and Poprawe R 2015 Laser additive manufacturing of ultrafine TiC particle reinforced Inconel 625 based composite parts: Tailored microstructures and enhanced performance *Mater. Sci. Eng. A* **635** 118–28

[12] Ma Q, Li Y and Wang J 2016 Surface modification of Q550 HSLA steel with Co–Fe–Cr composite coatings manufactured by fibre laser cladding *Surf. Eng.* **32** 934–42

[13] Bi G, Sun C N, Nai M L and Wei J 2013 Micro-structure and Mechanical Properties of Nano-TiC Reinforced Inconel 625 Deposited using LAAM *Phys. Procedia* **41** 828–34

[14] Zhang Y, Yu G, He X, Ning W and Zheng C 2012 Numerical and experimental investigation of multilayer SS410 thin wall built by laser direct metal deposition *J. Mater. Process. Technol.* **212** 106–12

[15] Dev S, Ramkumar K D, Arivazhagan N and Rajendran R 2018 Investigations on the microstructure and mechanical properties of dissimilar welds of inconel 718 and sulphur rich martensitic stainless steel, AISI 416 *J. Manuf. Process.* **32** 685–98

[16] Zhang B, Bi G, Wang P, Bai J, Chew Y and Nai M S 2016 Microstructure and mechanical properties of Inconel 625/nano-TiB_2 composite fabricated by LAAM *Mater. Des.* **111** 70–9

IOP Conf. Series: Materials Science and Engineering **744** (2020) 012032 doi:10.1088/1757-899X/744/1/012032

Atomic structures and scanning tunnelling microscopy of nitrogen-doped carbon nanotubes

Yoshitaka Fujimoto[1] and Susumu Saito[1,2,3]

1 Department of Physics, Tokyo Institute of Technology, Meguro, Tokyo 152-8551, Japan
2 Advanced Research Center for Quantum Physics and Nanoscience, Tokyo Institute of Technology, Meguro, Tokyo 152-8551, Japan
3 Materials Research Center for Element Strategy, Tokyo Institute of Technology, Yokohama, Kanagawa 226-8503, Japan
E-mail: fujimoto@stat.phys.titech.ac.jp

Abstract. We report on atomic structures, energetics, and scanning tunnelling microscopy images of nitrogen defects in carbon nanotubes (CNTs) based on first-principles density-functional study. As defect configurations in CNTs, not only the substitutional nitrogen defect but also the nitrogen-vacancy complex defects (pyridine-type defects) are considered. The substitutional nitrogen defect is found to be energetically the most stable defect. On the other hand, for the pyridine-type defects, the tetramerized and the trimerized configurations are found to be the possible configurations in energy. The scanning tunnelling microscopy (STM) images for defect configurations in N-doped CNTs are studied, and the various N defects in CNTs are expected to be identified by STM experiments.

1. Introduction

Carbon nanotubes (CNTs), which are composed of the quasi-one-dimensional cylinders, possess unique properties such as long mean free path and high current density [1,2]. Specifically, due to their high carrier mobility, CNTs are expected to be promising device materials to be used in nanoelectronics such as field effect transistors and sensors [3-13].

Substitutional doping with heteroatoms is known to be an effective way to tune the electronic properties of the CNTs. Nitrogen is one of promising dopants for carbon-based materials since it is a neighbouring element to carbon in periodic table. In the x-ray photoelectron spectroscopy observations, it has been suggested that there exist two major nitrogen-defect configurations in N-doped CNTs as well as N-doped graphene: One is the substitutional nitrogen defect where one carbon atom is replaced by one nitrogen atom. The other is the pyridine type where nitrogen atoms are located around the atomic vacancy [14-20]. Furthermore, it is reported that the transport properties of the substitutionally N-doped CNTs and graphene are much different from those of the pyridine-type defective ones: The substititionally N-doped CNTs show n-type doping property, whereas the pyridine-type defective CNTs have p-type one [20-25].

Scanning tunnelling microscopy (STM) method is a powerful tool to observe the local electronic structures of the semiconductor and metal surfaces at atomic levels. The STM images of nitrogen-doped as well as boron-doped graphene have been examined theoretically and experimentally, and the theoretical and experimental STM images are confirmed to show good agreement with each other [24,26-28]. However, our knowledge as to the pyridine-type defects in N-doped CNTs is still limited.

In this work, we report on the atomic structures, energetics, and STM images of N-doped (10,0) CNTs using the first-principles density-functional study. From our results of the calculated formation energy, it is found that the substitutional N defects have the lowest formation energy, whereas the tetramerized and the trimerized configurations are energetically plausible configurations among the pyridine-type defects. The STM images of the substitutional N, the trimerized pyridine-type, and the tetramerized pyridine-type defects in CNTs are simulated. It is found that the STM images of those N defects are different considerably from one another.

2. Computational method

The supercell models used in this work consist of two unit cells for a zigzag (10,0) CNT having 80 C atoms. Figure 1 shows the optimized atomic structures of N-doped (10,0) CNTs. In Fig. 1(a), substitutional N-doped (10,0) CNT is depicted [$C_{79}N$]. For the pyridine-type defects in N-doped (10,0) CNTs, the monomeric [$C_{78}N$], the dimerized [$C_{77}N_2$], the trimerized [$C_{76}N_3$], and the tetramerized [$C_{74}N_4$] configurations are shown in Figs. 1(b)-(e), respectively.

Our calculations are carried out using the first-principles total-energy methods within the density-functional theory [29]. The interactions between the ions and the valence electrons are described by the norm-conserving Troullier-Martins pseudopotentials [30], and exchange-correlation effects are treated using the local density approximation (LDA) parameterized by Perdew and Zunger [31,32]. The wave functions are expanded in a plane-wave basis and the kinetic-energy cutoff is taken to be 50 Ry [33]. The atomic positions are optimized until the Hellmann-Feynman force on each atom becomes less than 0.05 eV/Å. One-dimensional Brillouin-zone integration is performed with four k-point sampling.

The formation energy is defined as

$$E_f = E_{tot} - m_C\mu_C - m_N\mu_N, \qquad (1)$$

where E_{tot} is the total energy of the N-doped (10,0) CNT, and the chemical potentials μ_C and μ_N are the energies per carbon atom in pristine (10,0) CNT and nitrogen atom in N_2 gas phase, respectively.

The STM images are obtained based on the Tersoff-Hamann approximation [34]. This method is known to be highly accurate for many systems despite its simplicity [35-37]. In this approximation, the tunnelling current $I(r)$ is assumed to be proportional to the local density of states (LDOS) of the surface at the tip position integrated over a range of energy E restricted by the applied bias voltage V:

$$I(r) \sim \int_{E_F}^{E_F+eV} \rho(r,E)dE, \qquad (2)$$

whrere E_F is the Fermi energy.

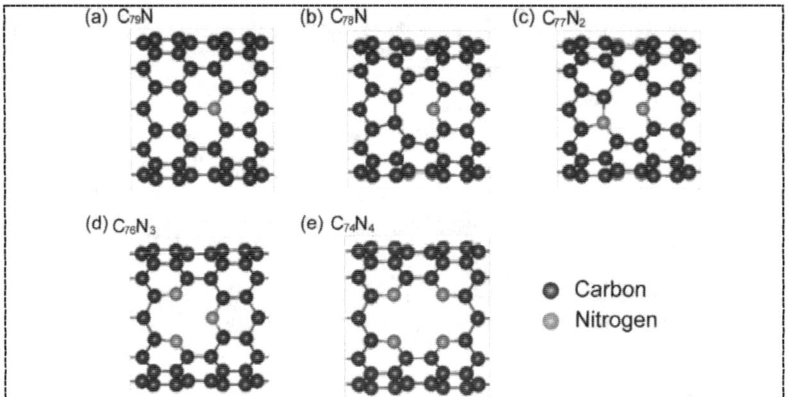

Figure 1. Optimized atomic structures for (10,0) carbon nanotubes with (a) substitutional nitrogen defect ($C_{79}N$), and (b) monomeric ($C_{78}N$), (c) dimerized ($C_{77}N_2$), (d) trimerized ($C_{76}N_3$), and (e) tetramerized ($C_{74}N_4$) pyridine-type defects.

3. Results and discussion

IOP Conf. Series: Materials Science and Engineering **744** (2020) 012032 doi:10.1088/1757-899X/744/1/012032

3.1. Energetics

We here examine the formation energies $E_f$calculated by Eq. (1). Table 1 lists the formation energies for various N-defect configurations in N-doped (10,0) CNTs. The formation energy of the substitutionally N-doped (10,0) CNT is found to be the lowest, 0.41 eV ($C_{79}N$). This implies that the substitutional N defect is the most plausible configuration when the N atom is doped into the (10,0) CNTs. For N-doped (10,0) CNTs with a vacancy, the tetramerized pyridine-type defects are found to be the most stable structure (E_f=1.78 eV), while the formation energy of the trimerized defects is 2.16 eV. The formation energies of the tetramerized and the trimerized configurations are much lower than those of the monomeric and the dimerized ones, implying the existence of the tetramerized and the trimerized configurations. It has been reported that the atomic vacancies are often observed experimentally in CNTs and graphene [38,39]. The pyridine-type defects are energetically favorable than the substitutional N defects in the presence of the atomic vacancy in the CNTs and the graphene [20,24]. Furthermore, the formation energies of the N-doped CNTs with the pyridine-type defects decrease as the number of the N atoms around the monovacancy in the N-doped (10,0) CNTs increases (see Figs. 1(b)-(d)). Therefore, the trimerized and the tetramerized pyridine-type defects would be plausible configurations if there exist the pyridine-type defects in the (10,0) CNT.

We here discuss the difference in energetics between the N-dope (10,0) CNTs and N-doped graphene. Interestingly, the formation energy of the substitutionally N-doped (10,0) CNT ($C_{79}N$) is higher than that of the substitutionally N-doped graphene, where the formation energy of the substitutionally N-doped graphene is reported to be 0.32 eV [24]. For the pyridine-type defects in the N-doped (10,0) CNTs, the formation energy of the N-doped (10,0) CNT with the tetramerized pyridine-type defect ($C_{74}N_4$) is lower than that with the trimerized one ($C_{76}N_3$). On the other hand, for the N-doped graphene, the formation energy of the N-doped graphene with the tetramerized pyridine-type defects is slightly higher than that with the trimerized one [24]. Thus, it showed be concluded that the behaviour of the formation energies for the N-doped (10,0) CNTs is considerably different from that for the N-doped graphene.

Table 1. Formation energies (E_f) for various N-defect formation in N-doped (10,0) CNTs. See Fig.1 for N-defect configurations.

Nitrogen-defect configuration	Formation energy (eV)
$C_{79}N$	0.41
$C_{78}N$	3.79
$C_{77}N_2$	2.91
$C_{76}N_3$	2.16
$C_{74}N_4$	1.78

3.2. Scanning tunnelling microscopy images

Figure 2(a) shows the STM image of the substitutional N defect in N-doped (10,0) CNTs. The STM image around the N atom in the substitutionally N-doped CNT is different from that at the C atoms since the STM image of the substitutional N defects mainly reflects the donor states below the conduction band [5]. This behaviour is also seen in the graphene and the hexagonal boron-nitride layers [26,40-47]. Therefore, the N atom in the substitutionally N-doped CNT could be identified in the STM images. Figures 2(b) and 2(c) also illustrate the STM images of the trimerized and the tetramerized pyridine-type defective (10,0) CNTs, respectively. It is found that the N atoms in the pyridine-type defects would appear to be individual regions in the STM images, which depends on the number of the N atoms in the pyridine-type defects [see Figs. 2(b) and 2(c)]. These STM images are similar to those of the N-doped graphene [24]. Thus, the STM images of the N-doped (10,0) CNTs with the substitutional N, the trimerized pyridine-type, and the tetramerized pyridine-type defects must be distinguished from one another and those three types of N-defect configurations would be observable by the STM measurements.

4. Conclusions

We have reported the atomic structures, energetics, and STM images of N-doped (10,0) CNTs using the first-principles density-functional study. From the results of the formation energy calculations, the substitutional N defect is found to be the most stable configuration. On the other hand, the tetramerized as well as the trimerized pyridine-type defects are the plausible configurations among the pyridine-type defects in the (10,0) CNTs. The STM images of (10,0) CNTs with the substitutional N, the trimerized pyridine-type, and the tetramerized pyridine-type defects are examined, and are found tobe considerably different from one another. Therefore, those N defects in the N-doped (10,0) CNTs could be identified by using the STM methods.

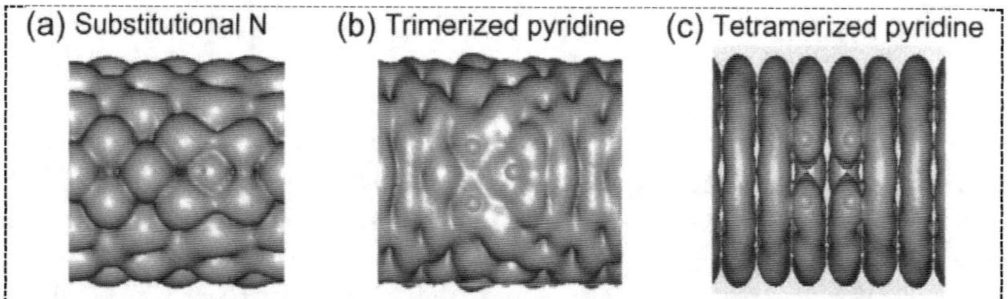

Figure 2. Simulated STM images of N-doped (10,0) CNTs with (a) substitutional nitrogen, (b) trimerized pyridine-type, and (c) tetramerized pyridine-type defects. The STM images are generated at applied bias voltage of -0.5 eV. The yellow balls denote the positions of N atoms in N-doped (10,0) CNTs.

Acknowledgments

This work was partly supported by MEXT Elements Strategy Initiative to Form Core Research Center through Tokodai Institute for Element Strategy, JSPS KAKENHI Grant Numbers JP19H01823 and JP17K05053. Computations were partly done at Institute for Solid State Physics, the University of Tokyo and at Global Scientific Information and Computing Center of the Tokyo Institute of Technology, and the Center for Computational Materials Science, Institute for Materials Research, Tohoku University for the use of MASAMUNE-IMR.

References

[1] Iijima S 1991 *Nature* **354** 56.
[2] Dürkop T, Getty SA, Cobas E and Fuhrer MS 2004 *Nano Lett.* **4** 35.
[3] Kong J *et al.* 2000 *Science* **287** 622.
[4] Schlapbach L and Zűttel A 2001 *Nature (London)* **414** 353.
[5] Fujimoto Y and Saito S 2011 *J. Phys.: Conf. Ser.* **302** 012006.
[6] Li E and Marzari N 2011 *ACS Nano* **5** 9726.
[7] Teerapanich P, Myint M, Joseph C, Hornyak G, Dutta J 2013 *IEEE Trans. Nanotechnol.* **12**, 255.
[8] Fujimoto Y and Saito S 2014 *J. Appl. Phys.* **115** 153701.
[9] Fujimoto Y and Saito S 2016 *Chem. Phys.* **478** 55.
[10] Nikolina A *et al.* 2016 *ACS Sens.* **1** 591.
[11] Yu L, Shearer C, and Shapter J 2016 *Chem. Rev.* **116** 13413.
[12] Fujimoto Y 2018 *New Polymer Nanocomposites for Environmental Remediation* ed C Hussain and A Mishra (Amsterdam: Elsevier) Chapter 9 pp 207-222.
[13] Fujimoto Y and Saito S 2019 *Jpn. J. Appl. Phys.* **58** 015005.
[14] Czerw R *et al.* 2001 *Nano Lett.* **1** 457.
[15] Ayala P *et.al.* 2007 *J. Phys. Chem. C* **111** 2879.

[16] Fang WC 2008 *Nanotechnol.* **19** 165705.
[17] Wei D, Liu Y, Wang Y, Zhang H, Huang L, and Yu G 2009 *Nano Lett.* **9** 1752.
[18] Lin YC, Lin CY, and Chiu PW 2010 *Appl. Phys. Lett.* **96** 133110.
[19] Susi T, Pichler T and Ayala P 2015 *Beilstein J Nanotechnol.* **6** 177.
[20] Fujimoto Y and Saito S 2011 *Physica E* **43** 677.
[21] Fujimoto Y and Hirose K 2003 *Phys. Rev. B* **67** 195315.
[22] Fujimoto Y, Asari Y, Kondo H, Nara J, and Ohno T 2005 *Phys. Rev. B* **72** 113407.
[23] Min YS *et al.* 2008 *Appl. Phys. Lett.* **93** 043113.
[24] Fujimoto Y and Saito S 2011 *Phys. Rev. B* **84** 245446.
[25] Yokoyama K *et al.* 2016 Carbon **142** 518.
[26] Shao L *et al.* 2013 *Nano Lett.* **13** 4659.
[27] Fujimoto Y and Saito S 2015 *Surf. Sci.* **634** 57.
[28] Fujimoto Y and Saito S 2017 *Springer Proc. Phys.* **186** 107.
[29] Kohn W and Sham L J 1965 *Phys. Rev.* **140** A1133.
[30] Troullier N and Martins JL 1991 *Phys. Rev. B* **43** 1993.
[31] Ceperley DM and Alder B J 1980 *Phys. Rev. Lett.* **45** 566.
[32] Perdew JP and Zunger A 1981 *Phys. Rev. B* **23** 5048.
[33] Computations have been performed using Tokyo Ab-initio Program Package (TAPP) which is developed by a consortium initiated at University of Tokyo: Yamauchi J, Tsukada M, Watanabe S, and Sugino O, 1996 *Phys. Rev. B* **54** 5586.
[34] Tersoff J and Hamann DR 1985 *Phys. Rev. B* **31** 805.
[35] Okada H, Fujimoto Y, Endo K, Hirose K, and Mori Y 2001 *Phys. Rev. B* **63** 195324.
[36] Fujimoto Y, Okada H, Endo K, Ono T, Tsukamoto S, and Hirose K 2001 *Mater. Trans.* **42** 2247.
[37] Fujimoto Y, Okada H, Inagaki K, Goto H, Endo K, and Hirose K 2003 *Jpn. J. Appl. Phys.***42** 5267.
[38] Hashimoto A, Suenaga K, Gloter A, Urita K, Iijima S 2004 *Nature (London)* **430** 870.
[39] Amorim R, Fazzio A, Antonelli A, Novaes F, daSilva A 2007 *Nano Lett.* **7** 2459.
[40] Terrones M et al. 2002 *Appl. Phys. A* **74** 355.
[41] Fujimoto Y and Saito S 2015 *J. Ceram. Soc. Jpn.* **123** 576.
[42] Ćirić-Marjanović G, Pašti I, and Mentus S 2015 *Prog. Mater. Sci.* **69** 61.
[43] Fujimoto Y and Saito S 2016 *Phys. Rev. B* **93** 045402.
[44] Fujimoto Y and Saito S 2016 *Phys. Rev. B* **94** 245427.
[45] Haga T, Fujimoto Y, and Saito S 2019 *Phys. Rev. B* **100** 125403.
[46] Haga T, Fujimoto Y, and Saito S 2019 *Jpn. J. Appl. Phys.* **58** SIIB03.
[47] Fujimoto Y and Saito S 2016 *J. Ceram. Soc. Jpn.* **124** 584.

IOP Conf. Series: Materials Science and Engineering **744** (2020) 012033 doi:10.1088/1757-899X/744/1/012033

Interlaminar fracture toughness analysis of fiber reinforced metal laminates based on APDL

Zhu Gong zhi[1], Zheng Chang liang[1], Xia Dong sheng[1] and Ren Ming fa[2]

1 Transportation Engineering College, Dalian Maritime University, Dalian 116026, China;
2 State Key Laboratory of Structural Analysis for Industrial Equipment, Dalian University of Technology, Dalian 116024, China
E-mail: dlmuzgz@126.com

Abstract. Fiber Reinforced Metal Laminates (FRMLs) are widely used in modern industry because of excellent comprehensive properties. However, due to the existence of the interface, there will be delamination damage in their use process. Interface is the weakest part of fiber reinforced metal laminates. The bonding strength at the interface is determined by the fracture toughness of the interface. It is an important mechanical property index of fiber reinforced metal laminates. It is usually measured by a single cantilever beam test. In this paper, the finite element model of fiber reinforced metal laminate was constructed based parametric design language APDL of ANSYS to simulate the fracture roughness tests of FRMLs using element birth-death technology. The results show that the numerical simulation is very close to the experimental results. The numerical simulation is a very effective method to determine the interlaminar fracture toughness of fiber reinforced metal laminates.

1. Introduction

The metal alloy sheets are bonded to the fiber/resin composite layer by alternating lamination, thus forming Fiber Reinforced Metal laminated plates(FRMLs). FRMLs combines the advantages of metal alloy and fiber/resin composite materials while overcoming their respective disadvantages:FRMLs has inherited the excellent impact resistance, corrosion resistance and mechanical processing performance of metal alloy, and at the same time, it has excellent fatigue resistance characteristics of fiber/resin composite materials due to the "bridging" effect [1-3]. Therefore, fiber-reinforced metal laminates are widely used in modern industry.

However, due to the unique structure of FRMLs itself, i.e. interface between the sheet metal and the composite material layer, the properties of materials on both sides of the interface are usually different. The essence of interface is a kind of internal restraint of material on one side of interface to material on the other side. Moreover, there are usually various defects at the interface, and stress concentration exists at the interface. Therefore, FRMLs is often damaged from their interfaces. when the stress in the thickness direction exceeds its strength limit, delamination fracture is initiated at the laminate interface [4].However, after the initiation of delamination fracture, the crack propagation is not controlled by the strength limit along the thickness direction of the laminates, but depends on the interfacial fracture toughness of the laminates. Therefore, interfacial fracture toughness is a very important mechanical property index of FRMLs.

Usually, the fracture toughness of the interface is determined by experiments. In this paper, based on the experiments, the parametric design language APDL of ANSYS is used to establish the model and the birth-death element technique is used to simulate the delamination fracture process of Carbon Fiber Reinforced Magnesium Laminates(CFRL/Mg) , and finally determine the interfacial fracture toughness.

2 Experimental determination of interfacial fracture toughness

The fracture toughness of CFRL/Mg interface was measured by single Cantilever Beam (SCB)structure. The SCB structure is shown in Figure 1. The loading rate is 0.5mm/s, and the crack propagation starts from the pre-crack and propagates along the interface between magnesium alloy and carbon fiber/resin composite plate.

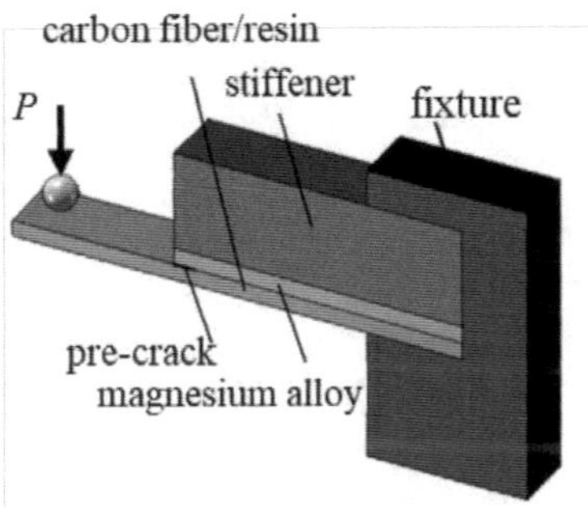

Figure 1 Schematic diagram of a single cantilever beam

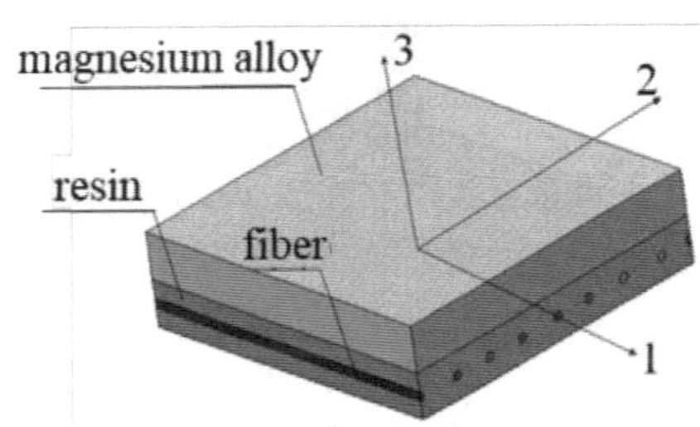

Figure 2 Material coordinate system of CFRL/Mg

Loads, displacements and crack lengths were recorded during the tests, and the interfacial fracture energy(Kc)was calculated by using the following equations.

$$K_C = \frac{P^2}{2\mathrm{B}}\frac{\mathrm{d}C}{\mathrm{d}a} \tag{1}$$

where

P: the applied force

B :the width of the specimen

C :the specimen compliance

a :the crack length.

The specimen compliance is expressed by:

$$C = C_0 + ma^3 \tag{2}$$

where

m is the slope of $C-a^3$ straight line, C_o is the constant of the given specimen, and it is determined by the ordinate of the intersection point of the straight line and the longitudinal axis.

Using equation (1) and(2) , fracture energy Kc can be determined:

$$K_C = \frac{3P^2ma^2}{2B} \tag{3}$$

The material coordinate system of CFRL/Mg is shown in Figure 2.

The preparation of the specimen, mechanical properties of magnesium alloys and carbon fibers/resins and the detailed test process are referred to in the references [5].

3 Failure criteria

During the CFRL/Mg fracture toughness test, the delamination failure of CFRL/Mg along the interface occurs as the crack propagates at the interface.

There are two main damage modes: matrix cracking of composite layers and interlaminar delamination(interlaminar damage) .

3.1 Failure Criteria for Matrix of Composite Layer

The modified Hashin [6] matrix failure criterion is adopted here. The failure degree of the matrix of the fiber/composite layer is determined by the matrix cracking failure factor e_m.

$$e_m^2 = (\frac{\sigma_{22}}{Y})^2 + (\frac{\sigma_{33}}{Y})^2 + (\frac{\tau_{23}}{S})^2 \tag{4}$$

$$Y = Y_C, \quad \sigma_{22} < 0$$

$$Y = Y_t, \quad \sigma_{22} .. 0$$

where

Y: the tensile and compressive strength of the laminate

S: the interlaminar shear strength of the laminate

Y_t: the tensile Strength of the laminate

Y_c: the compressive strength of the laminate

When $e_m \geq 1$, the matrix of the pavement is damaged.

3. 2 Interface Layering Criterion

The degree of interfacial delamination failure is determined by the interfacial delamination failure factor e_d:

$$e_d^2 = (\frac{{}^M\tau_{23}}{{}^M S})^2 + (\frac{{}^F\tau_{13}}{{}^F S})^2 + (\frac{{}^F\sigma_{22}}{{}^F Y})^2 \tag{5}$$

$$Y = Y_C, \quad \sigma_{22} < 0$$

$$Y = Y_t, \quad \sigma_{22} .. 0$$

where

The superscript M indicates magnesium alloy, F indicates carbon fiber/resin composite layer.

When $e_d \geq 1$, the interface is damaged.

3.3 The numerical simulation

The model of CFRL/Mg was established by APDL. The thickness of magnesium alloy plate and carbon fiber/resin layer was 5mm, and the width of the specimen was 20mm. In order to reduce the workload, according to the symmetry of the model and load, half of the entity is taken to build the model. The length of pre-crack is 10 mm. The SOLID92 entity unit is used to mesh the model. The tensile strength, shear strength and elastic modulus of magnesium alloy sheet are 174.3 MPa, 133.8 MPa and 42.1 GPa respectively. The tensile strength, compressive strength, shear strength and elastic modulus of carbon fiber/resin along the fiber direction are 630.2 MPa, 516 MPa, 42.2 MPa and 48.4 GPa.

Before the solution, the upper surface of magnesium alloy and the right end of the specimen are restrained. In the process of solving the problem, the load is divided into two stages. In the first stage, the load starts from zero and increases monotonously with the increase of head displacement. The load step is 1N. At the same time, the displacement and the magnitude of the load after the end of each load step were recorded and the matrix cracking failure factors e_m of each matrix element and the interface stratification failure factors e_d of the element at the interface were calculated. If $e_m \geq 1$, the single stiffness matrix is multiplied by a small number before the total stiffness matrix is introduced by using the element birth-death technology, so that the stiffness of the element is set to zero. If $e_d \geq 1$, the same method is used to set the stiffness of the failure element to zero. The appearance of CFRL/Mg delamination means the end of the first stage. In the second stage, the load does not increase monotonously with the increase of displacement. The load increases from zero, and the load step size is also set to 1N. Unlike the first step, it is not necessary to record the displacement of the head and the magnitude of the load at each load step. Displacement, load and crack length are recorded only when $e_d \geq 1$. The load increases from zero until the crack length reaches 55 mm, and the solution is completed. The meshing of the model and the shape of the deformed model are shown in figure 3.

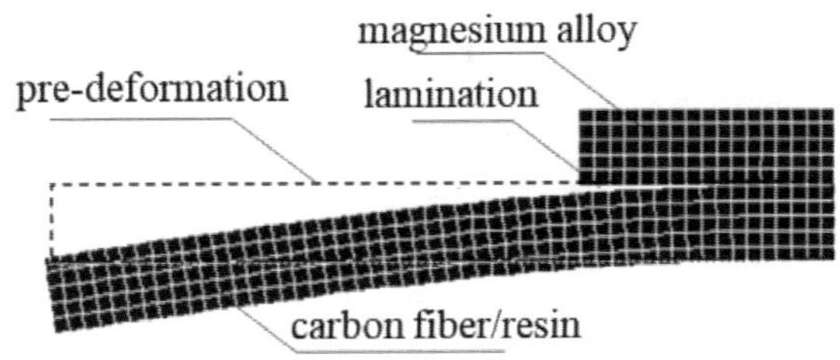

Figure 3 Mesh generation, lamination and deformation of CFRL/Mg

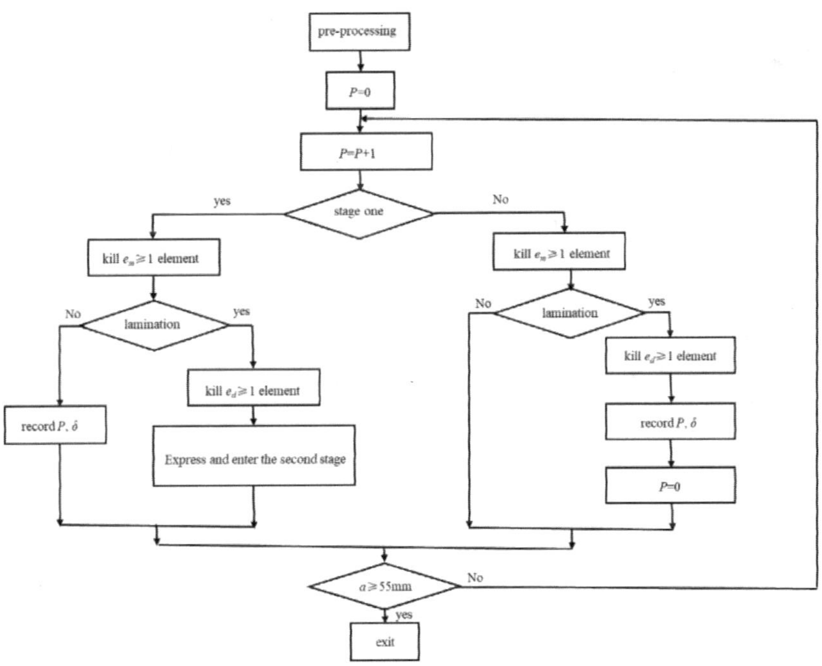

Figure 4 Schematic diagram of calculation process

The calculation flow diagram is shown in figure 4.

where

P: the applied force

δ: the displacement of head
a: the crack length

4 Comparison between numerical simulation and experimental results

The comparison of load-displacement curves between numerical simulation and test is shown in figure 5. It is very clear from the graph that the two curves are very close. Overall, the numerical simulation curve is slightly lower than the experimental curve. Moreover, the curve of numerical simulation is smoother than that of experiment. The reason for this result lies in the difference between the two constraints.The stiffness of the specimens is increased by using 45 # steel stiffener in the test, while the upper surface of the magnesium alloy plate is completely restrained while in the numerical simulation, which is equivalent to infinite stiffness. Therefore, the stiffness of the test specimen is less than that of the numerical simulation. As a result, the numerical simulation curve is slightly lower than the experimental curve, but there is little difference between them.

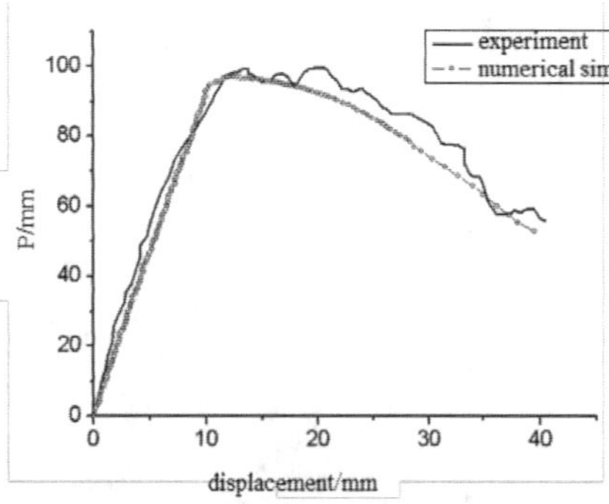

Figure5 Comparison of load-displacement curves between numerical imulation and test

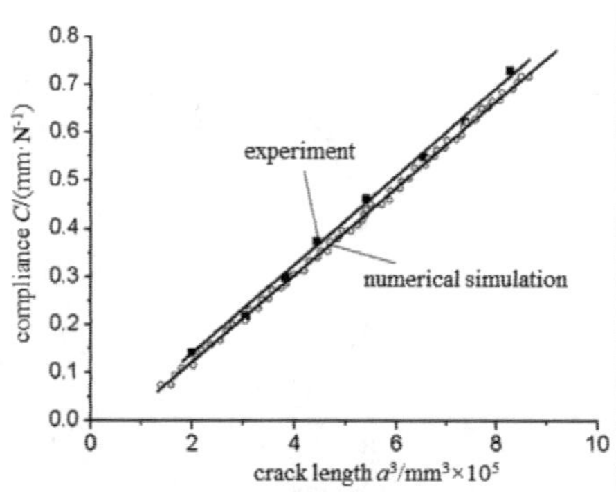

Figure 6 Comparison between numerical simulation and experimental C-a^3 curves

Figure 6 is the comparison of numerical simulation and experimental C-a^3 curves. It can be seen from the figure that the relationship between the compliance C of numerical simulation and test and the cubic of crack length a is approximately linear. This is completely consistent with the conclusion(equation(2)) obtained by theoretical deduction.

The compliance of the test specimen is slightly greater than that of the numerical

Figure 7 shows the comparison of numerical simulation and experimental *Kc-a* curves. It can be seen that before the crack length reaches 65 mm, the *Kc* of numerical simulation and test increases with the increase of the crack length, and both curves are relatively smooth. When the crack length exceeds 65 mm, the simulated curve approaches the level and remains smooth, while the experimental curve fluctuates. Overall, the two curves are very close.

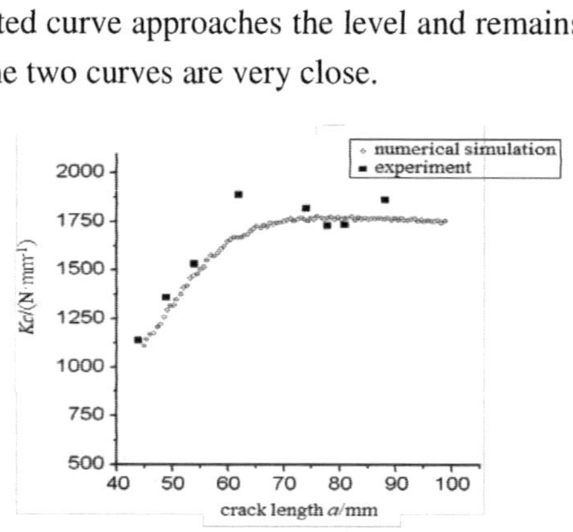

Figure 7 Comparison of numerical simulation and experimental *Kc-a* curves

5 Conclusion

Based on the platform provided by ANSYS, the finite element model of CFRL/Mg is established by using APDL, and the program is written. The growth process of interface crack is simulated by using birth-death element technology. Finally, the interface fracture toughness of laminates is determined. The results show that the load-displacement curves, C-a^3 curves and Kc-a curves obtained by numerical simulation are very close to those obtained by experiments. It is proved that the simulation of fracture delamination process of fiber reinforced metal alloy laminates based on APDL and life-and-death elements technology is an effective way to determine the interfacial fracture toughness.

6 Reference

[1] ZHAO Zu hu. *Fiber reinforced metal laminates for aerospace applications. Space Return and Remote Sensing*, 1996; **17**(1):42-51

[2] CHEN Shao jie. *Composites and A380 Aircraft. Aviation Manufacturing Technology*, 2002; (9): 27~29

[3] L.B. Vogelesang, A. Vlot. *Development of fibre metal laminates for advanced aerospace structures* . Journal of Materials Processing Technology, 2000; **35**(2):1-5

[4] S Mespoulet, J M Hodgkinson, F L Matthews et al. *Al/GFRL novel test method to determine the through-thickness tensile properties of long fibre reinforced composites.* Proceedings of Seventh European Conference on Composite Materials(ECCM-7), Institute of Materials London, UK, Woodhead Publishing, Cambridge, 1996; (2):131-150

[5] Zhu Gong zhi, Zheng Chang liang, Deng Yang bo. *Tensile properties and interlaminar fracture toughness of carbon fiber reinforced magnesium alloy laminates*. Science, Technology and Engineering, 2011; **11**(7): 1527-1531

[6] Hashin. *Failure Criteria for Unidirectional Fiber Composite*. Journal of Applied Mechanics, 1980; 47: 329-334

Design of Vacuum Pump Control System for Negative Pressure Shale Shaker Based on PLC

Xiangqian Yang*[1], Chao Zhou[2], Hua Guo[3] and Qinghua Zheng[4]

CNOOC Research Institute Co., Ltd.Beijing China
*[1]E-mail:979997829@qq.com;[2]E-mail:zhouchao@cnooc.com.cn;[3]E-mail:21346571@qq.com;[4]E-mail: zhengqh@cnooc.com.cn

Abstract. In order to improve the solid-liquid separation effect of drilling fluid in drilling operation, a new type of negative pressure vibrating screen was developed. The vacuum pump is its main component. At present, the degree of vacuum has a significant impact on the screening effect and working efficiency of the negative pressure vibrating screen. The value of the vacuum is based on experience, the setting parameters are not adjustable, or changing via the manual adjustment, and the control precision and the degree of intelligence are low. The system of the negative pressure vibrating screen vacuum pump control was designed with the moisture content of the filter cake as the controlled variable. The mathematical model between the moisture content and the vacuum degree of the filter cake was established. The fuzzy PID was studied by using Siemens S7-300 PLC as the controller. Control the law, build a control system based on the inverter as the actuator, and control the vacuum degree according to the drilling fluid quality in real time. The field application shows that the steady state error of the system is 11.1%, which meets the drilling operation requirements.

1.Introduction

The vibrating screen equipment is a kind of mechanical equipment that uses the vibration principle to perform the screening process. In the oil drilling industry, the shale shaker is the first stage solid control equipment, mainly used to clean the harmful solid particles in the drilling fluid. In order to obtain clean and reusable drilling fluid, secondary and tertiary solids control procedures is often required. The negative pressure shale shaker can meet the requirements of large processing capacity, high screening efficiency and high integration shale shaker. As a component of the apparatus, the vacuum pump has an important influence on the screening efficiency of the negative pressure shale shaker. The important parameter of the vacuum pump is the degree of vacuum, Most enterprises change the vacuum degree by experience, and the value is determined by experience, which make vacuum pump work in low efficiency, high energy consumption, and even unable to meet the work demands.

In order to improve the solid-liquid separation effect and work efficiency, real-time control of the vacuum pump is required. In this paper, moisture content of filter cake is the final control target, the mathematical model between the solid phase moisture content and the vacuum degree is established, and the vacuum pump vacuum degree control system is designed by using S7-300 PLC and frequency converter. The system intelligently adjusts the vacuum degree of the vacuum pump according to the moisture content of the filter cake, so as to improve the screening efficiency of the negative pressure shale shaker, reduce the energy consumption and the labor cost.

2. Principle of negative pressure shale shaker

The negative pressure shale shaker is mainly composed of frame, circulating screen mechanism, excitation system, negative pressure system, and cleaning mechanism, as shown in figure 1.The flexible screen is detachably attached to a chain, the sprocket drives the chain to circulate under the action of the drive motor, the chain drives the flexible screen fixed on it to circulate together. The excitation system drives the vibration bracket to vibrate through the pneumatic piston reciprocating vibrator, so that the chain transmission device vibrates in the vertical direction and drives the screen to vibrate. When the screen is circulated to the upper side of the chain drive, the drilling fluid is screened, and when the screen is circulated to the underside of the chain drive, the self-cleaning is realized under the action of the vibration and the cleaning mechanism. The negative pressure chamber is located under the upper screen, and a negative pressure is formed under the action of the vacuum pump, so that a pressure difference is formed on the surface of the screen to enhance the screening effect of the drilling fluid.

When the negative pressure vibrating screen is working, the drilling fluid containing cuttings flows from the inlet to the circulation screen. Circulating screen moves forward with drilling fluid and cuttings through chain transmission device. At the same time, under the action of negative pressure and vibration, the liquid phase is recovered by the drainer. The solid phase remains on the screen surface and continues along the screen until the outlet falls into the tank to achieve solid-liquid separation.

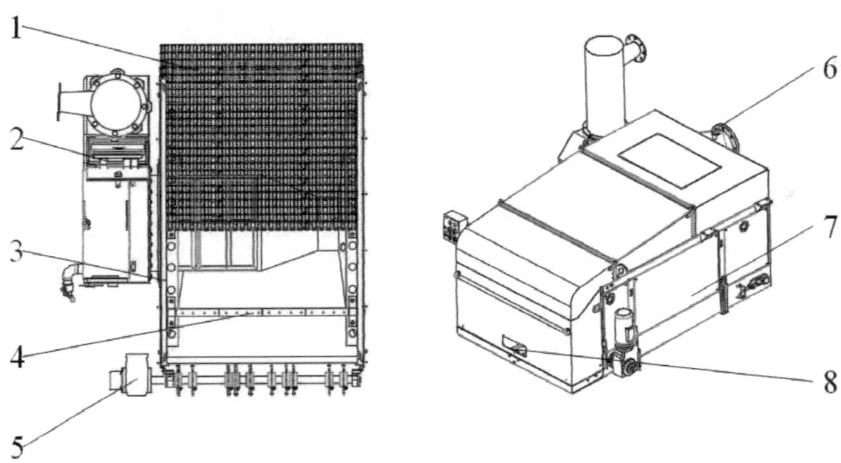

Circulating screen;2-Vacuum pump;3-Vibrating bracket;4-Cleaning system;
Motor;6-Inlet;7-Outer shell;8-Outlet

Figure 1. Negative pressure shale shaker structure diagram.

Compared with the traditional shale shaker, the negative pressure shale shaker greatly increases the screening efficiency by adding a negative pressure system. The negative pressure system is mainly composed of a vacuum pump and a negative pressure tank. The vacuum pump draws air, so that the pressure of the negative pressure tank under the screen is lower than atmospheric pressure. The solid phase diameter in the drilling fluid is larger than the sieve pore diameter, and is trapped and deposited to form a filter cake. The liquid passes through the screen and the filter cake under pressure difference and gravity to realize solid-liquid separation. At the same time, the suction action of the vacuum pump will cause the gas to pass through the filter cake into the negative pressure tank, take away part of the moisture in the filter cake, and enhance the screening ability of the negative pressure shale shaker.

3. Mathematical model of vacuum degree and moisture content of filter cake

IOP Conf. Series: Materials Science and Engineering **744** (2020) 012034 doi:10.1088/1757-899X/744/1/012034

The solid phase moisture content after solid-liquid separation is often used to evaluate the treatment effect. The moisture content of the filter cake is the mass ratio of the moisture in the filter cake to the solid content after drying, the lower the moisture content, the better the treatment effect is. The degree of vacuum refers to the degree of gas thinness under vacuum, for the negative pressure tank, the value is the difference between the atmospheric pressure and the internal pressure of the tank.

The moisture content of the filter cake is related to the flow speed of the drilling fluid. To simplify the study, it is assumed that the drilling fluid has no lateral flow along the width of the screen, the flow rate of the drilling fluid in the x direction is consistent with the screen surface and the effects of entrained gases in the drilling fluid is ignored. That is, only the flow through the screen is studied. Establish the coordinate system shown in figure 2, Dong's discrete element method is used to select the unit body $d_x d_y$ as the research object.

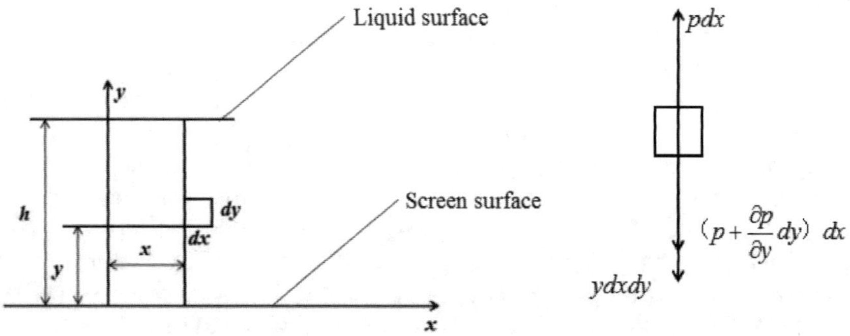

Figure 2. Screen unsteady flow and fluid unit.

When considering the application of negative pressure, there is a pressure difference between the upper and under surfaces of the drilling fluid, in the y direction, according to Newton's second law:

$$p\mathrm{d}x-(p+\frac{\partial p}{\partial y}\mathrm{d}y)\mathrm{d}x-\gamma\mathrm{d}y\mathrm{d}x=\frac{\gamma}{g}\mathrm{d}x\mathrm{d}y(\ddot{y}+\frac{\mathrm{d}v_y}{\mathrm{d}t}) \tag{1}$$

Where p is the pressure, $\ddot{y}=-\omega^2\lambda\sin\omega t$ is the acceleration of the screen along the y direction, γ is the gravity density of the drilling fluid.

introducing the relative inertia head H_{yi} and the flow loss H_{wy}. After calculation, it can be obtained that:

$$\frac{p_1-p_2}{g}+\frac{v_{y_1}^2-v_{y_2}^2}{2g}=\frac{1}{g}(\ddot{y}+g)(y_2-y_1)+H_{yi}+H_{wy} \tag{2}$$

In the above formula, $H_{yi}=\int_{y_1}^{y_2}(\frac{\partial v_y}{\partial t}+\frac{\partial v_y}{\partial x}v_x)\frac{\mathrm{d}y}{g}$ is the relative inertia head, $H_{wy}=\frac{\eta}{\gamma c}u_y+\beta\frac{\rho B}{\gamma\varepsilon^2 D}u_y^2+\frac{c_0}{\gamma}\sigma_0$ is the flow loss, P_1 is the pressure at the top of the microelement, P_2 is the pressure at the negative pressure tank, c is the screen liquid passing rate, σ_0 is static shear stress of drilling fluid, ε, B, D are respectively the porosity, thickness and size of the mesh, , β, c_0 are experimental constant.

At $y_1=h$, because the vertical flow rate of the drilling fluid is small, assuming its velocity is 0; then equation (2) can be simplified as:

$$\frac{u_y^2}{2g}-\frac{p}{\gamma}=\frac{h}{g}(\ddot{y}+g)-H_{yi}-H_{wy} \tag{3}$$

When $\ddot{y} > -g$, the drilling fluid passes through the screen, and the flow rate of the sieve is uy; when $\ddot{y} < -g$, there is no drilling fluid through the screen, and the drilling fluid movement equation is:

$$\begin{cases} \dfrac{u_y^2}{2g} - \dfrac{p}{\gamma} = \dfrac{h}{g}(\ddot{y}+g) - H_{yi} - H_{wy}, \ddot{y} > -g \\ u_y = 0, \ddot{y} \le -g \end{cases} \tag{4}$$

The vibration of the negative pressure vibrating screen is periodic, and the flow rate of the through-screen is also cyclically changed. In order to reflect its comprehensive performance, the average flow rate of the drilling fluid through the screen per unit time is required.

$$\overline{u_y} = \frac{1}{T}\int_{t_1}^{t_2} u_y \,\mathrm{d}t \tag{5}$$

Where t1 and t2 are the upper and under limit of integration, they are determined by equation (4), $t_1 = -\dfrac{\arcsin\dfrac{g}{\lambda\omega^2} - \pi}{\omega}$, $t_2 = -\dfrac{\arcsin\dfrac{g}{\lambda\omega^2}}{\omega}$.

The amount of treatment of the negative pressure shale shaker in unit time is expressed as :

$$Q_S = \frac{C_S}{C+C_S} I \int_0^{z_s} k\,\mathrm{d}z \int_0^{x_s} u_y \,\mathrm{d}x \tag{6}$$

In the formula, k is the proportion of screen perforation relative to the screen area, x_s is the length of the drilling fluid can effectively pass through the screen, z_s is the width of the drilling fluid can effectively pass through the screen, cs is the liquid passing rate conductivity of the cuttings layer at position x, I is the solid phase interference coefficient of the screen which is determined by experiment.

The inlet flow of the negative pressure shale shaker vibrating screen is set to G, The density of unscreened debris-containing drilling fluid can be measured as ρ_l, the mass percentage λ of solids in the drilling fluid can be obtained by experiment, and the volume percentage δ of fine particles passing through the screen relative to the through-screen drilling fluid is also obtained by experiment. So the moisture content H of the filter cake is

$$H = \frac{\rho_l G(1-\lambda) - \rho_l(1-\delta)Q_s}{\rho_s \lambda G - \delta\rho_s Q_s} \tag{7}$$

Mathematical model of vacuum degree and moisture content of filter cake can be obtained from equations (5), (6) and (7).

4. Design of control system

Negative pressure shale shaker vacuum degree control system is shown in figure 3. Part of the system parameters are determined by experiment, which makes the mathematical model is not accurate enough, it needs to change the control parameters with time, so the control system uses fuzzy PID control.

IOP Conf. Series: Materials Science and Engineering **744** (2020) 012034 doi:10.1088/1757-899X/744/1/012034

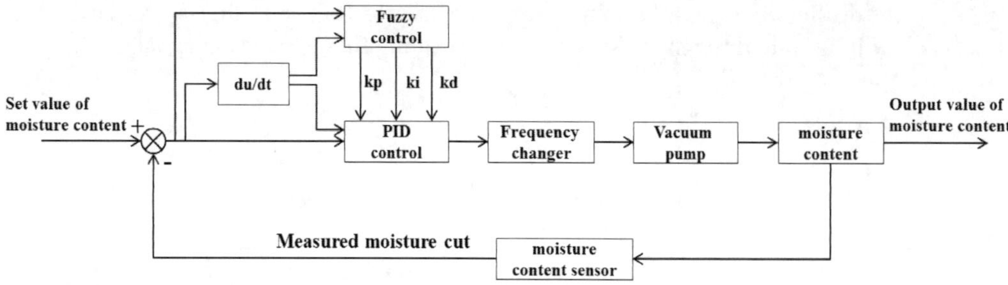

Figure 3. The block diagram of moisture content control program.

In practice, the density of drilling fluid is 1.015kg/m3, funnel viscosity is 48.8s, the density of pure drilling fluid is 1.009kg/m3, the density of filter cake is 2.34kg/m3, the vibration frequency is 96Hz. The transfer function established according to the mathematical model of the degree of vacuum and the moisture content of the filter cake is

$$G(s)=1-\frac{0.08e^{-4.02s}}{(0.5s^2+s)(2s+1)(0.4s+1)} \tag{8}$$

The simulation result of the filter cake moisture content fuzzy PID control process based on the transfer function formula (8) is shown in figure 4.

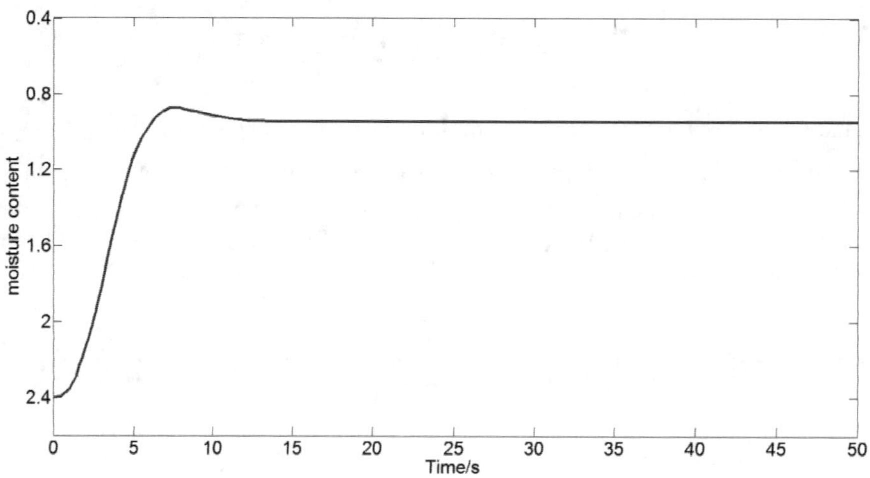

Figure 4. The simulation curve of Water content control process.

The set value of moisture content is 0.9, as can be seen from the figure, after 15s, the moisture content of the filter cake reaches the set value. The system has some overshoot, but the duration is short, and the system has almost no oscillation, which can meet the requirements of negative pressure shale shaker.

The system uses Siemens S7-300 PLC as controller to realize the real-time control of moisture content. When the system is working, the moisture content sensor feeds back the detected moisture content to the PLC, PLC compares the real-time moisture content returned from the feedback with the set value. The program uses fuzzy PID control algorithm to calculate and output the required amount of control. PLC converts the control amount into a voltage value, the frequency of converter is changed through the RS485 communication protocol. The PLC control flow is shown in figure 5.

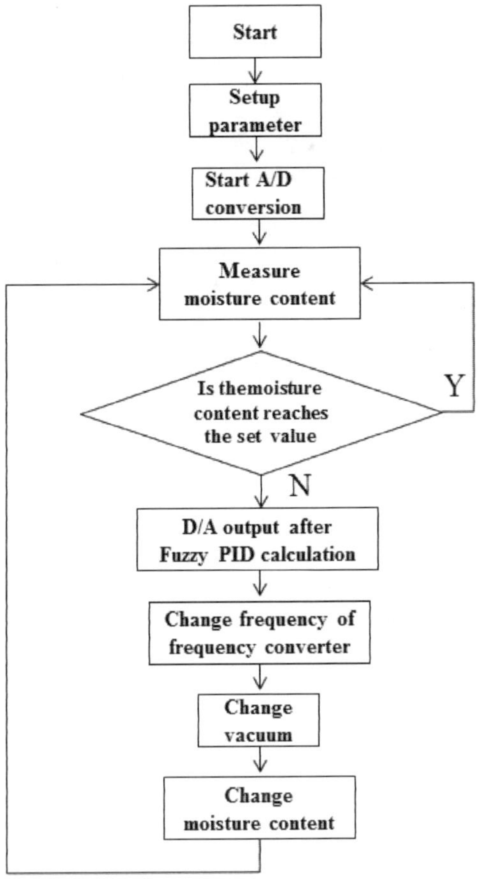

Figure 5. The flow diagram of PLC control process.

5. Experimental verification

Experiment is implemented to verify the rationality of the system, the moisture content control process can be expressed as the distribution of drilling fluid on the screen surface, as shown in figure 6.

(a)

(b)

Figure 6. Distribution of drilling fluid on the screen surface.

The set value of moisture content is 0.9,figure 6 (a) shows the distribution of drilling fluid on the screen surface when the moisture content is 2.4, figure 6 (b) shows the distribution of drilling fluid in the final stable condition, the adjustment time of the system is 16.5s, which is almost identical to the simulation result. The moisture content collected by the sensor is 0.8, and there is an error of 11.1% from the set value, it can meet the working needs of negative pressure shale shaker. The actual moisture content control process curve is shown in figure 7.

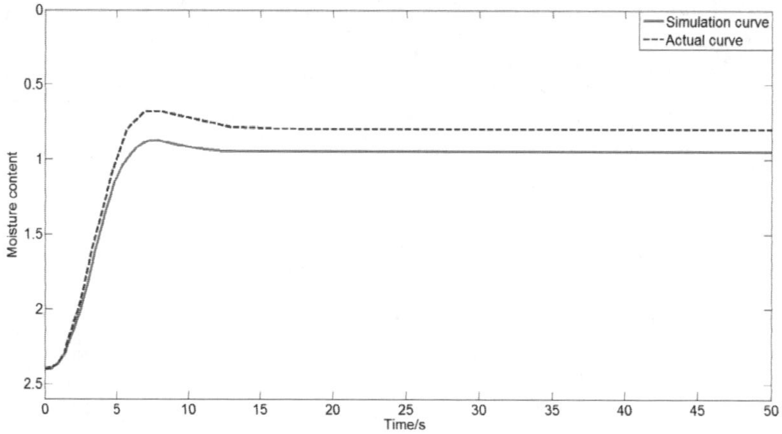

Figure 7. The curve of control.

6. Conclusion

The negative pressure shale shaker vacuum pump control system of this design is based on PLC, and the vacuum pump motor speed is changed by means of fuzzy PID control. The vacuum of the vacuum pump is adjusted. A mathematical model of vacuum and moisture content of the filter cake is established, automatic control of the moisture content of the filter cake is achieved. The experimental results show that the response time of the system is short and the steady-state error is 11.1%, which can meet the working requirements of the negative pressure shale shaker.

7. References

[1] M.H. Zhang, R. Deng and Q. Xu 2013 *WORKING THEORY AND TESTING TECHNOLOGY OF DRILLING VIBRATING SCREEN* (Petroleum industry press)

[2] Q.Y. Shi and J.H. Zhang 2019 Design of intelligent control system for negative pressure casting process based on PLC *Thermal processing technology* vol **13** 67-69+73

[3] Q. Zeng 2017 Analysis on dehydration mechanism of vacuum belt micro-vibrating mud screen (Yangtze university)
[4] Dong K J and Yu A B 2012 Numerical simulation of the particle flow and sieving behavior on sieve bend/low Head Screen Combination *Minerals Engineering*, vol **31** pp 2-9
[5] HoberockL L 1980 A Study of Vibratory Screening of Drilling Fluids *Journal of Petroleum Technology* vol **32(11)** pp 1889-1902
[6] W.X. Li 2018 Research on the theory of negative pressure vibrating screen(Southwest petroleum university)
[7] S. Guo 2019 Study on coordinated control strategy of multiple motors of solid liquid separator (Southwest petroleum university)
[8] A.W. Wen 2016 Development of fuzzy control algorithm based on SIEMENS s7-300 PLC (North China electric power university)
[9] M.M. Fang 2019 Application of PLC automatic control technology in frequency converter *Electronic technology and software engineering* vol **13** p 120

IOP Conf. Series: Materials Science and Engineering **744** (2020) 012035 doi:10.1088/1757-899X/744/1/012035

Synthesis and Photocatalytic Activity of Submicron Strontium Titanate in Soluble Polymer System

Yingnan Dong[1], Yan Zhao[1], Jian Tang[1] and Xiaoguo Bi[1]

1 College of New Energy, Shenyang Institute of Engineering, Shenyang, P. R. China
E-mail: dyn4870@qq.com

Abstract. Recently, micro-nano-strontium titanate photocatalytic materials have been one of the research hotspots because they can degrade organic pollutants. In the work, 200-400 nm strontium titanate powder was synthesized by soluble polymer (PEG) system and low-temperature direct precipitation method. Studies showed that the concentration of PEG and the degree of polymerization had a significant effect on the particle size of the powder. At the same time, the work studied the photocatalytic properties of the obtained strontium titanate powder, discussing the effect of particle size of strontium titanate on its photocatalytic activity.

1. Introduction

These Strontium titanate, a kind of perovskite-type composite oxide, is a new type of high-quality electronic ceramic material and photocatalytic material[1-6]. The low-temperature precipitation method was used to obtain strontium titanate powder with good uniformity of crystal particles between 200and 400 nm with $SrCl_2 \cdot 6H_2O$ and $TiCl_4$ as precursors in polyethylene glycol (PEG) system at different polymerization degrees. We investigated the effect of PEG concentration and degree of polymerization on the particle size of strontium titanate powder and its photocatalytic activity.

2. Experiment

2.1. Synthesis of strontium titanate powder

A certain amount of saturated cerium chloride hexahydrate solution and a certain amount of PEG were added to the reactor, and a certain amount of $TiCl_4$ solution was slowly added under 75 °C water bath and stirring. The pH of the mixed solution was controlled at pH=13 with sodium hydroxide, and the powder was washed with deionized water after heating in a water bath for 4 hours. Meanwhile, the circulating water multi-purpose vacuum pump was used for suction filtration. Repeat washing and suction filtration until there was no Cl^- in the filtrate. The powder was dried for 3 h, and then calcined in a muffle furnace at 750 °C for 3 h. The PEG polymerization degree was 2,000, and the strontium titanate powder obtained under the condition of a concentration of 5 g/L was marked as $STO_{PEG2000-5}$.

The surface morphology of the sample was observed using a Japanese Hitachi S-4800 scanning electron microscope, and the surface of the sample was sprayed with gold to eliminate charge accumulation. Set the emission current to 10 μA, and the acceleration voltage to 3.0 kV.

2.2. Photocatalytic performance evaluation of strontium titanate powder

IOP Conf. Series: Materials Science and Engineering **744** (2020) 012035 doi:10.1088/1757-899X/744/1/012035

The photocatalytic properties of strontium titanate powder were evaluated by methyl orange. A certain amount of methyl orange standard solution and our synthesized strontium titanate sample were added to the quartz reactor. After ultrasonic mixing, the photocatalytic experiment was carried out with TU1202 UV transillumination station UVC (254 nm) as the ultraviolet light source under stirring and a certain air flow rate. The reaction time was 0.5, 1, 1.5 and 2 h. After centrifugation, the solution was analyzed by colorimetric method. The concentration of methyl orange was determined by absorbance at 468 nm of ultraviolet-visible spectrophotometer, calculating the photocatalytic degradation rate of methyl orange solution.

The reagents used in this experiment were of analytical grade, and water was obtained by two-stage reverse osmosis.

3. Results and discussion

3.1. Effect of PEG concentration on the particle size of strontium titanate powder

We synthesized strontium titanate powder with PEG2000 at concentrations of 1, 5 and 10 g/L, respectively. Fig. 1 shows the SEM of the powder. The particle size statistics of 1,000 powder particles are counted by SEM image, and Fig. 2 shows the particle size distribution statistics. The particle sizes of $STO_{PEG2000\text{-}1}$, $STO_{PEG2000\text{-}5}$ and $STO_{PEG2000\text{-}10}$ are 325, 240 and 415 nm, respectively. The particle sizes are relatively uniform in normal distribution. The synthesized strontium titanate belongs to submicron powder.

Figure 1. SEM of $STO_{PEG2000\text{-}1}$, $STO_{PEG2000\text{-}5}$ and $STO_{PEG2000\text{-}10}$

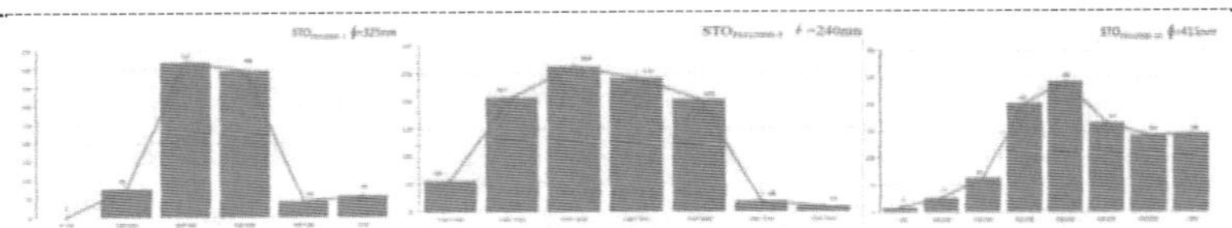

Figure 2. Particle size distribution of $STO_{PEG2000\text{-}1}$, $STO_{PEG2000\text{-}5}$ and $STO_{PEG2000\text{-}10}$

The experimental results showed that the strontium titanate powder particles decrease first and then increase with the increased PEG concentration. We speculate that when the PEG concentration is low, the formed strontium titanate crystal particles are incompletely encapsulated. The protection effect is not good, so the particle size is relatively large. When the amount of PEG increases to an appropriate concentration, PEG is ideal for the protection of strontium titanate grains, which can effectively prevent grain growth and agglomeration. When the PEG concentration rises, a large amount of hydrogen bonds can be generated due to the hydroxyl groups among the polyol compounds[7]. The bridging action between the hydrogen bonds of the PEG causes them to aggregate, which makes the size of the strontium titanate particles become large.

3.2. Effect of PEG degree of polymerization on the particle size of strontium titanate powder

Under the condition of PEG concentration of 5g/L, PEG1000, PEG2000 and PEG4000 were used as protective agents, respectively. Fig. 3 shows the SEM of the synthesized strontium titanate powder, and Fig. 4 shows the particle size distribution statistics. The experimental results showed that the crystals of the three samples were spherically monodisperse, and the average particle size of $STO_{PEG1000\text{-}5}$ was 293 nm. The average particle size of $STO_{PEG2000\text{-}5}$ was 240 nm, with good uniformity of the particle sizes of the two samples in normal distribution. The average particle size of $STO_{PEG4000\text{-}5}$ was 332 nm, but the uniformity of particle sizes was significantly reduced. The effect of PEG concentration on the particle size of strontium titanate was similar. Under the same concentration conditions, the polymerization degree of polyethylene glycol had no linear relationship with the particle size of strontium titanate powder. Rather, as the degree of polymerization of PEG increased, the average particle size of strontium titanate powder increased after decrease.

Figure 3. SEM of $STO_{PEG1000\text{-}5}$, $STO_{PEG2000\text{-}5}$ and $STO_{PEG4000\text{-}5}$

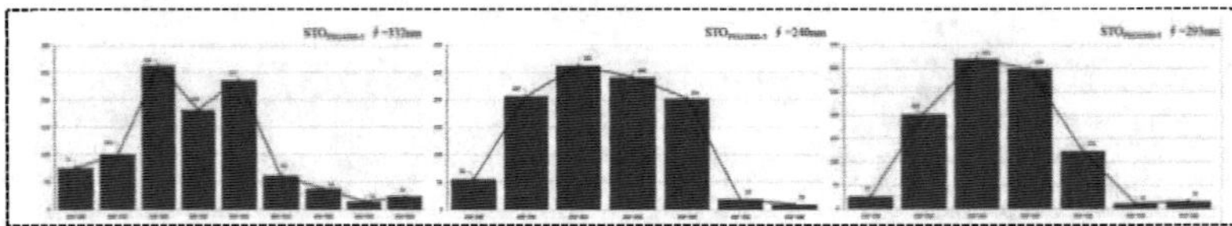

Figure 4. Particle size distribution of $STO_{PEG1000\text{-}5}$, $STO_{PEG2000\text{-}5}$ and $STO_{PEG4000\text{-}5}$

This phenomenon can be explained by the increase in the number of -(CH_2CH_2O)- units in the PEG molecule and the in situ formation of pseudocrown structure encapsulated by -(CH_2CH_2O)n- with the high degree of polymerization. As the -(CH_2CH_2O)n-chain is short in PEG1000 molecules with low degree of polymerization, a single molecule cannot form a complete pseudocrown ether structure. When pluralities of molecules encapsulate the pseudocrown ether, the interaction between the molecules is small with incomplete encapsulation of the strontium titanate particles. It cannot prevent the entropy effect of the high surface energy of the strontium titanate particles, thereby causing particle agglomeration and growth[8]. When the degree of polymerization of PEG is large, it may be due to the decrease in PEG monomer at the same mass/volume concentration and uneven encapsulation of strontium titanate. It makes the decreased distribution uniformity of particle size and the increased particle size. Thus, the molecular configuration of the protective agent is an important factor affecting the size of the strontium titanate particles.

3.3. Photocatalytic performance of strontium titanate powder

Methyl orange was used as a photocatalytic evaluation of strontium titanate. Fig. 5 and Table 1 show the photocatalytic experimental data of the synthesized strontium titanate powder. At the same time, as a comparison, we prepared a strontium titanate powder (φ=24 μm) obtained by oxalic acid coprecipitation without macromolecular polymer protection and a strontium titanate powder synthesized by a sol-gel method of acetic acid chelating agent (φ=301 nm) for photocatalytic experiments[9-11]. The results were grouped in Fig. 5 and Table 1.

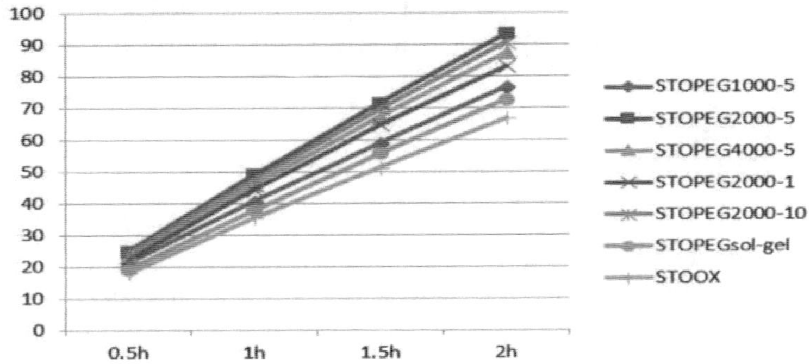

Figure 5. Photocatalytic experimental data of STO

Table 1. Photocatalytic experimental data of STO

		0.5h	1h	1.5h	2h
1	STOPEG1000-5	21.6	40.7	59.1	76.8
2	STOPEG2000-5	25.1	49.2	71.8	93.6
3	STOPEG4000-5	23.8	46.1	67.4	87.5
4	STOPEG2000-1	22.6	44.3	64.9	83.3
5	STOPEG2000-10	24.5	47.8	69.9	90.8
6	STOPEGsol-gel	19.4	37.8	55.8	72.7
7	STOOX	18.2	35.5	51.4	66.9

According to the experimental results, the photocatalytic activity of the strontium titanate powder obtained under the PEG system was significantly better than that of the oxalic acid method and the sol-gel method. The synthesis of strontium titanate photocatalyst with PEG as a protective agent is more promising.

The experimental results showed that $STO_{PEG2000-5}$ had the highest photocatalytic degradation rate and the best activity, and the degradation rate exceeded 93% at 2h.The photocatalytic degradation rate of $STO_{PEG1000-5}$ was relatively poor, and the degradation rate did not exceed 80% at 2h. The photocatalytic degradation rate of $STO_{PEG4000-5}$ could reach about 90%. The degree of polymerization of polyethylene glycol had a significant effect on the photocatalytic activity of strontium titanate powder, and it was not linear. As the degree of polymerization of PEG increased, the photocatalytic activity of strontium titanate powder first increased and then decreased.

Fig. 6 shows the relationship between the particle size of strontium titanate powder and photocatalytic activity. $STO_{PEG2000-5}$ has the smallest average particle size and the highest photocatalytic degradation rate. The average particle size of $STO_{PEG4000-5}$ is larger than that of $STO_{PEG1000-5}$, and the photocatalytic degradation rate of the former is higher than that of the latter. It indicates that there is no simple linear relationship between the average particle size and the photocatalytic degradation rate. The photocatalytic degradation rate of $STO_{PEG2000-5}$ is the highest because of its small crystal size[12-14]. The specific surface area is large, and more active sites are reacted, thereby improving the photocatalytic activity. Although the average particle size of $ST_{OPEG4000-5}$ is larger than that of $STO_{PEG1000-5}$, the degradation rate of the former is nearly 10% higher than that of the latter.

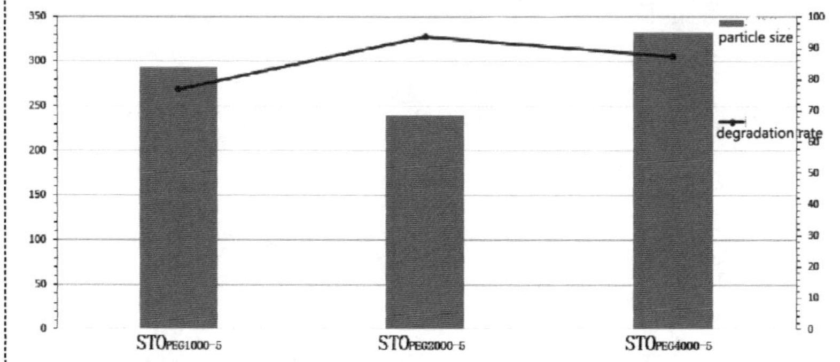

Figure 6. Relationship between the particle size of STO and photocatalytic activity

Some of the PEG is adsorbed on the surface of the strontium titanate crystal during the washing through the low-temperature precipitation method. In the subsequent calcination process, PEG evaporates at a high temperature, so there are many pores on the surface of the crystal, and the porous material is favorable for the improvement of photocatalytic performance. Since the addition of large polymerization degree PEG makes the pore density larger, $STO_{PEG4000-5}$ has a higher average particle size than $STO_{PEG1000-5}$, but the photocatalytic performance of the former is still better than the latter.

For the strontium titanate samples synthesized under different PEG concentrations, the same particle size and photocatalytic activity do not show a linear relationship. When the PEG concentration is large, the obtained strontium titanate powder has a large particle size, which is caused by the agglomeration of the strontium titanate grains from the bridging action of hydrogen bonds[15]. The agglomerated strontium titanate crystal grains have more pores, thus with better photocatalytic activity.

4. Conclusions

(1) A series of spherical and monodisperse strontium titanate powders were synthesized by low-temperature precipitation method using soluble polymer (PEG) as protective agent. The particle sizes were mainly distributed between 200 and 400 nm, rare for crystals smaller than 200 nm and larger than 400 nm. The strontium titanate powder synthesized by the method has good photocatalytic activity.

(2) As the concentration of polyethylene glycol increased, the particle size of the sample crystal increased after decrease. We briefly analyzed the factors affecting the particle size of strontium titanate powder by PEG concentration and degree of polymerization.

(3) Studies showed that PEG concentration and degree of polymerization were critical to the photocatalytic performance of the synthesized strontium titanate, which has been discussed in the work.

Acknowledgments

The study is supported by the fund from the Education Department of Liaoning Province (2019-ZD-0531) and the part regarding material characterization is partially supported by Taihong Chemistry, Science and Technology Co.Ltd. of Shandong province, China.

References

[1] Yingnan Dong, Jin Liu, Jian Tang et al 2017 *Boletín Técnico* (vol 55) pp 603-608
[2] Imai D, Dabwan A, Kaneco S, et al 2009 *Chemica Engineering Journal* (vol 148) pp 336-341
[3] Uyguner C, Bekbolet M 2005 *Catalysis Today* (vol 101) pp 267-274
[4] Wiszniowski J, Robert D, Surmacz-Gorska J 2002 *J PHOTOCH PHOTOBIO A* (vol 152) pp 267-273
[5] Kang K, Shinb H, Park H 2002 *Water Research* (vol 36) pp 4023-4032
[6] Salati S, Papa G, Adani F 2011 *Biotechnology Advances* (vol 29) pp 913-922
[7] Hoffmannmr, Martinst, Choiw 1995 *Chemical Review* (vol 95) pp 69-96
[8] Serikawa T, Shirai S 1998 *Vacuum* (vol51) pp 781- 783
[9] Wu W, Chiou B 1996 *Applied Surface Science* (vol 99) pp 237- 243
[10] Shouan Dong, Shiping Zhou 2007 *Materials Science and Engineering B* (vol 140) pp 153-159
[11] Masaya Matsuoka, Masaaki Kitano, Masato Takeuchi, et al 2007 *Catalysis Today* (vol 122) pp 51-61
[12] Dong Shouan, Zhou Qiang, Tang Chun, et al 2010 *Precious Metals* (vol 31) pp 1-6
[13] Yonggang Chang, Jian Xu， Yunyan Zhang, et al 2009 *Journal of Physical Chemistry C* (vol 113) pp 18761-18767
[14] Jian Xu， Yonggang Chang， Yunyan Zhang, et al 2008 *Applied Surface Science* (vol 255) pp 1996-1999.
[15] Wein-Duo Yanga, Kun-Ming Hunga, Chen-Feng Kao 2000 *Ceramics International* (vol 26) pp 475- 484.

IOP Conf. Series: Materials Science and Engineering **744** (2020) 012036 doi:10.1088/1757-899X/744/1/012036

Study on Mechanical Properties of Wall Panel with Insulation Decoration Structure

Jusong Zhang[1],Wangqiang Li[1] and Zhenyu Yang[2]

[1] College of materials science and engineering,Shenyang jianzhu University, Shenyang liaoning,China

[2] Yingkou Yishi energy-saving building Co. Ltd,Yingkou Liaoning,China

Email: 410194188@qq.com

Abstract. In order to explore the safety performance of the wall panel with insulation decoration structure in the actual application, some test were made.The test results of this paper show that: The rock wool with a density of 120 kg/m^3 and the polyurethane with a bonding density of 50 kg/m^3 are selected as the core material for the temperature retention, whose average tensile strength can reach 139.33 KPa; The mean value of single point anchoring force of 12 mm thick ceramic plate and back bolt as decorative surface is 2088.29 N, and the mean value of the structural adhesive strength between the ceramic plate and keel is 3092.07 KPa; When the uniformly distributed load on the wallboard reaches 6.53 KPa, its deflection reaches 1/180 of itself.

1. Introduction

Nowadays, due to the global energy shortage, all walks of life are promoting ecological civilization, energy saving and environmental protection.Building energy conservation is an important part of the energy-saving industry. Energy-saving buildings and industrialization of buildings have become the main theme of the development of contemporary construction industry.This article explores the mechanical properties of insulating core and keel in wallboard,and explores the bending resistance of wall panel by simulated wind pressure test.

2. Raw materials, test equipment and methods

2.1.Raw materials

The test block of rock wool composite polyurethane of different density;The rock wool composite ceramic plate;type I wall plate and type II wall plate:Among them, the keel in the I type wall panel is

No. 5 FRP and No. 10 channel steel,the keel in the II type wall panel is No. 10 FRP No. 6.3 channel steel,The length and width of the wallboard are 3000 mm and 1200 mm;The back bolt;Bagged cement.

2.2.Test equipment

The drawing instrument of ZQS6-2000A model;Tensile testing machine;driller.

2.3.Methods

Tensile strength test of rock wool composite polyurethane:Make the adhesive composite between the test block of vertical silk rock wool with the densities of 100 kg/m^3, 120 $kg/m^3$and 140 $kg/m^3$and sizes of 100mm×100mm×40mm and the rigid foam polyurethane test block with the densities of 50 $kg/m^3$and 60 kg/m^3 and sizes of 100mm×100mm×40mm.The drawing instrument can be started and the reading displayed on the sensor can be observed, and the reading can be recorded when the test block is broken.

Tensile strength test of rock wool composite ceramic plate:According to the first test,the vertical wire rock wool with a density of 120 kg/m^3 is selected as the core material for the temperature retention. 100mm×100mm×40mm vertical silk rock wool test block is bonded with a 100mm×100mm×8mm ceramic plate.the bonding strength is tested with the drawing instrument and the test data are recorded.

Connection performance test of back bolt and ceramic plate:The back bolt of is closely connected with the ceramic plate. Fix the connector of the puller on the expansion back bolt and test the connection strength between the back bolt and the ceramic plate.

Adhesion test of ceramic plate and keel:The structural adhesive is used to bond the ceramic plate with the glass fiber reinforced plastic component with the bonding area of 50 mm × 50 mm and the adhesive force of the two is tested and the data is recorded.

Bending test of wall plate:When the integrated wall panel of thermal insulation and decoration structure is used as the wall enclosure structure, the main load is the wind load. In this experiment, the average wind pressure once in 50 years at the height of 100 m in the open area of Shenyang City is the most unfavorable wind pressure. According to the standard JG/T432-2014 and execute it.

3. Results and analysis of test

3.1. Bonding performance of core material and decorative panel

Table 1.The tensile strength of the composite test block of different density

Density combination (kg/m^3)	100×50	100×60	120×50	120×60	140×50	140×60
Tensile strength (KPa)	114.48	122.37	139.82	169.59	218.94	237.75
	116.79	119.64	103.56	209.78	185.04	197.24
	102.46	102.36	156.32	204.68	198.74	197.77
	83.36	92.36	134.45	189.47	176.52	212.24

	97.48	122.76	156.32	134.65	201.10	163.36
	98.95	136.48	130.34	172.08	149.33	189.36
	67.54	136.07	171.25	153.57	215.27	231.28
	67.65	68.44	118.75	223.99	185.76	224.87
The average tensile strength (KPa)	95.56	116.78	139.33	183.96	197.34	208.79

According to the data in table 1, Combined with the requirements of external insulation regulations for external walls and economic costs, the density of rock wool in the core material for temperature preservation is selected as 120 kg/m^3,and the density of polyurethane is selected as 50 kg/m^3.

Table 2.Tensile strength of the rock wool composite ceramic plate

Number	1	2	3	4	5	6	7	8	Average value
Tensile strength (KPa)	152.53	170.38	134.02	150.36	142.64	125.57	136.69	143.22	140.72

According to the data in table 2, the experimental results meet the requirements of the 100 KPa specified value of the external insulation regulation of the outer wall.

3.2. Decorative panel and keel connection performance

Table 3.Tensile force of the back bolt connected with ceramic plate

Number	1	2	3	4	5	6	7	8	9	10	Average value
Tensile force(N)	1520 .22	1860 .47	2120 .95	1245 .39	1583 .45	2452 .72	2966 .19	1932 .34	2547 .73	2653 .44	2088.29

Table 4.Tensile strength of ceramic plate bonding FRP

Number	1	2	3	4	5	6	7	8	9	10	Average value
Tensile strength (KPa)	2662 .00	2844 .83	3091 .17	3551 .25	3306 .17	3029 .84	3136 .83	2780 .33	3381 .17	3137 .08	3092.07

According to the data in table 3 and 4 , the average tensile strength of the back bolt connected with 12mm ceramic plate can reach 2088.29 N and the average value of bonding strength of ceramic plate and FRP is 3092.07 KPa.In the insulation decoration structure integrated wall plate, 5 ceramic boards are laid flat, and 4 back bolts are mounted on each ceramic board, meet the requirements of the external insulation regulation of the outer wall.The bonding area between the ceramic plate and the keel is 0.3 m^2 ,therefore, the connection strength between the ceramic plate and the keel reached 269.27 KPa. And together with the bonding strength of rock wool and ceramic plates, which means that the strength of the connection among the ceramic plate,insulation materials and the keel in a wall panel can reach 397.61 KPa,meet security needs.

3.3. Bending resistance of wall plate

Table 5.Load and deflection

Load(kg) \ Deflection(mm) \ Model	Type I wall plate	type II wall plate
0	1	1
600	2	2
1200	4	3
1800	9	6
2400	19	Destroyed

In experiment, the load of type I wall plate reaches 2400 kg,but type II wall plate is damaged,the reason is that channel steel is small.The theoretical calculation of this part is carried out below.

First, I type wall panel is used as the analysis object.keel is a two-phase composite,assuming that its modulus of elasticity is E_c;FRP elastic modulus is E_m;Volume rate is V_m;The modulus of elasticity of the channel is E_p;Volume rate is V_p;Without considering Poisson's ratio,And both are in line with the tandem model,then the calculation formula of the keel elastic modulus is:

$$E_c = \frac{E_m E_p}{E_m V_p + E_p V_m} \tag{1}$$

$E_m = 2.2\times10^4 MPa$; $E_p = 2.06\times10^5 MPa$; $V_m = 33\%$; $V_p = 67\%$;Substituting values to formulas: $E_c = 5.48\times10^4 MPa$.The keel and ceramic plate combination also conforms to the tandem model,use the above formula again to calculate,Elastic modulus is $E_d = 2.12\times10^5 MPa$.After calculation, the elastic modulus of type II wall plate is $E_d = 1.68\times10^5 MPa$.

Next,calculating the mid-span deflection of a single-span simply supported plate under the uniform load of I wall plate:

$$f = \frac{5W_k L^4 B}{384 E_f I_0}(1+k)(1-\beta) \tag{2}$$

$$k = \frac{9.6 E_f I_0}{L^2 G_e A_e} \tag{3}$$

$$\beta = \frac{I_f}{I_f + \frac{I_0}{1+k}} \tag{4}$$

The model is in the form of a sandwich,the keel volume is included in the panel volume.The volume of keel is $2.16\times10^7 mm^3$,after the panel volume is included, the thickness of the panel is increased by $6\,mm$.According to the formula: L =3000mm; B =1200mm; e =75mm; t =15mm; $G_e=1.7\times\rho/100$; $\rho=120$kg/m^3; $I_0=Bte^2/2=5.06\times10^7$ mm^4; $I_f=Bt^3/6=6.75\times10^5$ mm^4; $W_k=6.67\times10^{-3}MPa$; $E_d=E_f=2.12\times10^5 MPa$; $A_e=Bd=1.8\times10^5$;Substituting values into formulas: $k=31.5$, $\beta=0.3$, $f=17.55$ mm.According to this algorithm, the value of type II wall plate is 16.95mm. The calculated result is very close to the test result,so the test is feasible.

The calculation formula of the maximum wind pressure of the insulation system of the external wall is:

$$\omega_k=\beta_{gz}\mu_s\mu_z\omega_0 \tag{5}$$

Take the coefficient according to the specification GB50009-2012 and substituting the coefficient of Shenyang City into the formula: ω_k =-2.54KN/m^2 (Reverse wind pressure).

This value is less than the value in the test,therefore, the wallboard has strong safety performance.

4. Conclusion

(1) The way to meet the economic cost requirements of mechanical requirements is that the density of rock wool is 120 kg/m^3 and the density of polyurethane is 50 kg/m^3.

(2) The connection strength among the ceramic plate,insulation materials and the keel in a wall panel can reach 397.61 KPa,meet the requirements of the regulations.

(3) Ability to withstand wind loads of type I wall plate is 6.67 KN/m^2 the load when II wall plate is destroyed is 5.55 KN/m^2,exceeding the standard value that is 2.54 KN/m^2.

5. Reference lists

[1] Lifang Jia 2019 *Summary of Research on Prefabricated Exterior Wall Panels* (Sichuan: Building Materials of Sichuan)

[2] Danmei Ye and Yuming Xiao 2019 *Application status of Prefabricated Building Wall Panels* (Guangxi:Guangxi Town Construction) p 163

[3] Ziyue Wang and Han Shen 2019 *Application Status and Development Trend of Assembled Composite Wall Panels* (Sichuan: Building Materials of Sichuan)

[4] Dingqiu Chen 2016 *Review of Research on Prefabricated Concrete Wall Structure System* (Magazine and Editorial of Building Structure)

[5] Dingfeng Xu 2017 *Lightweight Composite Energy-saving Wallboard* (Chinese Silicate Society)

3D Numerical Simulation of Expendable Bathythermograph Probe Dropping

Wenlong Lv [1,2], Yongjie Wang [1,2] and Fang Li [1,2]

1 State Key Laboratories of Transducer Technology, Institute of Semiconductors, Beijing, China
2 College of Materials Science and Opto-Electronic Technology, University of Chinese Academy of Sciences, Beijing, China
E-mail: lwl@semi.ac.cn

Abstract. In order to study the motion of Expendable Bathythermograph (XBT) probe, the transient dropping process is numerically simulated by using Fluent software. In the paper, the dynamic mesh method, the 6DOF solver and the volume of fluid (VOF) model are used in Fluent. The trajectory, speed and rotating speed of the probe are investigated, and we also analyze the influence of the tail fin's shape on the rotating speed. The numerical results show that the trajectory and key velocity of XBT probe are in accordance with the relevant research results. And we could get more detailed information than theory or experiment methods. The work can provide a research basis for the research on the underwater motion, stability and structure design of the XBT probe.

1. Introduction

The Expendable Bathythermograph (XBT) is a kind of instrument for measuring the temperature of seawater at different depths. There are important applications in scientific research, marine environment investigation and military purposes. It is rapid, easy and inexpensive in measuring the temperature of seawater. The XBT probe is a disposable device, it is usually released from the ship and quickly measures the temperature of seawater during navigation. Figure 1 shows the shape of XBT probe. The probe is a torpedo-shaped detector with a thermistor inside the diversion chamber, which measures seawater's temperature when dropping in water. The data of the temperature is transmitted to an on-board data acquisition system by fine coated wire. Since it doesn't contain pressure sensors, the depth of the XBT probe is not directly measured, but determined by the fall-rate equation (FRE) obtained by CTD calibration [1].

Most of the researches on XBT probe are mainly around the fall rate equations. Green (1984) firstly proposed the motion equations of XBT falling [2], taking into account the probe's mass, drag coefficient, wire loss, etc. Later, some scholars started using the CFD methods to study the flow state and motion state of XBT in water; In order to obtain the drag coefficient, Abraham (2012) numerically simulated the XBT probe rotating at a certain speed by CFX software [3]; Hong Xiao (2012) simulated the falling process of XBT probe with the VOF method [4]. Early numerical simulations on XBT were mainly steady flow analysis, which were based on the quasi-steady state assumption of flow field. Few people have numerically simulated the whole sinking process of the probe and analyzed the rotation.
This paper not only simulates the dynamic process of XBT probe dropping, but also considers the rotation. Here, we analyze the trajectory, velocity and rotating speed with Fluent software. At the same time, the influence of the tail fin's shape on the rotation is also analyzed.

IOP Conf. Series: Materials Science and Engineering **744** (2020) 012037 doi:10.1088/1757-899X/744/1/012037

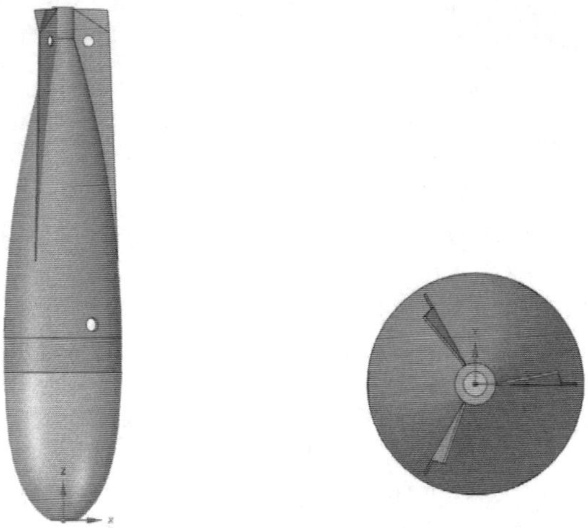

Figure 1. Front and elevation view of an XBT probe.

2. Methodology

2.1. Governing equations for rigid body in motion

According to the working mechanism of XBT, we need to determine the correspondence between depth and temperature. Therefore, it is necessary to study the speed of the probe before practical application.

XBT probe is released freely in the air, because the air density is very low, the drag and buoyancy force of the probe in air can be ignored. When XBT probe hits the water, the speed reaches V_{ff}:

$$v_{ff} = \sqrt{2gh_0} \tag{1}$$

Where $g = 9.81\ m \cdot s^{-2}$ is the gravitational acceleration and h_0 is the drop height.

Then the momentum of probe will reduce due to the impact on water, which means the velocity of the probe is reduced to v_0 after impacting, and the velocity loss of probe in water can be obtained:

$$\Delta v = v_{ff} - v_0 \tag{2}$$

Finally, the probe is submerged in water, we can get the velocity v_{new} of probe at each moment according to the velocity v of the previous time step:

$$v_{new} = v + \frac{\Delta t}{m_p}\left[(m_p - m_w)g - C_d \frac{1}{2}\rho v^2 A - v^2 \frac{dm_p}{dz}\right] \tag{3}$$

The model is account for variations of the buoyancy force, the drag force and the mass reduction due to wire payout. The term m_p is the probe mass, m_w is the mass of the water discharged by the probe, C_d is the drag coefficient, and ρ refers to the seawater density [5].

2.2. Governing equations for fluid

The dropping of the probe involves some interactions with water, and the basic control equations for fluid include: mass conservation equation, momentum conservation equation and energy conservation equation [6]. However, the energy conservation equation is not considered here.

2.2.1. Mass conservation equation

$$\frac{\partial \rho}{\partial t} + \rho div(\vec{u}) = 0 \tag{4}$$

Where the term ρ is the density of fluid, t is time, div is divergence, and $\vec{u}$ refers to the velocity vector [4].

2.2.2. Navier-Stokes equation

$$\frac{\partial(u_i)}{\partial t}+\frac{\partial(u_iu_j)}{\partial x_j}=-\frac{1}{\rho}\frac{\partial p}{\partial x_i}+\frac{1}{\rho}\frac{\partial \tau_{ij}}{\partial x_j}+f_i \tag{5}$$

Where the terms u_i and u_j represent the velocity vectors. p represents the static pressure. f_i is the other momentum forcing components. τ_{ij} is the viscous term, is shown as the following:

$$\tau_{ij}=\mu\left(\frac{\partial(u_i)}{\partial t}+\frac{\partial(u_j)}{\partial x_j}\right) \tag{6}$$

2.2.3. Turbulence model

The moving environment of the probe is viscous, incompressible and unsteady flow field. Here we use the realizable k-ε model. The transport equations for k (J) and ε (%) are as following:

$$\frac{\partial(\rho k)}{\partial t}+\frac{\partial(\rho k u_i)}{\partial x_i}=\frac{\partial}{\partial x_i}\left[\left(\mu+\frac{\mu_t}{\sigma_k}\right)\frac{\partial k}{\partial x_j}\right]+G_k-\rho\varepsilon \tag{7}$$

$$\frac{\partial(\rho \varepsilon)}{\partial t}+\frac{\partial(\rho \varepsilon u_i)}{\partial x_i}=\frac{\partial}{\partial x_i}\left[\left(\mu+\frac{\mu_t}{\sigma_\varepsilon}\right)\frac{\partial \varepsilon}{\partial x_j}\right]+\rho C_1 E\varepsilon-\rho C_2\frac{\varepsilon^2}{k+\sqrt{v\varepsilon}} \tag{8}$$

$$C_1=MAX\left(0.43,\frac{\eta}{\eta+5}\right) \tag{9}$$

$$E_{ij}=\frac{1}{2}\left(\frac{\partial u_i}{\partial x_j}+\frac{\partial u_j}{\partial x_i}\right) \tag{10}$$

Where k is the turbulent kinetic energy, and ε is the turbulent dissipation rate. G_k represents the generation of turbulent kinetic energy due to the average velocity gradient. σ_k and σ_ε is the turbulent Prandtl numbers of the turbulent kinetic energy and its dissipation rate respectively, $\eta=(2\,E_{ij}\cdot E_{ij})^{1/2}\frac{k}{\varepsilon}$, σ_k=1.0, σ_ε=1.2, C_2=1.9 [6].

2.3. Boundary conditions and initial conditions

In order to ensure the full development of the flow field, we set a large cylinder as the computational domain, its length is 15.5 meters, and its diameter is set as 10 times the diameter of XBT probe. The boundary of the computational domain is shown in figure 2. The top edge of the computational domain is defined as the pressure outlet, the outlet pressure is atmospheric pressure, and the remaining boundaries and probe surface are defined as no-slip walls. Others were set default.

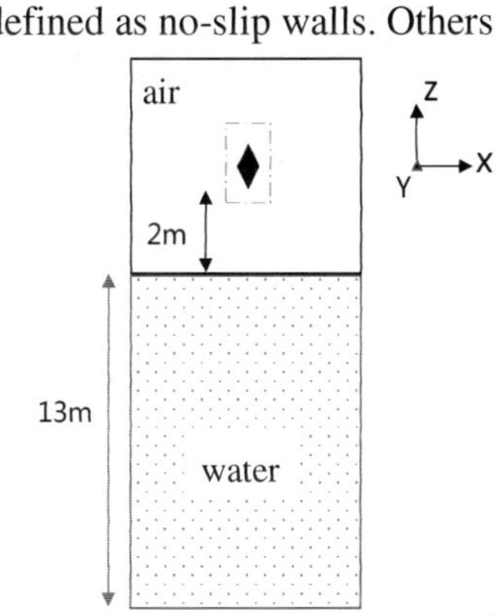

Figure 2. Schematic diagram of the initial position of the XBT probe

IOP Conf. Series: Materials Science and Engineering **744** (2020) 012037 doi:10.1088/1757-899X/744/1/012037

The upper part of the flow field is defined as the air , the lower part is defined as the water. At first, the XBT probe is 2 meters high above the water surface. The probe is released freely in the air. And the direction of the gravity acceleration is -Z direction.

2.4. Meshing

Based on the structure of the XBT probe, we build a simulation model by SpaceClaim software and divide it into lots of tetrahedral mesh. In order to ensure the quality of mesh, the mesh near XBT probe is encrypted. The encrypted area can move together with the XBT probe to ensure that there is no negative volume mesh. They are shown in figure 3. The total number of mesh is 4499538 after completion.

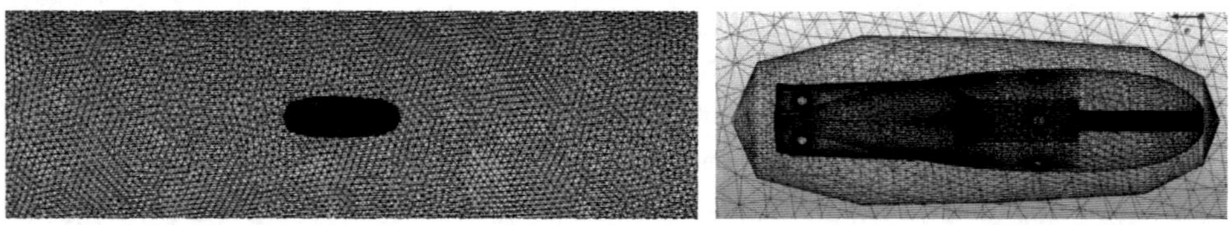

Figure 3. Tetrahedral Mesh of the computational domain and probe

2.5. Dynamic mesh update

Compared to the steady state simulation of XBT probe by some previous scholars [7], we make a 3-D transient simulation about the probe's falling process, which is more realistic to show the dynamic process of falling and rotating. Here, we make use of dynamic mesh method in consideration of the mesh's motion and deformation. Specifically, the transient simulation uses the methods of Spring Smoothing and Local Remeshing.

3. Simulation results and analysis

3.1. Moving position of XBT probe

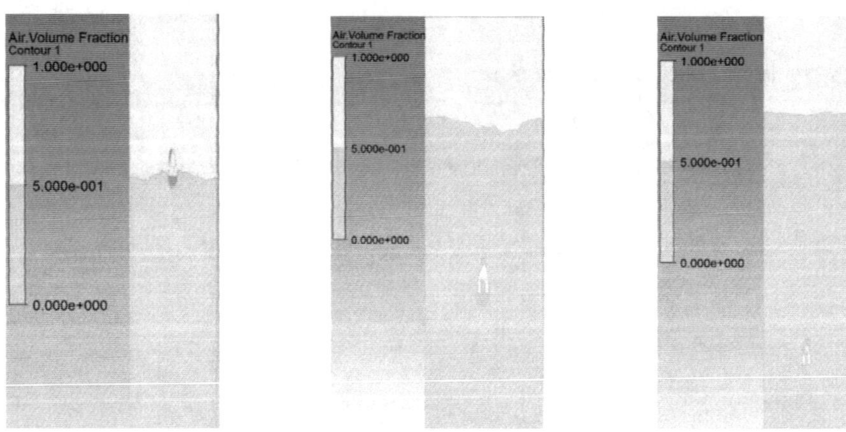

(a) t=0.648s (b) t=0.773s (c) t=0.898s

Figure 4. Position profile of the XBT probe at different time (contours were cut through the vertical axis of the computational domain).

Figure 4 shows the results obtained by using the multiphase flow model in Fluent software. The VOF model is utilized to track the free-surface of the probe, corresponding to the falling position at 0.648s, 0.773s and 0.898s respectively. Air is selected as the primary phase and water is the secondary phase. It can be seen that the XBT probe is farther and farther away from the water surface over time,

IOP Conf. Series: Materials Science and Engineering **744** (2020) 012037 doi:10.1088/1757-899X/744/1/012037

indicating that the simulation is successful. Moreover, the section of the probe becomes smaller and smaller, which indicates that a certain deviation occurs when the probe falls. Figure 5 shows the trajectory of the whole movement process of the probe dropping. The six degree of freedom (6DOF) solver is utilized to track the trajectory of the probe by calculating the buoyancy force and the drag force. After impacting the water, the probe gradually rotates. The trajectory is kind of like a spiral line. And Stark J had noted that the XBT probe motion in near surface layer is slow rotation with helicoidal trajectory [8]. While on the whole, the trajectory is a straight line.

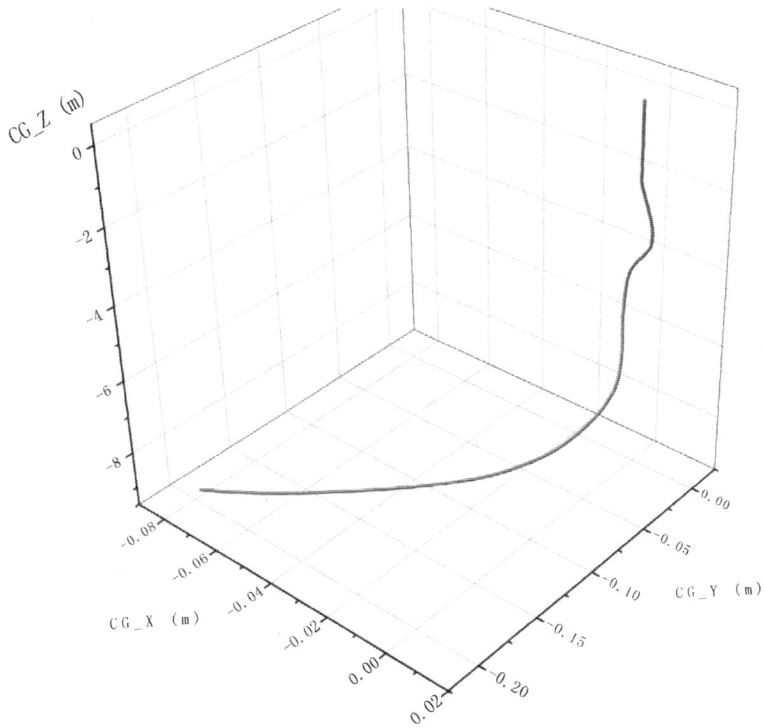

Figure 5. 3D trajectory of the XBT probe

3.2. Vertical velocity of XBT probe

Using the 6DOF solver of Fluent, we can obtain the displacement and rotating angle of XBT probe in the X, Y and Z directions. Figure 6 is the speed of the probe in the direction of gravity, roughly corresponding to the initial movement process of the probe. The probe is freely released 2 meters above the water surface. When it hits the water at about 0.63s, it reaches the maximum speed of 6.18m/s. Then there is a loss of momentum due to the impact on water surface. At about 0.66 s, the probe is submerged in water. As a result of the fluid force, XBT probe will rotate in the water, and the falling speed tends to be stable.

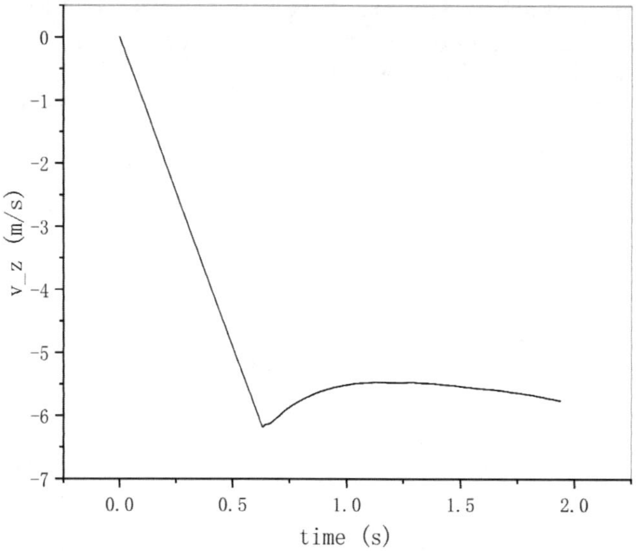

Figure 6. Vertical velocity of XBT probe in numerical simulation

A similar experiment was conducted abroad in 2015 [9]. The American scholar Bringas conducted the falling test of the XBT probe in a 11.12 meters deep water tank, and captured the position of the probe at different moments by a digital camera. Table 1 compares the experiment and the simulation results. We can see that the initial speed v_0 in water and the loss of speed Δv are close to respectively.

Table 1. Simulation results and experiment results.

	Drop height (m)	v_0 ($m{\cdot}s^{-1}$)	Δv ($m{\cdot}s^{-1}$)
Numerical simulation	2	6.137	0.123
Tank experiment	2	5.73±0.42	0.53±0.42

3.3. Rotating speed of XBT probe

There are three triangular tail fins at the end of the XBT probe. The tail fins are bent at a certain angle, so the lift force that is perpendicular to the probe axis generates torque on the probe. The torque makes the probe rotate at a certain speed, so as to maintain the attitude stability of XBT probe. At the same time, it is beneficial to loosen the thin copper wires connected to the launch ship.

Since the rotation of the probe in the water depends on the curved tail, the bending angle of the tail fin may also affect the effect of rotation. We conduct numerical simulations on three XBT probes with different shapes, and the bending angle of the tail fins are respectively 20°, 30° and 40°. From the simulation, we obtain the 6DOF information of the probe with time going. Due to the limitation of the computing domain, the case was only calculated about 1.4s. The change of its rotating displacement about the z-axis was specially analyzed, as shown in figure 7. It can be seen from the figure that the probe hardly rotates in the air. After entering the water at about 0.66s, the rotating displacement of probe around the z axis significantly increases, which conforms to a certain actual situation. According to the rotating angle of different probes in figure 7, the larger the bending angle of the tail fin, the higher the corresponding rotating speed. The final rotating speed of 20°, 30° and 40° probe in water is 6.4r/s, 7.8r/s and 10.4r/s, respectively.

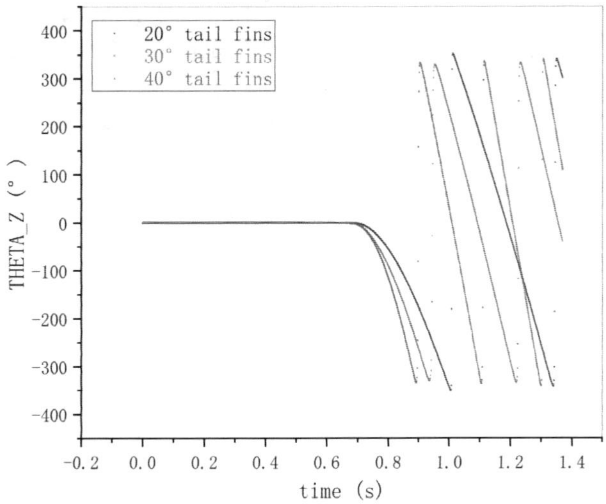

Figure 7. Rotating angle of XBT probes with different tail fins around z axis

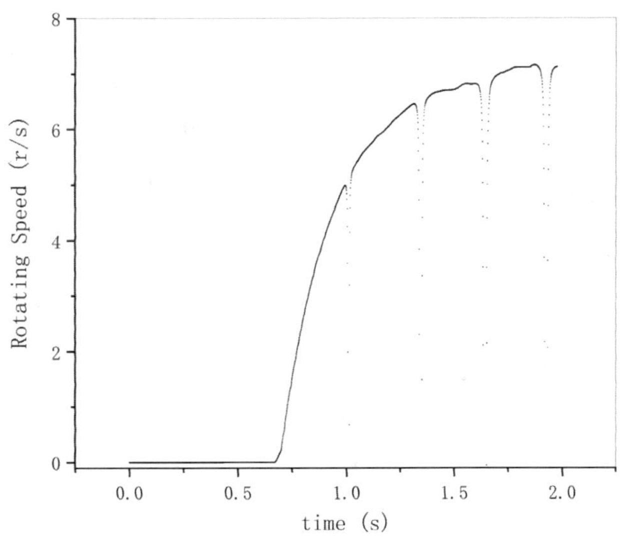

Figure 8. Rotating speed of the probe with the 20° tail fins

Figure 8 shows the rotating speed of the probe with the 20° tail fins within 2s, including the process of free falling in air, impacting the water, rotating and falling in water. After entering the water, the probe rotates faster and finally reaches a stable rotating speed of about 7.1r/s, which was somewhat different from the 10r/s observed by foreign scholars [3]. In addition, there are discontinuities in the figure, which may be related to the selection of the time step.

4. Conclusion

The falling process of the XBT probe was simulated by using the VOF model and the 6DOF solver in software Fluent. The trajectory and key velocity are close to the relevant experiments respectively. The falling trajectory of XBT probe is a straight line with slight deflection in water.

Here we use the VOF method in the Fluent software to track the XBT probe's motion interface at different time. The curved tail fins help the probe reach a high rotating speed after entering the water and maintain the posture stability in the seawater. The larger the angle of the tail fin, the easier it is to obtain a higher rotating speed.

The numerical simulation in this paper provides a research basis for further research on the underwater motion and the attitude stability of the XBT probe, which is of great significance for improving the accuracy of temperature measurement. Due to the limitation of the computational domain, the simulation results can't fully reflect the falling process of the probe. This process will be further studied in later work.

Acknowledgments
This project is supported by The Senior User Project of RV *KEXUE* (KEXUE2018G08).

References

[1] Kizu S, Sukigara C and Hanawa K 2011 Comparison of the fall rate and structure of recent T-7 XBT manufactured by Sippican and TSK *Ocean Science* **7** pp 231-244
[2] Green and A. W 1984 Bulk dynamics of the expendable bathythermograph (XBT) *Deep Sea Research* **31** pp 415-426
[3] Abraham J, Gorman J, Reseghetti F and Sparrow E 2012 Drag coefficients for rotating expendable bathythermographs and the impact of launch parameters on depth predictions *Numerical Heat Transfer Applications* **62** pp 25-43
[4] Hong X and Xinhua Z 2012 Numerical investigation of the fall rate of a sea-monitoring probe *Ocean Engineering* **56** pp 20-27
[5] Abraham J, Gorman J, Reseghetti F, Sparrow E and Minkowycz W 2012 Turbulent and Transitional Modeling of Drag on Oceanographic Measurement Devices *Modelling and Simulation in Engineering* **2012** pp 1-8
[6] Fujun W 2004 *Calculation Hydrodynamic Analysis* (Beijing: Tsinghua University Press) p 268
[7] Zhengbao L, Libin D, Jie L and Haijing H 2013 Hydrodynamic Characteristics Analysis of Expendable BathyThermograph Probe *Communications in Computer and Information Science* **418** pp 231-238
[8] Stark J, Gorman J, Hennessey M, Reseghetti F, Willis J, Lyman J, Abraham J and Borghini M 2011 A computational method for determining XBT depths *Ocean Science* **7** pp 733-743
[9] Bringas F and Goni G 2015 Early dynamics of deep blue XBT probes *Journal of Atmospheric and Oceanic Technology* **32** pp 2253-63

Effect of annealing on recrystallization of GAP particles of Fe-Nd-B system alloy

V Chaika[1], V Savin[1], M Sorokovikov[1], A Marukhin[1] and A Martyushev[1]

[1]Immanuel Kant Baltic Federal University, Kaliningrad, Russia
E-mail: chayka8@rambler.ru

Abstract. The magnetic and mechanical properties magnets of the Fe-Nd-B system depend more on the size and shape of the grain structure. The influence of annealing on the microstructure of gas atomized powder (GAP) particles of Fe-Nd-B alloy was investigation. Conditional three types of microstructure are allocated and their changes in the process of recrystallization are analyzed. It is shown that the process of recrystallization is accompanied by an increase in the intergranular phase and the formation of grains in it. Increasing the annealing time enhances these processes.

1. Introduction
Magnetic materials of alloys of the iron-neodymium-boron system have unconditional very high practical significance. Operating conditions and high heat sometimes lead to irreversible processes and loss of material properties. The theory of boride ferromagnetism of stoichiometry $TM_{14}REM_2B$ does not have a complete accurate quantitative model, requires verification, refinement and further development. The magnetic properties of the Fe-Nd-B alloy material and products made of such an alloy are sensitive to the type of grain structure and the size of the grain size. Annealing of the gas atomization powders (GAP) causes recrystallization and changes in crystallographic parameters of grain structure, size and shape, which is associated with changes in magnetic properties [1-5].

2. Experimental
Gas atomized powders (GAP) were obtained by gas atomization a melt jet with an inert gas stream at the URZHMV-3 liquid metal atomization unit (Ukraine). In this case, the pressure of the energy carrier gas reached 12 atm. and the nozzle "Laval nozzle" was used. The modes of the gas atomization process are described in detail in [4]. Fractures of composition $Fe_{77.5}Nd_{14.5}B_{8.0}$ were investigated in the work.

Powders with a fraction from 0.0 to 0.630 mm were taken for the investigation. Separation into fractions was performed by sieves with numbers: 0.020, 0.056, 0.100, 0.160, 0.400 and 0.630 mm. The analysis of grain size, shape and structure was performed by quantitative metallography at the SIAMS-800 scanning station for panoramic applications in reflected light, including the OLYMPUS BX-51 metallographic microscope and SIAMS Drive System.

3. Results and discussion
The determination of the fractional composition by sieve analysis of the investigated compositions shows that the maximum mass fraction falls on the fraction of 0.160 - 0.400 mm (Fig.1).

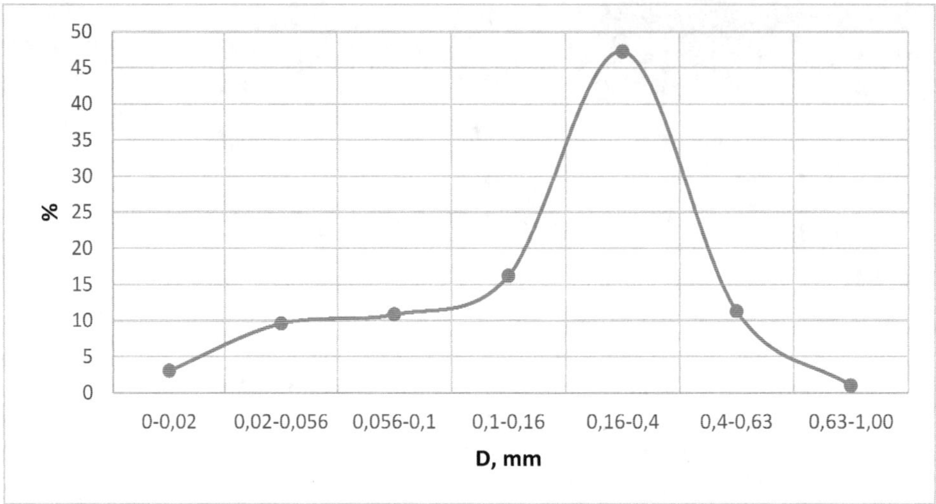

Figure 1. Distribution by fractions of fracturing alloy $Fe_{77.5}Nd_{14.5}B_{8.0}$.

To identify the microstructure, several types of etchants based on sulfuric acid and nitric acid with the addition of glycerol and phosphoric acid were used. Analysis of the microstructure of the studied alloys revealed some features. The main feature of particle formation during gas spraying is the difference in crystallization rates of particles of different sizes. Fracturing is characterized by the formation of particles into which other smaller particles were embedded during solidification. During the preparation of micro-grinding and etching, the boundary between the embedded particle and the base is destroyed, and the embedded particle falls out, leaving a cavity. Since the cooling rate of the embedded particle and the substrate differ, the microstructures also differ. Due to this feature, there is an uneven distribution of the neodymium containing phase, which causes differences in the microstructure of the particles. Three conventional structures are accepted for consideration: 1) coarse-grained structure with anisotropy of grain shape greater than three (Fig.2 a); 2) fine-grained structure with grain shape anisotropy less than three (Fig.2 b); grain block structure with radial gradient of grain growth direction (Fig.2 c).

The formation of the block structure is explained by the excellent solidification temperature of the different phases of the Fe-Nd-B system. when the fracturing particles cool down, the regions enriched with Nd with a fine-grained structure first harden. For figure 2c these areas look like small rounded nucleation centers of elongated grains with a radial direction of growth. The centers themselves are a collection of a large number of equiaxial grains with an average size of about 0.5 microns.

When annealing in air for 1 hour at 600 °C in all fractions there is a change in the shape of the grains. On the surface of the particles, a region saturated with oxygen is formed, which is colored with a characteristic light gray color (Fig. 3). Inside the particles, the intergranular space is also highly oxidized and expanded.

Analysis of the microstructure of particles after annealing for 1 hour at 600 °C in argon protective medium shows that different structures change differently. In the coarse-grained structure, grain associations and an increase in the intergrain space are manifested (Fig.4 a), in which there is the origin and crystallization of a new phase. In the fine-grained structure, the grains are crushed until the grain boundaries disappear, and the intergranular phase is formed in the form of fine particles (Fig. 4b). The dissolution of grain boundaries is observed in the block structure (Fig. 4c).

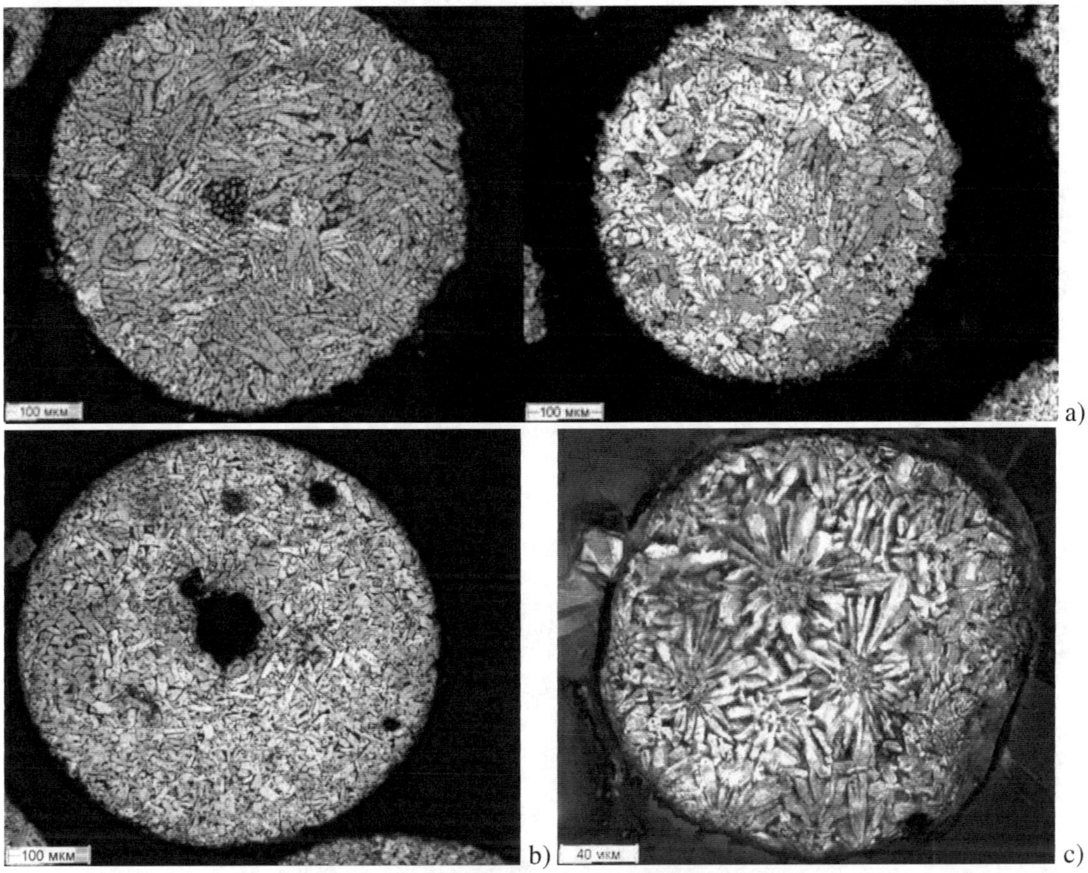

Figure 2. Microstructure of fracturing alloy particles системы $Fe_{77.5}Nd_{14.5}B_{8.0}$.

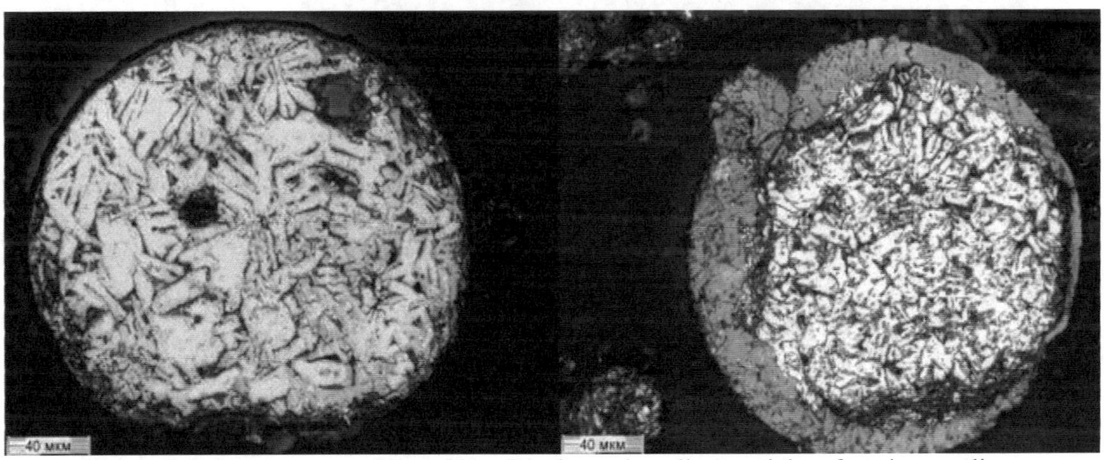

Figure 3. Microstructure of $Fe_{77.5}Nd_{14.5}B_{8.0}$ fracturing alloy particles after air annealing.

Increasing the annealing time to 2.5 hours increases the phenomenon of grain fusion and phase crystallization in the intergranular space in all fractions (Fig. 5)

IOP Conf. Series: Materials Science and Engineering **744** (2020) 012038 doi:10.1088/1757-899X/744/1/012038

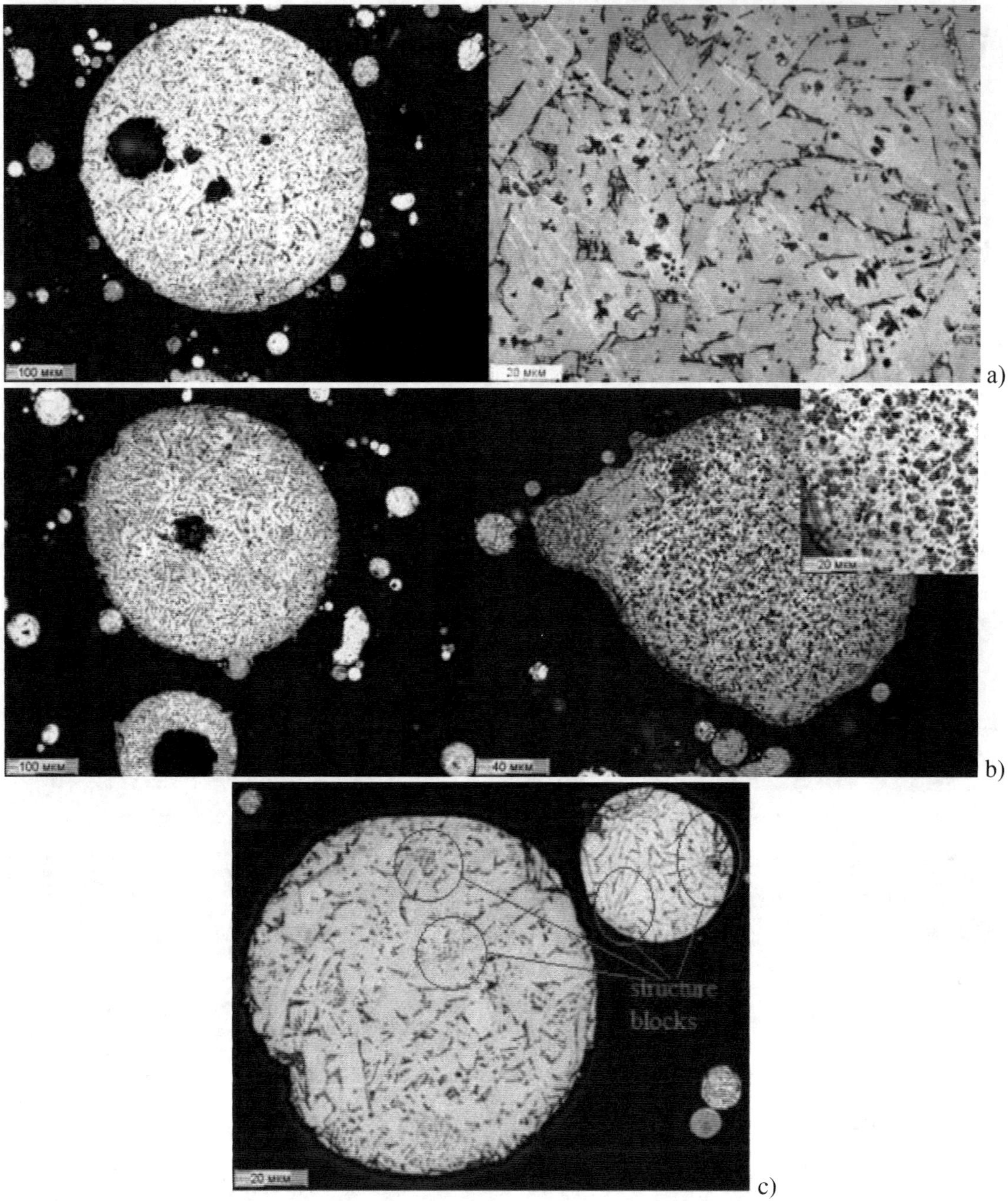

Figure 4. The microstructure of particle fracturing of the alloy $Fe_{77.5}Nd_{14.5}B_{8.0}$ after annealing in argon.

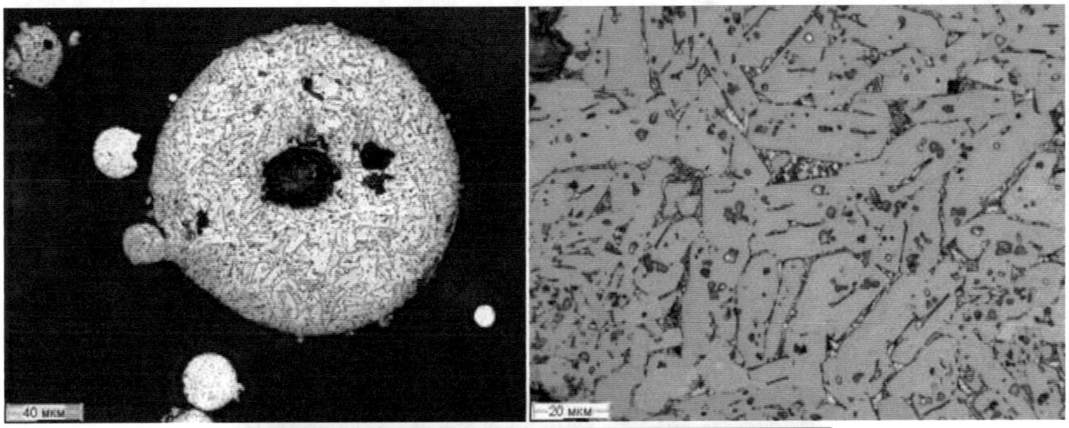

a)

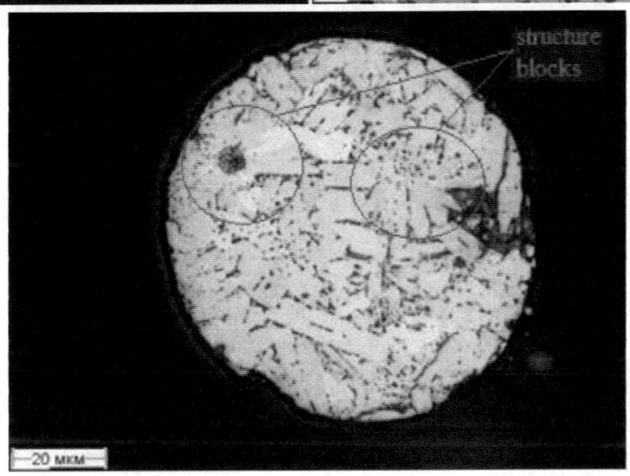

b)

Figure 5. Microstructure of $Fe_{77.5}Nd_{14.5}B_{8.0}$ 0 fracturing particles after 2.5 hours annealing in argon.

4. Conclusions

1. The recrystallization process has a different mechanism for different initial microstructure.
2. During annealing, there is always a change in the morphology and size of the intergranular structural component in the Fe-Nd-B fracturing alloy particles.
3. In the intergranular structural component for all fractions of the studied powder, the separation of new structural formations is observed.

Acknowledgments
This work was supported by the Russian Foundation for Basic Research (RFBR) project № 19-48-390002\19.

References

[1] X. D. Fan, et al. 2015 Influence of Nd Doping on the Magnetic Properties of $Nd_2Fe_{14}B/\alpha$-Fe Nanocomposite Magnets *Materials Science Forum* **809-810** 88
[2] B. B. Li, et al. 2017 Effects of Co, Zr Additions on the Magnetic Properties of Nanocomposite $Nd_2Fe_{14}B/\alpha$-Fe Magnets *Materials Science Forum* **898** 2128
[3] L.Q. Yu, et al. 2009 Effect of Boron on Magnetic Properties and Corrosion Resistance of High Energy Magnets *Advanced Materials Research* **79-82** 1043
[4] S. Sugimoto and D. Book 2006 HDDR process for the production of high performance rare-earth magnets *Handbook of Advanced Magnetic Materials* 977
[5] K. L. Alvarez, J. M. Martín, M. Ipatov, and J. Gonzalez 2018 Soft magnetic amorphous alloys (Fe-rich) obtained by gas atomization technique *J. Alloys Compound* **735** 2646

Photoelasticity Ray-casting for Residual Stresses in the Axisymmetric Glass Cylinder

Xun Zhang[1], **Quan Tian**[2], **Donghai Chen***[1], **Licheng Huang**[3]

[1]Changjiang institute of survey, planning, design and research, Wuhan, Hubei 430081, China
[2]Zhixing college of Hubei University, Wuhan, Hubei 430081, China
[3]Key Laboratory of Metallurgical Equipment and Control Technology of Ministry of Education, Wuhan University of Science and Technology, Wuhan, Hubei 430081, China
*Corresponding author Email: chendonghai@cjwsjy.com.cn

Abstract. Residual stresses are important index to evaluate the precision molded glass plane. Residual stresses inside a glass cylinder was simulated by Finite Element Method. The circular polariscope was used to compare retardation and isoclinic angle between the simulated residual stresses and the thermal treated glass cylinder. The retardation and isoclinic angle of simulated residual stresses were calculated by Jones calculus. The comparison results show a good agreement.

1. Introduction

As an alternative fabrication method to traditional glass devices fabrication process, precision compression molding of glass optics has attracted more and more attention because of its high output and net molding fabrication capability [1]. Glass molding is a kind of hot forming process. The heated glass blank is pressed into required geometry by a pair of optical polishing molds. The heat pressed glass is held in the mold for a predetermined time at the molding temperature to release the stress caused by the pressing. During the cooling stage, cooling rate is controlled to keep residual stresses and thermal shrinkage at or below the required levels.

Residual stresses are important index to evaluate the precision molded optical glass devices. When the glass solidifies completely in the cooling process, the residual stresses freezes in the molded glass plane. The residual stresses will cause the change of refractive index, which will result in unnecessary light deviation and intensity change. The unnecessary light deviation and intensity change will lead to the deterioration of image quality. Photoelastic tomography is the most commonly used method to measure the residual stresses inside glass pieces[2-4]. This method utilizes the photoelasticity of a glass sample when polarized light pass through, and is based on the birefringence of the glass when it is stressed.

In this research, the birefringence characteristics of the molded glass cylinder are simulated by Jones calculus. First, Finite element method (FEM) was employed to simulate the thermal treatment process[5] and then the residual stresses were calculated by the numerical simulation. Then the simulated residual stresses were used as source to simulate the circular polariscope outputs. The isoclinic angle and optical relative retardation calculated. By comparing isoclinic angle and retardation between the simulation and experiment, the results showed that FEM simulations are confirmed with the residual stresses inside glass lenses by glass molding.

IOP Conf. Series: Materials Science and Engineering **744** (2020) 012039 doi:10.1088/1757-899X/744/1/012039

2. Numerical Simulation of residual stresses of a Glass Cylinder

2.1. FEM Model of a Glass Plane

Numerical simulation of the heat treatment process of BK7 glass cylinder was performed by using the FEM software MSC/MARC[1]. The glass cylinder selected in this research is high 7 mm, and diameter of 7 mm. In the heat treat experiment, the glass cylinder sit on a ceramic plane with 4 mm height and 15 mm diameter. Then the glass cylinder and the ceramic plane were placed in an oven to perform heat treatment. The two-dimensional (2D) axisymmetric model is adopted by using the rotational symmetric geometry of the mold and cylinder, as shown in Fig. 1. There are two steps in the numerical simulation: (1) heating the glass cylinder and mold to the molding temperature, (2) at a certain cooling rate, cooling the molded glass plane and mold to room temperature.

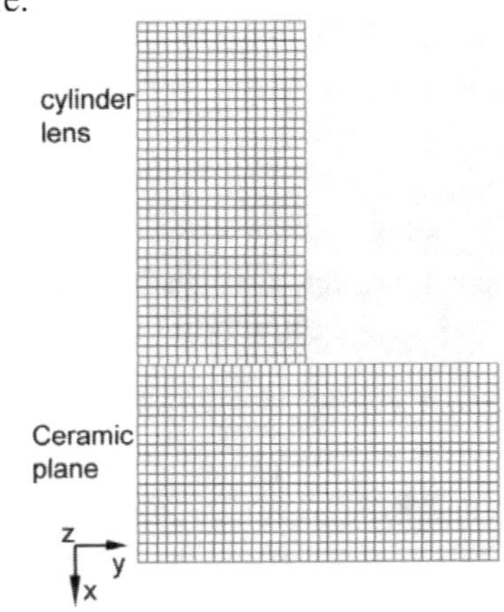

Figure 1 FEM model of the glass cylinder

2.2. Simulated Residual Stresses

Fig. 2 shows the distribution of residual stresses in the glass cylinder in cylindrical coordinates. The figures display half of the normal cross section of the cylinder. In the stress distribution figures, compressive stress is represented by negative value and tensile stress is represented by positive value.

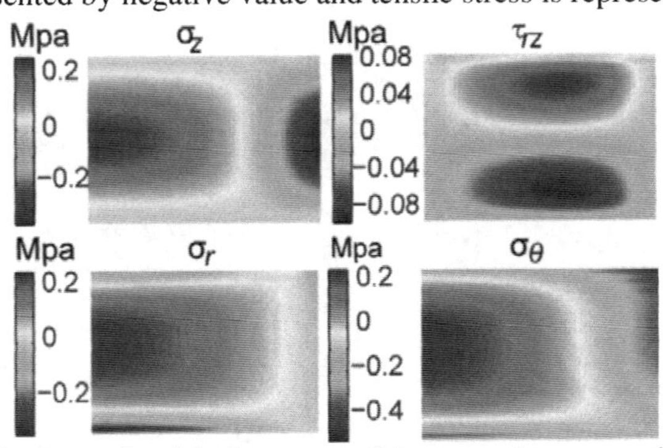

Figure 2 Simulated distributions of residual stresses of the normal cross section of the glass cylinder in cylindrical coordinates: (a) ${}_{z}$ axial stress, (b) ${}_{rz}$ shear stress , (c) ${}_{r}$ radial stress , (d) circumferential stress

3. Photoelasticity Ray Casting

According to Jones Calculus, the light exporting from the circular polariscope can be expressed as [2, 6]:

$$\begin{pmatrix} E_x \\ E_y \end{pmatrix} = \begin{pmatrix} \cos^2\beta & \cos\beta\sin\beta \\ -\cos\beta\sin\beta & \sin^2\beta \end{pmatrix} \times \begin{pmatrix} i\cos^2\gamma+\sin^2\gamma & (i-1)\sin\gamma\cos\gamma \\ (i-1)\sin\gamma\cos\gamma & i\sin^2\gamma+\cos^2\gamma \end{pmatrix}$$
$$\times J(\Delta,\varphi)\times \begin{pmatrix} i\cos^2\varepsilon+\sin^2\varepsilon & (i-1)\sin\varepsilon\cos\varepsilon \\ (i-1)\sin\varepsilon\cos\varepsilon & i\sin^2\varepsilon+\cos^2\varepsilon \end{pmatrix} \times \begin{pmatrix} 0 \\ 1 \end{pmatrix} \quad (1)$$

$J(\Delta,\varphi)$ is the Jones matrix of the measured medium, Δ and φ are optical retardation and isoclinic angle of the specimen, respectively. ξ and γ are the direction of the fast-axis of quarter wave plates before and after the specimen. The polarizer's orientation is fixed at $\pi/2$, β is the direction of the analyzer. According to the equivalence principle, $J(\Delta,\varphi)$ can be expressed as:

$$J(\Delta,\varphi) = \begin{pmatrix} \cos\varphi & -\sin\varphi \\ \sin\varphi & \cos\varphi \end{pmatrix} \begin{pmatrix} e^{i\Delta} & 0 \\ 0 & 1 \end{pmatrix} \begin{pmatrix} \cos\varphi & \sin\varphi \\ -\sin\varphi & \cos\varphi \end{pmatrix}$$
$$= \begin{pmatrix} e^{i\Delta}\cos^2\varphi+\sin^2\varphi & (e^{i\Delta}-1)\sin\varphi\cos\varphi \\ (e^{i\Delta}-1)\sin\varphi\cos\varphi & e^{i\Delta}\sin^2\varphi+\cos^2\varphi \end{pmatrix} \quad (2)$$

Along the light path, the object can be divided into n layers with a thickness of dz. The relation between the outgoing light E_{out} and the incident light E_{in} is expressed as [3, 4]:

$$E_{out} = \sum_{k=1}^{n}(1+A_k dz)E_{in} \quad (3)$$

where, $E = \begin{pmatrix} E_x \\ E_y \end{pmatrix}$, $A = -\frac{1}{2}iC_0 \begin{pmatrix} \sigma_{xx}-\sigma_{yy} & 2\tau_{xy} \\ 2\tau_{xy} & -(\sigma_{xx}-\sigma_{yy}) \end{pmatrix}$. σ_{xx}, σ_{yy} and τ_{xy} are stress components in XY plane, i is imaginary.

Then, the Jones matrix of the photoelastic object is obtained as:

$$J_{stress} = \sum_{k=1}^{n}(1+A_k dz) = J(\Delta,\varphi) \quad (4)$$

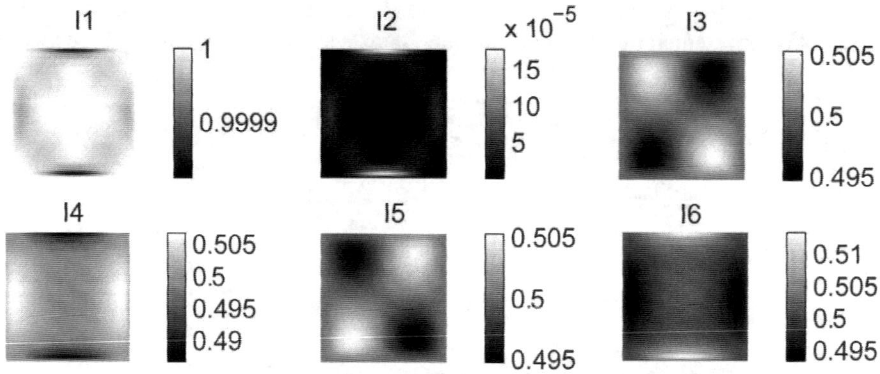

Figure 3 The intensity images of the glass cylinder from simulated stresses by Jones calculus

With the simulated intensity images of the glass cylinder shown in Fig.3, the isoclinic angle and optical retardation of the glass cylinder can be obtained as follows:

$$\varphi = \frac{1}{2}\arctan\left(\frac{I_5-I_3}{I_4-I_6}\right) \quad (5)$$

$$\Delta = \arctan\left(\frac{(I_5 - I_3)\sin 2\varphi + (I_4 - I_6)\cos 2\varphi}{I_1 - I_2}\right) \tag{6}$$

Therefore, the optical retardation and isoclinic angle of the glass cylinder can be calculated through intensity images in Fig. 4. In Fig. 4, the isoclinic angle retardation are expressed in radian.

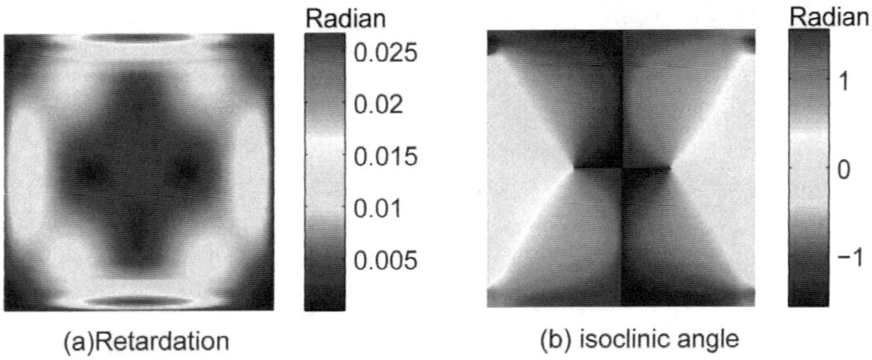

Figure 4 Optical retardation and isoclinic angle of the glass cylinder expressed in radian, and calculated from intensity images: (a) optical retardation, (b) isoclinic angle.

4. Results and Discussion

Due to the containment and the residual stress effect of the glass tank, intensities shown in Fig. 5 are hard to tell by eyes. According to Eqs. 5 and 6, the retardation and isoclinic angle can be calculated by the figures in Fig. 5. The isoclinc angle in radian and the retardation in nm, calculated from the intensity images in Fig.5, are shown in Fig. 6 (a) and Fig. 6 (b) respectively [4].

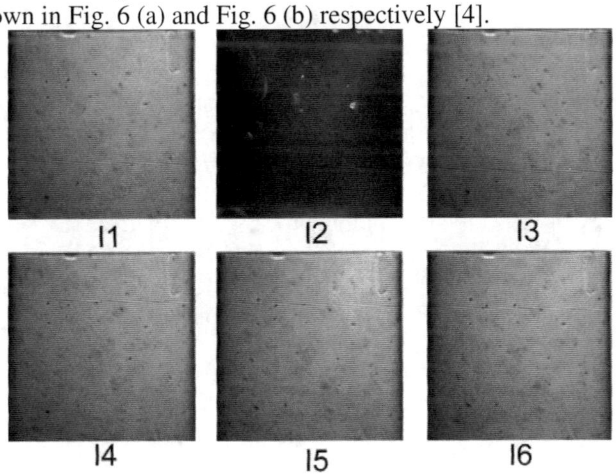

Figure 5 Experiment intensity images of the glass cylinder

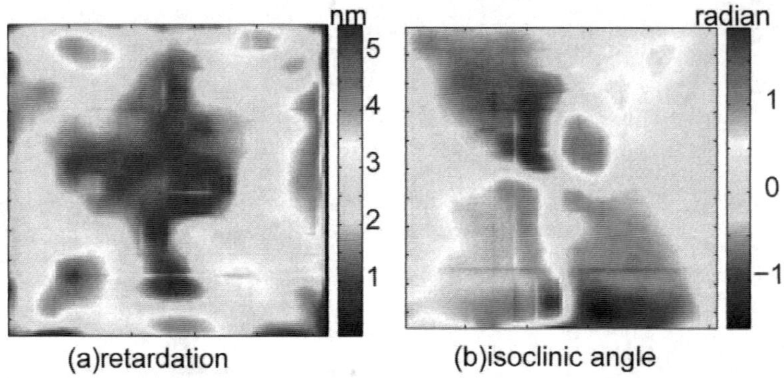

Figure 6 Experimental calculated retardation and isoclinic angle of the glass cylinder: (a) retardation (nm), (b) isoclinic angle (radian)

5. Conclusions

Residual stresses of a glass cylinder were simulated by using FEM with a structural relaxation model. At the same time, the light intensity images of glass cylinder from the circular polariscope under the same experimental conditions are measured. The comparison of the retardation and isoclinic angle from simulation by simulated residual stresses and experiments demonstrates that residual stresses calculated in numerical simulation have a reasonable conformance with experimental results.

References

[1] A. Y. Yi and A. Jain, “Compression molding of aspherical glass lenses–a combined experimental and numerical analysis,” *J. Am. Ceram. Soc.*, **88** [3] 579–586.

[2] H. Aben and C. Guillemet, *Photoelasticity of Glass*, 97-101, Springer-Verlag.

[3] Yi, A. Y., et al. "Residual stresses in glass after molding and its influence on optical properties." *Procedia Engineering* **19**: 402-406.

[4] Bo. T., Peng H., Lianguan S., Allen Y. “Quantitatively measurement and analysis of residual stresses in molded aspherical glass lenses”. *Int. J. Adv. Manuf. Tech.*, **74**(9-12), 1167-1174.

[5] A. Jain, G. C. Firestone, and A. Y. Yi, “Viscosity measurement by cylindrical compression for numerical modeling of precision lens molding process,” *J. Am. Ceram. Soc.*, 88 [9] 2409–2414.

[6] Bo. T., Shuang X., and Honghui Y., “Numerical analysis of residual stresses reconstruction for axisymmetric glass components.” In 2017 International Conference on Optical Instruments and Technology: Optical Systems and Modern Optoelectronic Instruments. International Society for Optics and Photonics.

Nonlinear algorithms of residual stresses reconstruction for axisymmetric problem

Donghai Chen[1], Quan Tian[2], Xun Zhang*[2], Juntong Yun[3]

[1]Changjiang institute of survey, planning, design and research, Wuhan, Hubei 430081, China
[2]Zhixing college of Hubei University, Wuhan, Hubei 430081, China
[3]Key Laboratory of Metallurgical Equipment and Control Technology of Ministry of Education, Wuhan University of Science and Technology, Wuhan, Hubei 430081, China
*Corresponding author Email: zhangxun@cjwsjy.com.cn

Abstract. A non-destructive testing method for three-dimensional stress of a glass cylinder based on photoelasticity is presented. Based on simulated stresses in the glass cylinder, intensity images output from a circular polariscope are calculated by Jones calculus. Therefore, the isoclinic angle and optical retardation can be simulated. Then the residual stresses in the glass cylinder can be reconstructed in cylindrical coordinates. By comparing the reconstructed stresses with numerical simulated stresses, there is good agreement but with some mismatches.

1. Introduction

With a large demand of the compression molded glass products, especially precision glass lenses, nondestructive method for stress measurement inside the glass components are needed. However, due to the nonlinear problem of the stress. Stress reconstruction in photoelastic material is a problem[1]. In order to study the reconstruction error of the stress reconstruction method, a residual stress reconstruction process of a glass cylinder is analyzed. Firstly, residual stresses in the glass cylinder are simulated by finite element method (FEM). Based on the simulated stresses, intensity images emerging from a circular polarizer can be calculated by Jones calculus. With six step phase-shifting technique (six step PST) [2], the isoclinic angle and optical retardation of the glass cylinder was then obtained. Therefore, the residual stresses were reconstructed through the isoclinic angle and optical retardation by integrated photoelasticity. The error of reconstruction method is presented by comparing the reconstructed stresses with the stresses through FEM.

2. Numerical simulation of a circular polariscope

Figure 1 shows a schematic of a circular polarizer. In Fig. 1, the polarizer and analyzer are identical. S represents the slow axis, F represents the fast axis. ξ and γ represents the fast axis of two quarter wave plates; φ is isoclinic angle of the specimen. The orientation of the polarizer is fixed at $\pi/2$, β is the orientation of the analyzer.

IOP Conf. Series: Materials Science and Engineering **744** (2020) 012040 doi:10.1088/1757-899X/744/1/012040

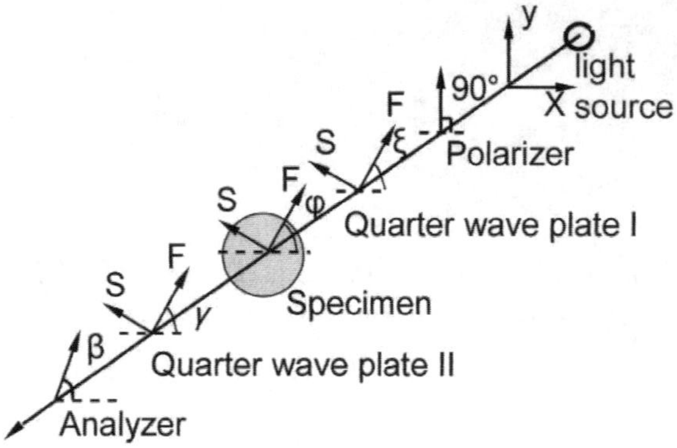

Fig. 1 Schematic of a circular polariscope

According to Jones Calculus, the light emitting from the analyzer can be described as [3]:

$$\begin{pmatrix} E_x \\ E_y \end{pmatrix} = \begin{pmatrix} \cos^2\beta & \cos\beta\sin\beta \\ -\cos\beta\sin\beta & \sin^2\beta \end{pmatrix} \times \begin{pmatrix} i\cos^2\gamma + \sin^2\gamma & (i-1)\sin\gamma\cos\gamma \\ (i-1)\sin\gamma\cos\gamma & i\sin^2\gamma + \cos^2\gamma \end{pmatrix}$$
$$\times J(\Delta,\varphi) \times \begin{pmatrix} i\cos^2\varepsilon + \sin^2\varepsilon & (i-1)\sin\varepsilon\cos\varepsilon \\ (i-1)\sin\varepsilon\cos\varepsilon & i\sin^2\varepsilon + \cos^2\varepsilon \end{pmatrix} \times \begin{pmatrix} 0 \\ 1 \end{pmatrix} \tag{1}$$

$J(\Delta,\varphi)$ is the Jones matrix of the specimen, and Δ and φ are optical retardation and isoclinic angle of the specimen, respectively. The $J(\Delta,\varphi)$ can also be expressed as:

$$J(\Delta,\varphi) = \begin{pmatrix} e^{i\Delta}\cos^2\varphi + \sin^2\varphi & (e^{i\Delta}-1)\sin\varphi\cos\varphi \\ (e^{i\Delta}-1)\sin\varphi\cos\varphi & e^{i\Delta}\sin^2\varphi + \cos^2\varphi \end{pmatrix} \tag{2}$$

Along the light path, the object can be divided into n layers with a thickness of dz. The relation between the outgoing light E_{out} and the incident light E_{in} is expressed as [3, 4]:

$$E_{out} = \sum_{k=1}^{n}(1 + A_k dz)E_{in} \tag{3}$$

where, $E = \begin{pmatrix} E_x \\ E_y \end{pmatrix}$, $A = -\frac{1}{2}iC_0\begin{pmatrix} \sigma_{xx} - \sigma_{yy} & 2\tau_{xy} \\ 2\tau_{xy} & -(\sigma_{xx} - \sigma_{yy}) \end{pmatrix}$. σ_{xx}, σ_{yy} and τ_{xy} are stress components in XY plane, i is imaginary. Then, the Jones matrix of the photoelastic object is obtained as:

$$J_{stress} = \sum_{k=1}^{n}(1 + A_k dz) = J(\Delta,\varphi) \tag{4}$$

2.1. Experiment simulation

The residual stresses in a thermal treated glass cylinder are simulated by MSC/MARC[5] based on a two-dimensional axisymmetric model. Figure 2 shows the residual stresses in cylindrical coordinates. The stress distribution displays half of the normal cross section of the cylinder. In the stress distribution figures, compressive stress is represented by negative value and tensile stress is represented by positive value.

Based on Fig. 2, light intensity images of six step PST can be calculated, shown in Fig. 3. In the calculation, refraction of light passing through the glass cylinder is not considered. And the light wavelength is 683 nm, the photoelastic constant of BK7 is C_0=2.77e-6/Mpa.

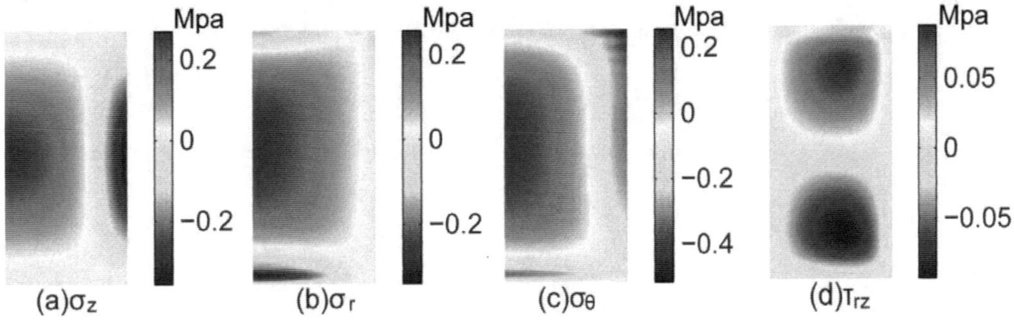

Fig. 2 Numerical simulation of residual stresses in a glass cylinder of the normal cross section in cylindrical coordinates: (a) axial stress $_z$, (b) radial stress $_r$, (c) circumferential stress , (d) shear stress $_{rz}$

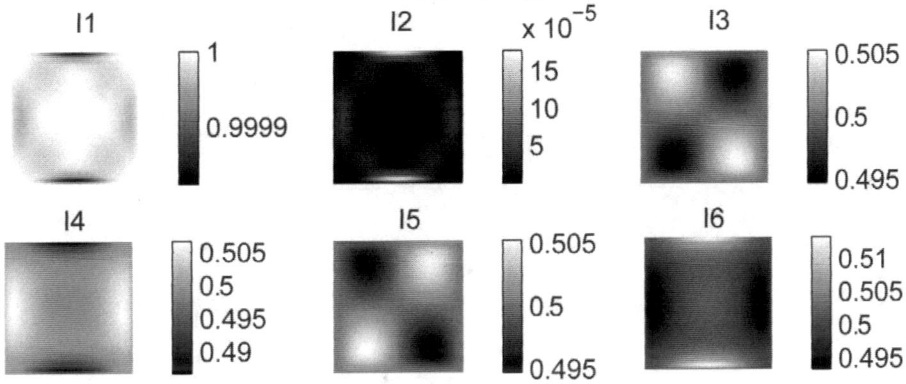

Fig. 3 Intensity images of the glass cylinder from FEM simulated stresses by Jones calculus

3. Residual stress reconstruction

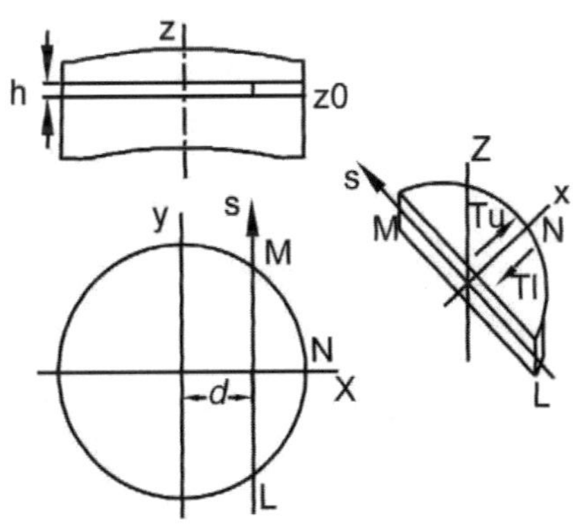

Fig. 4 Diagram of force balance of a glass cylinder[6]

As shown in Fig. 4, along the light ray s, in the 3D region LMN formed by plane $z=z_0+h$ and plane $z=z_0$, the shear forces of plane $z=z_0+h$ and plane $z=z_0$ can be expressed as:

$$T_u = \int_d^R \left(\int \tau_{zx} dy\right) dx = \frac{1}{2C_0}\int_d^R V_2' dx \quad (5)$$

$$T_l = \int_d^R \left(\int \tau_{zx} dy\right) dx = \frac{1}{2C_0}\int_d^R V_2 dx \quad (6)$$

IOP Conf. Series: Materials Science and Engineering **744** (2020) 012040 doi:10.1088/1757-899X/744/1/012040

where, R is the outer diameter of this area; d is the distance to the axis of symmetry of light s. In the auxiliary plane $z=z_0+h$, V_1 and V_2 are denoted as V'_1 and V'_2 , respectively.
Based on the equilibrium condition of the 3D region LMN:

$$C_0\int_L^M \sigma_z dy = \frac{1}{2h}\left(\int_d^N V_2' dx - \int_d^N V_2 dx\right) - V_1 \tag{7}$$

The shear stress τ_{rz} and axial stress σ_z can be calculated, through Eqs. (6) and (7), respectively [6, 7].

3.1. Circumferential stress and radial stress
The circumferential stress σ and radial stress σ_r, can be calculated through the equation of equilibrium:

$$\frac{\partial \sigma_r}{\partial r} + \frac{\sigma_r - \sigma_\theta}{r} + \frac{\partial \tau_{rz}}{\partial z} = 0 \tag{8}$$

and the generalized sum rule:

$$\sigma_\theta + \sigma_r = \sigma_z - 2\int_0^r \frac{\partial \tau_{rz}}{\partial z} dr + C_1 \tag{9}$$

where, C_1 is an constant and depends on boundary conditions. Therefore, the circumferential stress σ and radial stress σ_r can be obtained as:

$$\sigma_r(r) = -\int_0^r \frac{\partial \tau_{rz}}{\partial z} dr + \frac{1}{r^2}\int_0^r r\sigma_z dr + \frac{1}{2}\frac{C_2}{r^2} + C_4 \tag{10}$$

$$\sigma_\theta(r) = \sigma_z - \int_0^r \frac{\partial \tau_{rz}}{\partial z} dr - \frac{1}{r^2}\int_0^r r\sigma_z dr - \frac{1}{2}\frac{C_2}{r^2} + C_4 \tag{11}$$

where, C_2 and C_4 are constants depending on C_1. In order to acquire C_2 and C_4, a cost function PF is introduced [4].

$$PF = \frac{1}{n}\sum_{i=1}^{n}\left[\left(V_{1c}(i) - V_{1m}(i)\right)^2\right] \tag{12}$$

Subscript m denotes experiment values and subscript c denotes calculated values. The relation between $V_{1c}(i)$ and $V_{1m}(i)$ can be expressed as:

$$C_2\left(C_0\int_L^M\left(\frac{x^2}{r^4} - \frac{1}{2r^2}\right)dy\right) + C_4\left(C_0\int_L^M dy\right) = V_{1m} - V_{1c}' \tag{13}$$

Therefore, the C_2 and C_4 on each layer can be obtained and optimized by using the least square method.

4. Results and discussion

Based on the method in section 3, the reconstructed stresses are shown in Fig. 5. Compressive stress is represented by negative value and tensile stress is represented by positive value.

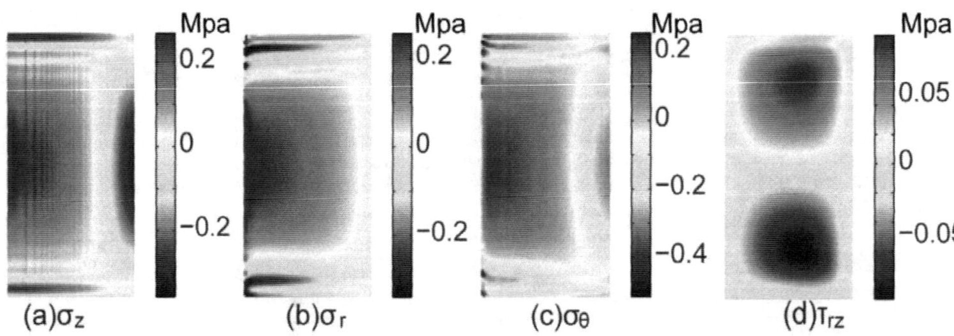

Fig. 5 Residual stresses calculation of the normal cross section in cylindrical coordinates: (a) $_z$, (b) $_r$, (c) , (d) $_{rz}$

Comparing Fig. 2 and Fig. 5, the reconstructed stresses shows the same magnitude and distribution with the FEM simulation but the axial stress, radial stress and circumferential stress have certain errors.

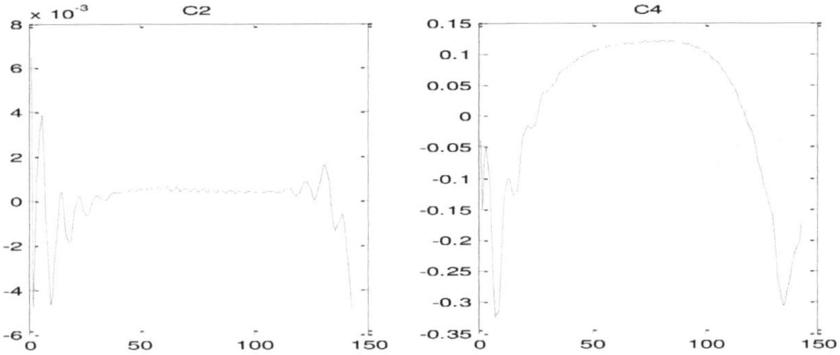

Fig. 6 C_2 and C_4

Figure 6 shows the C_2 and C_4 along the layer. It can be seen that the C_2 is a very small value ,and can be ignored.C_4 is the mainly factor.

5. Conclusion

In this paper, the residual stress distribution in a thermal treated glass cylinder is calculated by FEM simulation. Based on the stress tensor, intensity images from the circular polariscope were simulated and the stresses reconstruction was carried out. By comparing the reconstructed stresses with the stresses by numerical simulation, it shows that the method can be used to reconstruct the 3D stresses but with some errors. In the calculation of circumferential stress and radial stress the constant, which depends on boundary conditions, in generalized sum rule plays an import factor.

References

[1] Doyle, J. F., and Danyluk, H. T., "Integrated photoelasticity for axisymmetric problems," *Exp. Mech.* **18**(6), 215-220.

[2] Tao, B., He, P., Shen, L., and Yi, A., "Quantitatively measurement and analysis of residual stresses in molded aspherical glass lenses," *Int. J. Adv. Manuf. Tech.* **74**(9-12), 1167-1174.

[3] Theocaris, B. P. S., Gdoutos, E. E., [Matrix Theory of Photoelasticity]. Springer-Verlag.

[4] Aben, H. K., Josepson, J. I., and Kell, K. J., "The case of weak birefringence in integrated photoelasticity," *Opt. Lasers Eng.* **11**(3), 145-157.

[5] Chen, Y., Allen, Y. Y., Su, L., Klocke, F., and Pongs, G., "Numerical simulation and experimental study of residual stresses in compression molding of precision glass optical components," *J. Manuf. Sci. E.-T. Asme* **130**(5), 051012.

[6] Aben, H., and Errapart, A., "Photoelastic Tomography with Linear and Non-linear Algorithms," *Exp. Mech.* **52**(8), 1179-1193.

[7] Aben, H., Ainola, L., and Errapart, A., "Application of the Abel inversion in case of a tensor field," *Inverse Probl. Sci. Eng.* **18**(2), 241-249.

IOP Conf. Series: Materials Science and Engineering **744** (2020) 012041 doi:10.1088/1757-899X/744/1/012041

Development of water leakage detection and localization system

Xun Zhang[1], Quan Tian[2], Yuting Liu[3]

[1]Changjiang institute of survey, planning, design and research, Wuhan, Hubei 430081, China

[2]Zhixing college of Hubei University, Wuhan, Hubei 430081, China

[3]Key Laboratory of Metallurgical Equipment and Control Technology of Ministry of Education, Wuhan University of Science and Technology, Wuhan, Hubei 430081, China

Email: taoboq@wust.edu.cn

Abstract: Detection and localization of pipeline leakage is one of the important means to save water resources. A set of leakage detection was set up and experiments were carried out. The system contains flowmeter, pressure gauges and vibration sensor. The signal was continuous collected to constantly detect and alert water leaks. The leakage location is based on GCC-PHAT algorithm. a leakage detection and location system based on GCC-PHAT algorithm is proposed. Experiments have verified that our system works well.

1. Introduction

With the continuous improvement of China urbanization level, the water supply network system is improved. It is understood that the length of water supply pipeline in China presents a trend of accelerating growth every year. Since 2008, the construction speed of water supply pipeline in China has increased at an average annual rate of 30,000 kilometers, with an average annual growth about 6.5%. However, there are some issues about pipe maintenance due to the rapid development of urban water supply network. Water leakage is one of the main issues.

Traditional leak detection methods include acoustic principle detection[1], Thermography[2], radar detection[3], current method and potential gradient method utilized based on metal properties of metal pipe. However, these methods cost a lot of labour and equipment, with low detection accuracy. Therefore, many experts and scholars have done further research on the principles, instruments and data processing of water leakage location.

Laucelli et al. [4]proposed an optimal DMA design in a real WDN(Water Distribution Networks) to reduce leakage. They refer to a two-step strategy for changing the hydraulic paths to reduce pressure and leakages through the WDN. Anuj Purwar et al. [5]used the Internet of things approach to design automated water management systems. Ismall et al.[6]employs an accelerometer sensor MPU6050 to measure the vibrations of the water pipeline of Acrylonitrille Butadience Styrene (ABS) pipe from three different axes, which are x-, y- and z-axis. An Arduino controller board was used to process the vibration data from the accelerometer sensor and the data was transmit over wireless ZigBee networks every ten milliseconds. M. JayaLakshmi et al. [7]proposed an enhanced underground pipeline water leakage monitoring and detection system using wireless sensor network.They use small Printed Circuit Boards (PCB) and different types sensors (acoustic, pressure, temperature, flow rate, etc.) to locate the leakage, the proposed system shown in Figure.1. Flow sensor is used to measure the flow of the water between two points. Transmitter and receiver module constructed using AT Mega 8 Microcontroller. Communication between the two modules is done by ZigBee wireless communication. The transmitter has the flow sensor which monitors the flow rate. The flow rate values are shown in LCD. If there is a change in the flow rate value, then it sends an alarm signal.

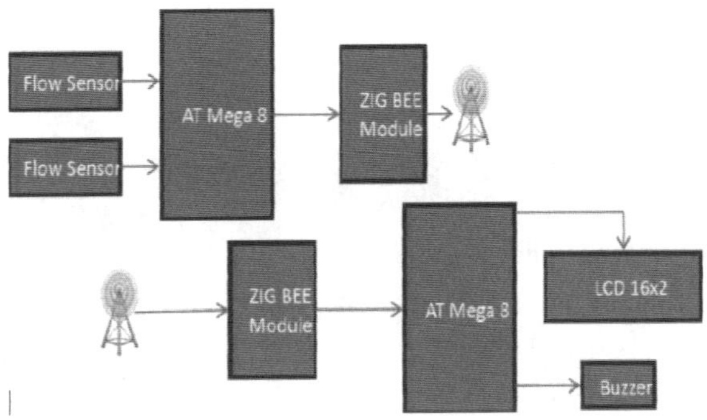

Figure 1. M. JayaLakshmi et al. proposed system [7]

2. Water leakage detection and localization system

2.1 GCC-PHAT algorithm

GCC is a generalized cross-correlation algorithm, which can process multiple signals and reduce or eliminate the influence of noise on the accuracy of estimation of time delay[8-9]. The schematic is shown in Figure 2. Firstly, the sensor signal is obtained by sensors 1 and 2, and then sampled by A/ D converter to get the original signal. Then by using a movable finite length window to frame and analyse signal with Fourier transform, we could get cross-power spectrum of the frame. After the cross-power spectrum is weighted in frequency domain according to a certain weight, the cross-correlation function of this frame is obtained by inverse Fourier transform. According to the principle of cross-correlation function, the time corresponding to the peak of the function is the delay time between two sensors. The weighting of the signal in the power spectrum is beneficial to highlight the relevant parts of the signal and suppress the disturbed parts by the noise. In this way, the peak value of the correlation function delay is more prominent. By comparing the quality of CC(Cross Correlation), Roth, SCOT(Smoothed Coherence Transform), improved SCOT, PHAT(Phase Transformation) and other weighting functions, we selected PHAT function as the weighting function in GCC algorithm. Figure 2 shows the schematic flow diagram of the GCC-PHAT algorithm[10].

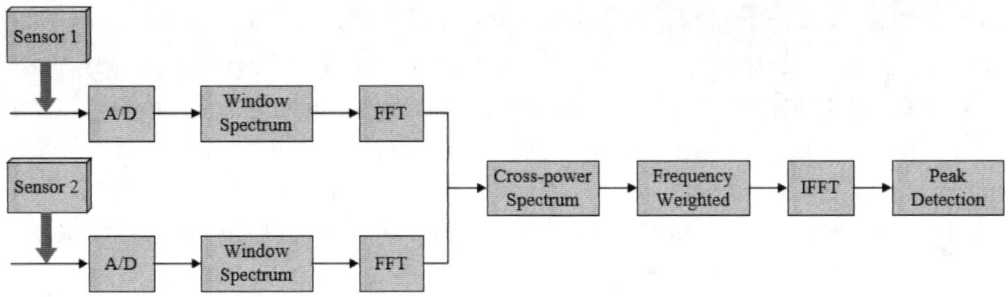

Figure 2. The schematic flow diagram of GCC-PHAT algorithm

2.2 Water leakage detection and localization system

This paper proposes a kind of system which can detect and locate the leaking point in real time The system use three different instruments: pressure gauges, flowmeters, vibrating sensors to get several environmental parameters. By adjusting the relevant parameters, analysing the data of the vibration sensor and calculating the leakage point, the position of the leakage point is finally located, and the positioning error is within 1 metre. Figure 3 shows the flow chart of the whole system. Two ends of the experiment pipeline are respectively equipped with pressure gauges, flowmeters and vibration sensors. The acquired signal is connected to PLC by RS485 communication line, and the data is transmitted to PC by network.

The PLC module is mainly used for data acquisition, preliminary judgment of pipeline leakage and preliminary processing of vibration signal。The specific process is as follows:

(1) PLC uses MODBUS-RTU protocol to collect the real-time data of six instruments from site and send them to PC.
(2) PC saves the data to SQL database.
(3) Computers make corresponding response, after the data processed and judged.
(4) Users could read the specific data in website.

In the data processing algorithm, we use the GCC-PHAT algorithm. The algorithm weights the power signal. In this way, the related signal is highlighted and the noise interference is suppressed. The peak value of the correlation function at the time delay is more prominent. The higher the GCC-PHAT function value is, the more reliable the sound source signal will be received.

IOP Conf. Series: Materials Science and Engineering **744** (2020) 012041 doi:10.1088/1757-899X/744/1/012041

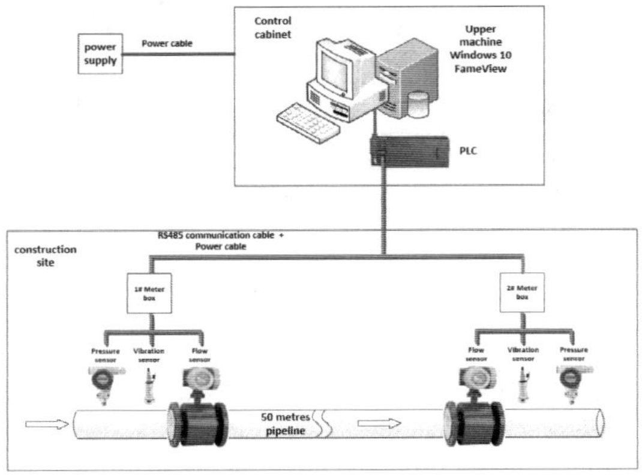

Figure 3. Flow chart of the water leakage detection and localization system

3. Experiment and results

Water leakage detection and localization experiment is based on the water saving project of Wuhan Changjiang Water Resource Commission Of the Ministry Of Water resource. We installed the aboved system on the 50 metres experimental road section, and observed and collected data for 15 days. During the experiment, a leakage valve was installed in a certain part of the pipeline. By adjusting the valve, the situation of whether the pipeline has leakage was simulated. The flowmeters at both ends of the pipeline change as shown in Figure 4. It shows that the indication of flowmeter is changed according to the amount of water leaked out. Figure 5 shows the change of corresponding pressure gauge value. As shown in Figure 6, the collected signals was calculated using GCC-PHAT algorithm. From the Figure 6, we could find that the results of the location of water leakage fluctuates to some extent, but the fluctuation range is not too much, which means it can locate the leakage point well.

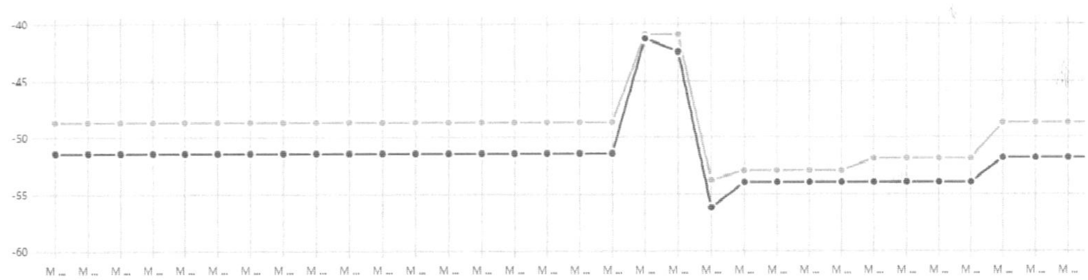

Figure 4. The changes of flowmeters

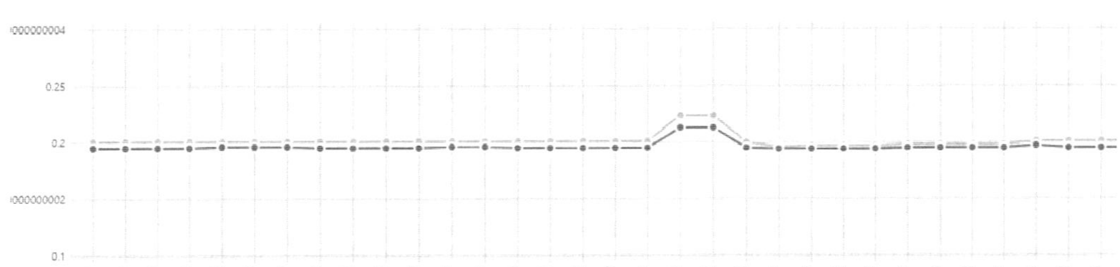

Figure 5. The changes of corresponding pressure gauge value

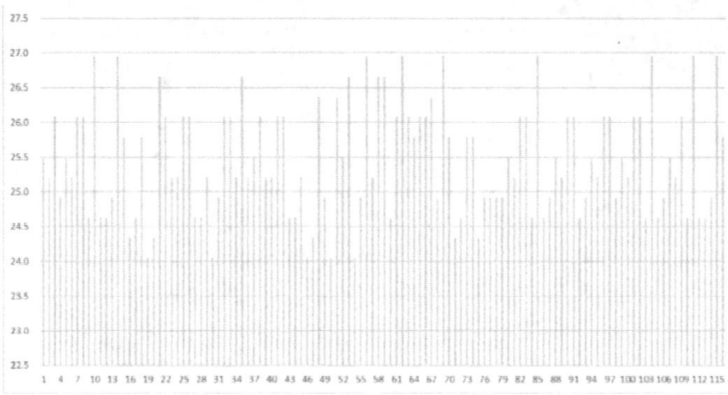

Figure 6. Results of water leakage location

4. Conclusion

Pipeline leakage detection and localization technology is one of the important means to save water resources. In this paper, a water leakage detection and location system was built and experiments were carried out. Experimental results have verified the effectiveness and the localization accuracy. Furthermore, the method is robust against the low frequency structural vibration introduced by other external forces. Meanwhile, we test the system by water saving project and get expected experimental result.

However, our system still needs improvement to adapt to large scale deployment, for examples: miniaturize equipments, which use current microcontrollers and wireless transmission to collect and control the sensor data, and algorithm optimization; and adopt better algorithms in data processing for more accurate positioning results.

References

[1] Fabrício C L Almeida, Brennan M J, Joseph P F, Dray S and Paschoalini A T 2015 *Journal of Sound and Vibration* **359** 40-55
[2] Schuetz T, and Weiler M 2010 *Egu General Assembly*
[3] Amran T S T, Ismail M P, Ahmad M R, Amin M S M, Ismail M A and Sani S 2018 *IOP Conference Series: Materials Science and Engineering* **298** 012002
[4] Laucelli D B, Berardi L, Giustolisi O and Andrei C G 2017 *Int. Conf. on ENERGY and ENVIRONMENT*
[5] Purwar A, Patel M, Garg M and K Ahuja 2018 *7th Int. Conf. on Reliability, Infocom Technologies and Optimization* p 373-7
[6] Ismail M, Dziyauddin R A and Salleh N A A 2016 *IEEE Int. Symposium on Robotics & Intelligent Sensors*
[7] Jayalakshmi M and Gomathi V 2015 *Int Conf on Soft-Computing and Networks Security*
[8] Kang H G, Graczyk M and Skoglund J 2016 *IEEE Inter. Workshop on Acoustic Signal Enhancement*
[9] An L. Wang P. Jiang G. Li G& Shen G 2012 *Zhongguo Dianji Gongcheng Xuebao/Proceedings of the Chinese Society of Electrical Engineering* **32** 16-23
[10] Kang H G, Graczyk M and Skoglund J 2016 *IEEE Int. Workshop on Acoustic Signal Enhancement*

Study on properties of sisal fiber modified foamed concrete

Yunxue Liu[1], Zhen Wang[1],Zhaorong Fan[1], Jin Gu[2]

1. School of Material Science and Engineering,Shenyang Jianzhu University,Shenyang110168,China;
2. China Railway No.9 Group Co., Ltd., Shenyang110044, China
Email: 1413472465@qq.com

Abstract. Sisal fiber modified foamed concrete was prepared with ordinary Portland cement and vegetable protein foaming agent as the base material and sisal fiber as the modifier. The distribution of fibers in the sample changes. The results show that when the fiber content is 0.75% and the fiber length is 5mm, the mechanical properties of foamed concrete are improved most obviously, and the respectively. Sisal fiber can effectively improve the shrinkage of foamed concrete. The longer sisal fiber is, the smaller the drying shrinkage of modified foamed concrete is. From the optical photos of concrete, it can be seen that at the same mixing amount (1%), fibers with lengths of 5mm and 10mm are more evenly dispersed than those with lengths of 15mm, which have little impact on the bubble holes.

1. Introduction

The use of plant fibers to prepare concrete construction materials can reduce the negative impact of traditional concrete materials on the environment to a certain extent.Plant fiber can enhance the mechanical properties of concrete, which is in line with the current concept of green and sustainable development .Due to the combination of cement and fiber, a mesh support system is formed inside the foamed concrete, thereby protecting the air holes.During the preparation of foamed concrete, the incorporation of fibers protects some of the bubbles and reduces the degree of damage, and at the same time reduces the solidification and sedimentation rate of the slurry, improves the overall stability of the foamed concrete, hinders the generation of cracks and Diffusion makes the strength and toughness of foamed concrete improved .

2. Test materials and schemes

2.1. Raw materials

The cement uses Shenyang Jidong Cement Plant Dunshi P · O42.5 grade cement, and the fiber uses 5mm, 10mm, 15mm sisal fiber, and the additives include foaming agent, water reducing agent, etc.

2.2. Hemp fiber processing

The sisal fibers with a length of 5mm, 10mm, and 15mm were washed and dusted, and then placed in a drying box at 80 ° C and dried. The dried fibers were immersed in a 10% NaOH solution for 30 minutes.

2.3. Sample preparation

First, the mother liquid of the foaming agent is mixed with water in proportion to play a dilution role. The diluted mother liquid is added to a foaming machine to prepare a foam. After that, the pure slurry is mixed, and the pure slurry is prepared according to the test proportion and blended with plant fibers, and then the prepared foam is added, and the foamed concrete is prepared after uniform mixing.

2.4. parameter test

For compressive and flexural strength testing, refer to "Testing Methods for Autoclaved Aerated Concrete Performance" (GB / T 11969-2008); For drying shrinkage performance, refer to "Foam Concrete" (JG / T 266-2011).The size of the test piece is 100mm ×100mm × 100mm, no less than three test blocks are selected for each group, and the dispersion of the test values meets the requirements of the specification.

3. Results and discussion

3.1. Hemp fiber processing method

Since the plant fiber contains lignin that has an inhibitory effect on cement hydration, sisal fiber must first be treated with alkali to remove a large amount of lignin. After the preliminary test, the sisal fiber was impregnated with NaOH solution at a concentration of 10-30%, and the immersion time was 10-60 minutes. It was found that with the increase of time and concentration, the mass loss of sisal fiber will tend to first A fixed value will then continue to decrease, as shown in Figure 1:

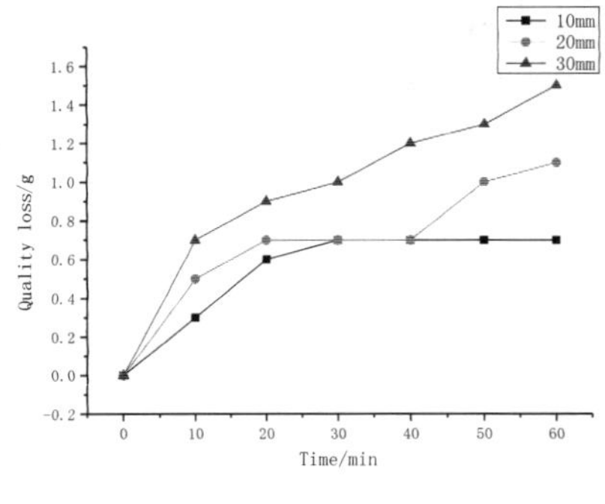

Fig.1 Changes of fiber quality loss

When the lignin in the fiber is basically removed, the mass loss of the fiber tends to a fixed value, and it will continue to decrease after the fixed value, because the internal structure of the fiber is also damaged. The ultra-deep microscope was used to observe the internal structural changes of 10% NaOH immersion for 30 minutes and 30% NaOH immersion for 60 minutes, as shown in Figure 2:

IOP Conf. Series: Materials Science and Engineering **744** (2020) 012042 doi:10.1088/1757-899X/744/1/012042

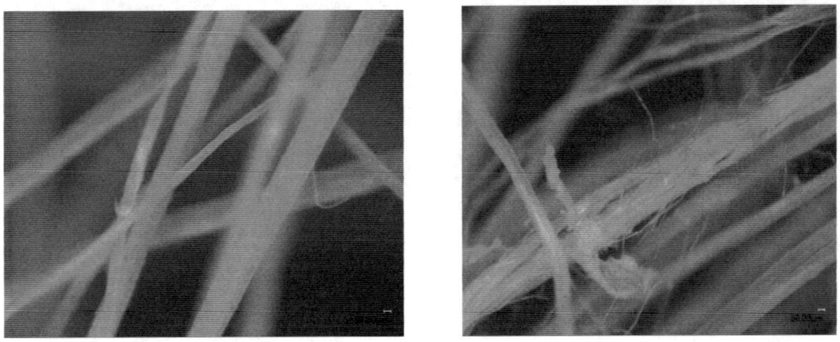

(a)10% NaOH for 30 minutes (b) 30% NaOH for 60 minutes

Fig.2 3D ultradeep micrograph of sisal fibers

It can be seen from Figure 2 (a) that the surface of hemp fiber is smooth without obvious splitting, and the fiber integrity is better.It can be seen from Fig. 2 (b) that the internal structure of hemp fiber has been damaged after prolonged soaking, the surface of the fiber is cracked, and the branches generated are crisscross.Such a fibrous structure can easily blister bubbles in the cement slurry, causing a collapse phenomenon. It can be seen that soaking for 30 minutes is more suitable.

3.2. Effect of sisal fiber on the mechanical properties of foamed concrete

The effect of sisal fibers with different amounts and different lengths on the 28-day compressive strength of foamed concrete is shown in Figure 3:

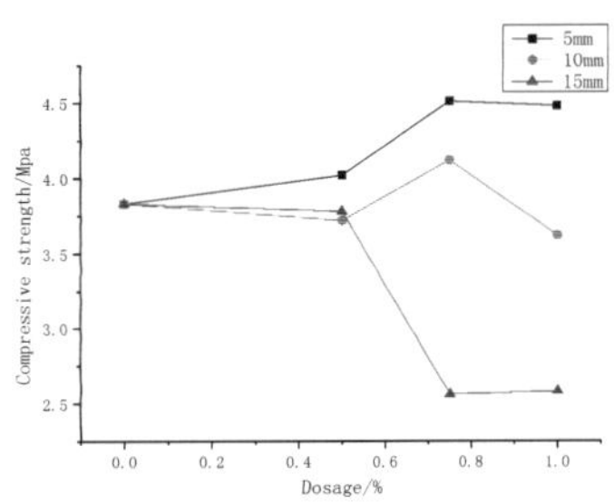

Fig.3 Effect of fiber content on compressive strength of foamed concrete

As can be seen from Figure 3 (a): when the fiber length is 5mm, the 28d compressive strength increases with the increase in the amount of the fiber, and when the fiber content exceeds 0.75%, the 28d compressive strength is basically unchanged;When the fiber length is 10mm, the compressive strength at 28d first increases and then decreases with the content. When the fiber content is 0.75%, the compressive strength reaches the maximum.When the fiber length is 15mm, the compressive strength decreases, and when the content exceeds 0.75%, the decrease becomes slower; when the content remains the same, the shorter the fiber length is, the more obvious the compressive strength is. The effect of sisal fibers with different amounts and different lengths on the flexural strength of foamed concrete is shown in Figure 4:

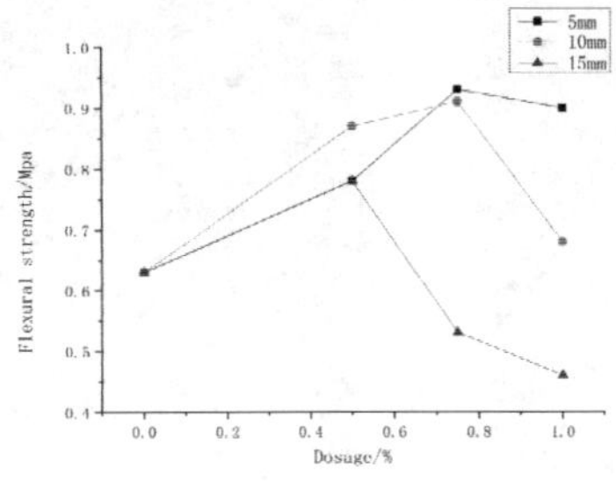

Fig.4 Effect of fiber content on the flexural strength of foamed concrete

It can be seen from Fig. 4 that as the length and amount of sisal fibers increase, the flexural strength of the foamed concrete tends to increase first and then decrease. When the content of sisal fiber was 0.75%, the flexural strength of the test pieces with a length of 5mm and 10mm reached the maximum. When the content of sisal fiber is 0.5% and the length is 15mm, the compressive strength reaches the maximum. When the fiber length is 5mm and the blending amount is 0.75%, the flexural strength of 7d and 28d reaches the maximum, which is higher than the other two specifications of hemp fiber foam concrete.

The reason for the above situation is that the incorporation of fibers reduces the occurrence of some small cracks in the foamed concrete, which improves the integrity of the foamed concrete. When the cracks appear, the fibers can bear a part of the external force, thereby improving the flexural strength.

3.3 Effect on drying shrinkage of foamed concrete

The effect of different amounts and different lengths of sisal fibers on the dry shrinkage of foamed concrete is shown in Figure 5:

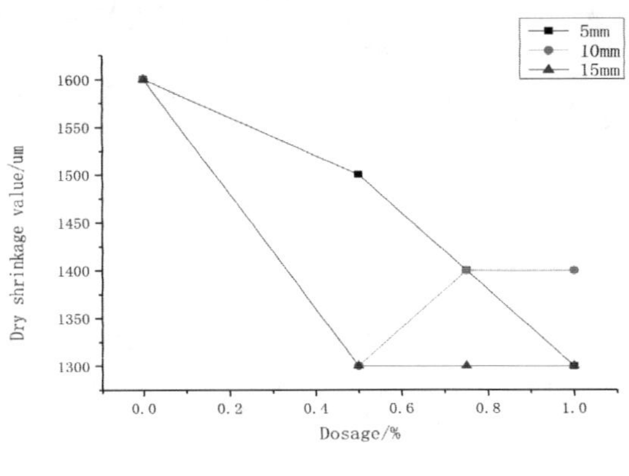

Fig.5 Effect of fiber content on dry shrinkage of foamed concrete

It can be seen from Figure 5 that with the increase in the amount of hemp fiber and hemp fiber in the four specifications, the dry shrinkage value of foam concrete decreases. When the content is 0.5%, the dry shrinkage value of 10mm and 15mm fiber concrete is the smallest;The dry shrinkage value of

15mm fiber-reinforced concrete is the smallest when the content is 0.75; the dry shrinkage value of 5mm and 15mm fiber-reinforced concrete is the lowest when the content is 1.0%.

4. Conclusion

- Modified hemp fiber with alkali concentration of 10% NaOH and soaking time of 30 minutes has the best effect on improving the performance of foam concrete.
- When the fiber content is 0.75% and the fiber length is 5mm, the mechanical properties of the foam concrete are most improved, and the compressive and flexural strengths are increased by 17.8% and 47.6%, respectively.
- Sisal fiber can effectively improve the shrinkage performance of foamed concrete. Under the same content, the longer the sisal fiber, the smaller the shrinkage value of the modified foamed concrete.

5. Reference

(1)Yi feng. Research progress of plant fiber concrete [J]. Building materials and decoration, 2017(34):125-126.
(2)Zhou yongsheng, zhang siyuan, du yuxin, li yuming. Study on improving foamed concrete performance by adding sugarcane leaf fiber with mastic film [J]. Concrete and cement products, 2017(10):56-59.
(3)Wang xue, zhai yanyu, guo yuanchen, wang zhi, zhao ting, liu jun. Mechanical properties of sisalfiber reinforced concrete [J]. Chinese journal of silicate, 2017, 36(07):2488-2
(4)Zhan binggen, guo jianlei, Lin xingsheng. Experimental study on the properties of glass fiber reinforced foam concrete [J]. Journal of hefei university of technology (natural science edition), 2009, 32(02):226-229.
(5)Zhang xuesong, zhu yanfeng, huang linchong. Influence of polypropylene fiber on properties of foam concrete [J]. New building materials, 2019, 46(03):140-142+147.

Compressive strength and flexural strength are increased by 17.8% and 47.6%

Study on Properties of Graphite Polystyrene Permeable Composite Insulation Board

Jusong Zhang[1], Yuwei Cao[1] and XinWang[2]

[1] College of materials science and engineering, Shenyang jianzhu University, Shenyang liaoning, China
[2] Liaoning super insulation technology Co. Ltd, liaoning dandong 118000
E-mail:739007960@qq.com

Abstract. In order to further optimize the properties of polystyrene permeable flame retardant composite insulation board, the composition, properties and flame retardant mechanism of graphite polystyrene permeable flame retardant composite insulation board were discussed in this paper. The properties of graphite polystyrene substrate and permeable flame retardant, the optimum ratio of permeable flame retardant. The tensile strength, compressive strength, thermal conductivity and softening coefficient perpendicular to the plate surface are discussed, and the endothermic flame retardant mechanism of osmotic flame retardant was discussed. The results show that graphite polystyrene permeable flame retardant composite insulation plate does not reduce its combustion and mechanical properties when reduces the density and thermal conductivity.

1. Introduction

With the economic development and social progress, people's awareness of environmental protection has increased, and the sustainable development of building energy conservation has received more and more attention. Polystyrene boards have the characteristics of excellent thermal insulation performance, small bulk density, low thermal conductivity, and convenient construction. It is easy to burn, and generates a lot of harmful gases when burning. Magnesium penetrating flame retardant is a halogen-free green flame retardant, and water and oxides generated during combustion will not cause pollution to the environment; and the flame retardant effect is relatively stable, but magnesium-based penetrating flame retardant and graphite polystyrene insulation board There have been no reports at home and abroad. For this reason, the properties of graphite polystyrene magnesium flame retardant composite panels have been discussed in this paper.

IOP Conf. Series: Materials Science and Engineering **744** (2020) 012043 doi:10.1088/1757-899X/744/1/012043

2. Raw materials, experimental equipment and methods

2.1.raw materials

Graphite polystyrene board and graphite polystyrene permeation flame retardant composite thermal insulation board are produced from the production line of Liaoning Chaoqiang Fireproof Thermal Insulation Technology Co., Ltd., the size is 1000mm ×600mm ×100mm.

Magnesium based penetrating flame retardant: magnesium oxysulfide based material.

2.2 Experimental methods

Test the basic performance of graphite polystyrene insulation board and inorganic penetrating flame retardant graphite polystyrene composite insulation board according to the national standard "Molded polystyrene board thin plaster external wall insulation system material" GBT29906-2013.

According to GB 8624 standard method, the burning performance of graphite polystyrene insulation board and inorganic permeable flame retardant graphite polystyrene (heterogeneous material) composite board was tested.

Optimization of magnesium penetration flame retardant:The molar ratio of magnesium oxide to magnesium sulfate (oxygen-sulfur ratio) is 3: 1, 5: 1, 7: 1, of which 5: 1 is added with 30% fly ash, and tested for strength and thermal analysis. The universal surface tensile tester (QL-5E) was used to test the tensile strength of the vertical surface. The intelligent thermal conductivity tester (DRCD-3030) was used to test the thermal conductivity. The high-precision rivet puller (MD60), the test calorific value uses the building material product combustion calorific value measuring device (JCRZ), the test flammability uses the building material non-combustibility experimental device (JCBR), the test temperature uses the infrared thermometer (AS882), and observes its microscopic morphology using scanning Electron microscope (S-4800).

3. Graphite polystyrene board

According to the method described above, the density of the graphite polystyrene board was 18 kg/ m³, the thermal conductivity was 0.030, the tensile strength perpendicular to the plate surface was 0.06 MPa.The flammability of graphite polystyrene has been tested to reach B1 level.The TG and DSC curves of the graphite polystyrene board are shown in Fig. 1.

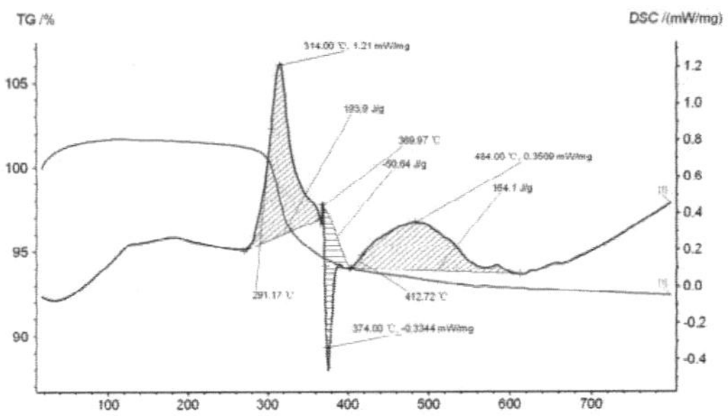

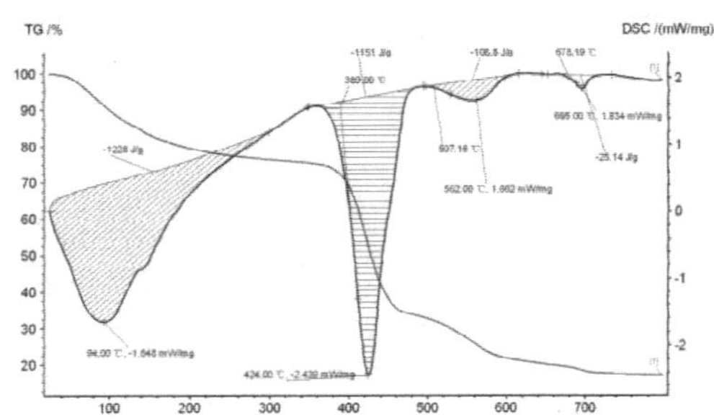

Fig. 1 TG/ DSC curves of the board Fig. 2 TG / DSC curve of magnesium penetrating flame retardant

The DSC curve in Fig. 1 shows that the first peak temperature appears at about 314℃, and the heat released is about 193.9J/g. At the temperature of 484℃, the second exothermic peak of the graphite polystyrene substrate exists. The heat is 164.1 J/g, and the total of the two is about 358 J/g. When the temperature is 374℃, an endothermic peak appears and the heat is 50.64J/g. It is 307.36 J/g. The TG curve shows that the temperature consistent with the starting point of the exothermic peak starts from 291.17℃ to 412.72℃, and the mass rapidly decreases from 100% to about 93%. With the second exothermic peak, the mass slowly decreases to 91% and stabilizes.

The exotherm of graphite polystyrene board should be the result of the combustion of polystyrene. The heat absorption may be caused by graphite and other flame retardant components.

4. Magnesium penetration flame retardant

Table 1 shows the experimental results of the strength of magnesium infiltration flame retardants with different oxygen-sulfur ratios.

Table 1 Strength of magnesium permeable flame retardants with different oxygen-sulfur ratios

Oxygen to sulfur ratio	3:1	5:1	5:1(Fly ash)	7:1
strength/ MPa	55.3	58.8	62.4	60.2

Experimental results of strength test based on the different oxygen-sulfur ratio of magnesium penetration flame retardant.With the increase of oxygen-sulfur ratio, the strength of magnesium penetration flame retardant is increased. When the oxygen sulfur ratio is 5:1 and fly ash is added, the highest strength is 62.4mpa. The main reason of high strength when adding fly ash may be that the two main components of magnesium material, magnesium oxide and magnesium sulfate, are the activators of pozzolanic material respectively, which makes the chemical activity of fly ash play a role.

The TG / DSC curve of the optimization of the proportion for magnesium penetration flame retardant is shown in Fig. 2.As can be seen in Fig. 2, there are four endothermic peaks in the DSC temperature curve. The total endothermic area is 2513 J / g, the first endothermic peak temperature is 94℃, the endothermic area is 1228 J / g, and the second endothermic peak The temperature is 434 ℃, the endothermic area is 1151J / g, the third endothermic peak temperature is 562 ℃, the endothermic area is 109J / g, the fourth endothermic peak temperature is 695 ℃, and the endothermic area is 25J / g. The TG curve shows that the mass loss corresponding to the first endothermic peak is about 25%, the mass loss corresponding to the second endothermic peak is about 40%, and the mass loss corresponding to the third and fourth endothermic peaks is about 15%.

Table 2 shows the DSC curve analysis results of magnesium-based penetrating flame retardants with different oxygen-sulfur ratios and optimized ratios.

Table 2 DSC curve analysis results of different oxygen-sulfur ratios

Oxygen to sulfur ratio	First	Second	Third	Fourth	Fifth	In total Calorific value(J/g)
3:1	183/1774	444/183	568/123	647/97.85	0/0	/2178
5:1	174/1664	442/361	566/172	640/37	704/8	/2242
5:1(Fly ash)	163/1639	441/349	556/139	644/36	709/7	/2170
7:1	160/1378	438/377	564/179	624/21	704/14	/1969
Optimized ratio	1228/94	1151/424	109/562	25/695	-	/2513

Note: temperature (℃) /Calorific value (J/g) of different peaks in table.

Magnesium penetration flame retardant with different oxygen-sulfur ratio and optimized ratio according to the relevant research of magnesium oxysulfide material [1-2], it can be concluded that the first endothermic peak is mainly caused by the desquamation of the adsorbed water and the loose binding water from the basic magnesium sulfate material $xMg(OH)_2 \cdot MgSO_4 \cdot yH_2O$ and magnesium hydroxide $Mg(OH)_2$, which will take the heat away. The second, third, fourth and fifth endothermic peaks are mainly caused by the desquamation of the closely bound water.

DSC curves of different oxygen-sulfur ratio and optimized ratio of magnesium penetration flame retardant have similar rules with Fig. 2. There are four or five endothermic peaks at similar temperature, with slightly different calorific value. The change rule of calorific value shows as: the calorific value of 5:1 is greater than 3:1 and 7:1, and the calorific value of 5:1 with a certain amount of fly ash is slightly lower than 5:1 without fly ash, and the maximum calorific value of optimized ratio is 2513 J/g.

According to the relevant research of magnesium oxysulfide material [3-5], the reason for the change of calorific value causes by water and the main crystal phase of the product as $5Mg(OH)_2 \cdot MgSO_4 \cdot 7H_2O$ (5·1·7). After adding fly ash, the basic magnesium sulfate with higher water content in the product is replaced by the magnesium silicate (aluminum) hydrate with lower water content due to the chemical excitation of magnesium oxide and magnesium sulfate, which is consistent with the test results of the above strength.The results of comprehensive strength and thermal analysis shows that the optimal one of magnesium penetration flame retardant in different oxygen sulfur-ratio is 5:1.

5. Graphite polystyrene penetration flame retardant composite insulation board

Fig.3 is the mechanical properties of the composite plate .

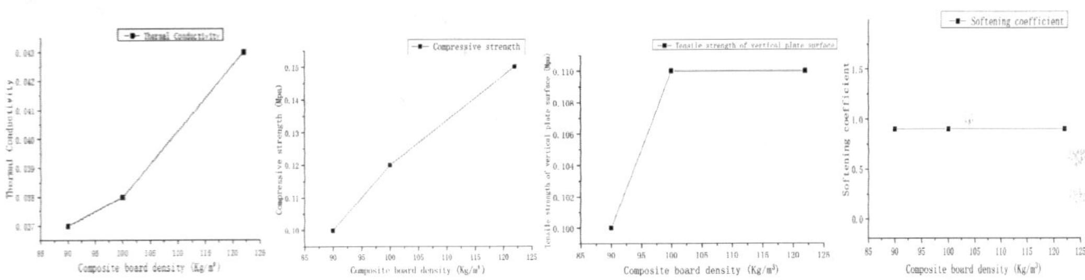

Fig. 3 Mechanical properties of composite plates

As can be seen in Fig. 3, as the density decreases from 122kg/m³ to 90kg/m³, the thermal conductivity of the composite plate decreases from 0.043 w/m.k to 0.037 w/m.k, and the compressive strength decreases from 0.15MPa to 0.10MPa. When the density of the composite plate is increased from 90 kg/m³ to 100kg/m³, the tensile strength of the vertical plate surface is increased from 0.10MPa to 0.11MPa, and when the density of the composite plate is increased from 100 kg/m³ to 122kg/m³, the vertical plate surface tensile strength is basically constant. Within the test density range, the softening coefficient of the composite board did not change significantly.

The change rule of thermal conductivity and compressive strength conforms to the rule of general composite materials. The change rule for tensile strength of vertical board should be caused by the brittle character of magnesium flame retardant penetrant as the inorganic cementitious materials. The increase of density has limited effected on the improvement of tensile capacity of materials.

The TG/DSC curve of the composite insulation board with the density of 90 Kg/m³.Produced with graphite polystyrene and the optimized ratio of penetrating flame retardant as the raw material is shown in Fig. 4.

Fig. 4 is visible, the endothermic peaks at 100℃ and 150℃ exactly match the endothermic peaks in Fig. 2, because the polystyrene has not been fully burned and exothermic at this stage shown in Fig. 1, and the characteristics of the endothermic peaks are determined by the flame retardant. The combustion exothermic peak at about 300℃ in Fig. 1 is not obvious due to the influence of the heat absorption of the flame retardant, but the characteristics of the peak temperature still exist. The characteristics of the endothermic peak at about 380℃ in Fig. 1 and the endothermic peak at about 420℃ in Fig. 2 still exist in Fig. 4, and the endothermic peak at 562℃ in Fig. 2 is still obvious, because the exothermic polystyrene has been completed at this stage.

As all known, the mechanism of flame retardancy of combustibles: oxygen, combustibles and ignition point are the three necessary conditions for combustion. Generally, there are three types of flame-retardant mechanisms studied: gas-phase flame-retardant, condensed phase flame-retardant and interruption of heat exchange: (1) Gas-phase flame-retardant mechanism: when the flame retardant is heated, it releases free radicals to interrupt the polymer combustion chain reaction, prevent combustion, release non-combustible gas, dilute the concentration of combustible gas and oxygen around the flame-retardant composite; some flame retardants decompose their own crystal water when heated, it are released in the form of water vapor, which can absorb or take away the heat of polymer surface, to reduce the surrounding temperature and achieve the flame retardant effect.

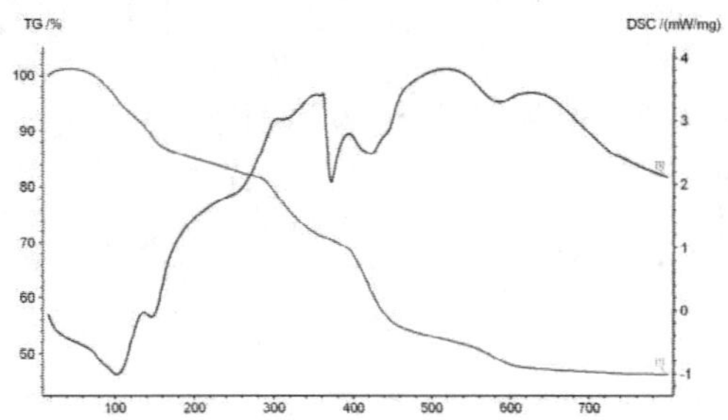

Fig. 4TG / DSC curve of composite board infiltration

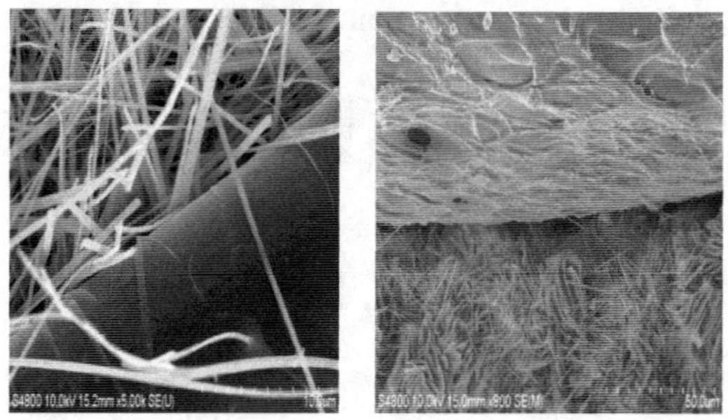

Fig. 5- SEM of graphite polystyrene composite insulation board

(2) Condensed phase flame retardant mechanism: it is also known as solid-phase flame retardant mechanism, the flame retardant decomposes into the incombustible substances when heated, and covers the polymer surface to form a protective layer, so as to isolate the outside air and heat from entering into the base material inside, and at the same time prevent the combustible gas inside from escaping, so as to achieve the flame retardant effect; or some flame retardant decomposes and absorbs a lot of heat when heated, which can reduce the surface temperature of polymer and prevent the thermal degradation of polymer. (3) The mechanism of heat exchange interruption: the flame retardant takes a lot of heat away during the combustion of the composite, so as to reduce the surface temperature of the composite and make it lower than the degradation temperature of the polymer, then the material will appear self extinguishing at time.

Through the above analysis, it can be seen that the action mechanism of magnesium permeating flame retardant may be all of the above mechanisms, but the basic characteristics are clear: heat absorption and decomposition dehydration change the atmosphere around the combustibles after heat exposure, and the decomposed non combustibles isolate the oxygen between the combustibles and the environment. In addition, there is no doubt that the matching between the exothermic temperature of polystyrene combustion and the endothermic decomposition temperature of flame retardant is very good Important.

SEM analysis of typical interface of graphite polystyrene infiltration composite insulation board is shown in Fig. 5.Shown by the SEM test, the typical characteristics of magnesium permeating and flame retardant materials hydrate of graphite polystyrene permeating composite board are needle-like, sheet-like and needle mesh-like shape, which should be closely related to the water material ratio and process parameters of raw materials. These hydrates are tightly wrapped around the polystyrene particles, that is the structure of inorganic isolation and bin separation is formed when the composite board encounters high temperature In the environment, on the one hand, when the magnesium permeating flame retardant material and polystyrene are in contact with the high temperature environment at the same time, which will be the complex interaction. On the other hand, even if a polystyrene particle burns in the fire and the fire will be blocked, due to the isolation of the magnesium permeating flame retardant material.

Tested by the JCRZ, the heat value of the graphite polystyrene infiltration composite insulation board is about 2.5MJ / kg, which meets the A2 level requirements.

6. Conclusion

1. Graphite polystyrene is used as the substrate, and sulfur-oxygen gelled material is used as the permeation flame retardant. Graphite polystyrene permeation composite insulation boards with a density of less than 100 Kg / m^3 and a thermal conductivity of less than 0.04 w / m.k can be prepared.
2. The tensile strength of the plate perpendicular to the surface is 0.1 MPa, the thermal conductivity is 0.039 w / m.k, and the softening coefficient is 0.9.
3. Graphite polystyrene board has Class B1 flame retardant ability, and can be compounded with Magnesium penetrating flame retardant material to achieve Class A2 heat value less than 3MJ / Kg. The flame retardant mechanism of magnesia penetrating flame retardant materials was analyzed as the endothermic flame retardant effect at appropriate temperature.

7. References

[1]Wu Chengyou, Yu Hongfa, Zhang Huifang. Extraction of aluminum by pressure acid-leaching method from coal fly ash. Transactions of Nonferrous Metals Society of China. 2012, 22 (9):2282-2288. (SCI)

[2] Wu Chengyou, Yu Hongfa, Zhang Huifang. The effects of aluminum -leached coal fly ash residue magnesium oxychloride cement. Advances in cement research, 2013, 25(5): 254-261. (SCI)

[3] Jusong Zhang Ruge Chen1 and Xin Wang.Study on Basic Properties of Inorganic Isolation Partitioned Polystyrene Insulation Board IOP Conf. Series: Materials Science and Engineering 544 (2019) 012055 doi:10.1088/1757-899X/544/1/012055.

[4] Tomce Runcevski, Wu Chengyou*, Yu Hongfa, Yang Bo, Robert E, Dinnebier. Structural Characterization of New Magnesium Oxyslufate Hydrate Cement Phase and Its Surface Reactions with Atmospheric Carbon Dioxide. Journal of the American ceramic Society, 2013, (9):1-8. (SCI)

[5]Wu CY, Zhang HF, Yu HF. Preparation and properties of modified magnesium oxysulfate cement derived from waste sulfuric acid [J]. Advances in cement research,2016.

Magnetic graphene oxide as sorbent for SPE of melamine from water

W Xu, J Huang and M Liang*

Tianhong Measurement Institute, Zhuzhou 412036, China
Email: ljweisz@163.com

Abstract. Melamine introduced into environment water through various industrial effluents can pose negative effects on ecosystem, thus it is required to establish suitable analytical methods for detection of melamine in environmental water samples. In this work, magnetic graphene oxide was synthesized and used as the sorbent for solid phase extraction (SPE) of melamine. The parameters that affect the extraction efficiency of melamine were studied. The optimum conditions, including sample pH, extraction and desorption time, and eluent volume were obtained. The analytical performance of the method was evaluated, and it was found that the limit of detection for melamine was 0.15μg/L and the relative standard deviation was 3.6 %. The proposed method was utilized in the detection of melamine in various water samples with recoveries of 92.0-95.8%.

1. Introduction

Melamine (1, 3, 5-triazine-2, 4, 6-triamine, $C_3H_6N_6$) has been produced in large amounts. Melamine-contaminated infant formulas have resulted in serious health problems [1-3]. Nowadays, melamine has also been present in the environment as a result of its widespread uses [4-6]. Consequently, it is necessary to develop suitable analytical methods for determination of melamine in environmental samples. Because melamine in environmental water samples is often at low levels, suitable sample preparation procedures are often needed before its determination. Among various sample preparation techniques for trace analysis, solid phase extraction (SPE) is the most often used one and has achieved satisfactory results [7-18].

Graphene oxide (GO) is the oxidized derivative of graphene, which has an ultrahigh-specific surface area, superior chemical property and physical stability. Thus, GO can be used as sorbent materials for SPE [19-29].

Traditional SPE is always a tedious process. However, magnetically separation technology can provide an easy and rapid way for separation of magnetic particles from solution by applying a magnetic field [30-36].

In this work, magnetic graphene oxide (MGO) was prepared, and applied for SPE of melamine from aqueous solution. The parameters that affect the extraction efficiency of melamine were investigated, including sample pH, extraction time and desorption time, and eluent volume. The analytical performance of the method was examined. Finally, the developed method was applied for the determination of melamine in environmental water samples, and satisfactory analytical results were acquired.

2. Materials and methods

2.1. Chemicals

Melamine (>99.9%) was obtained from Aladdin Reagent (Shanghai) Co., Ltd (Shanghai, China).Nature graphite powder (99.95%) was purchased from Sinopharm Chemical Reagent Co., Ltd (Shanghai, China). Other chemicals and materials were purchased from Tianjin Damao Chemical Reagent Factory (Tianjin, China).

2.2. Apparatus

A HPLC system (SHIMADZU, Kyoto, Japan) equipped with diode array detection (DAD, SPD-M20A) and a Shim-pack VP-ODS column (150 × 4.6 mm I.D.) was utilized for analysis.

2.3. Preparation of MGO

Graphite oxide (GO) was prepared from nature graphite powders by a modified Hummers method, and magnetic graphite oxide (MGO) was synthesized by a one-pot solvothermal reaction [19].

2.4. Solid phase extraction

50 mg of MGO was added to the 200 mL melamine containing water sample, and the solution pH was adjusted to pH 6. After shaking for 20 min, the MGO was separated from solution using a magnet. The adsorbed melamine was eluted with 3 mL of 6% ammonia-methanol solution.

3. Results and discussion

3.1. Selection of sample pH

The influence of pH of aqueous solution on the extraction efficiency of melamine was explored by varying pH value from 3 to 10. As shown in Fig.1, the recovery increased with the increase of pH from 3 to 5, and then dropped slightly up to 8. When the pH value was larger than 9, the recovery decreased significantly. Therefore, pH 6 was selected.

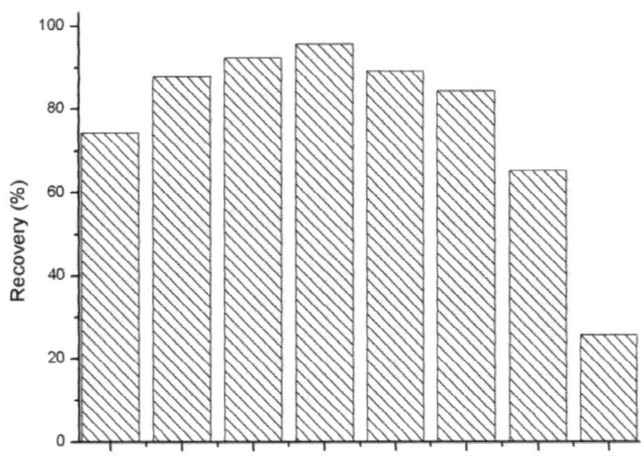

Figure 1. Effect of solution pH.

3.2. Selection of extraction time

The effect of extraction time on extraction efficiency of melamine was investigated. As can be seen from Fig. 2, the recovery increased drastically at the initial adsorption stage from 1 to 10 min. After 20 min, the recovery showed very little changes since the sorption equilibrium was achieved. Based on these results, 10 min was employed as the optimal extraction time.

IOP Conf. Series: Materials Science and Engineering **744** (2020) 012044 doi:10.1088/1757-899X/744/1/012044

3.3. Selection of eluent volume

In this study, 6% ammonia-methanol solution was used as the eluent for stripping of melamine from the MGO. To examine the effect of eluent volume on the extraction recovery of melamine, different volumes of the eluent (1, 2, 3, 5, 8 and 10 mL) were tested. The results showed that 3 mL of this solution could efficiently elute the adsorbed melamine.

3.4. Selection of desorption time

In order to investigate the optimum desorption time, various time periods were examined in the range of 1 to 20 min. The results were shown in Fig.3. According to Fig.3, 5 min was found to be sufficient for quantitative elution of melamine from MGO.

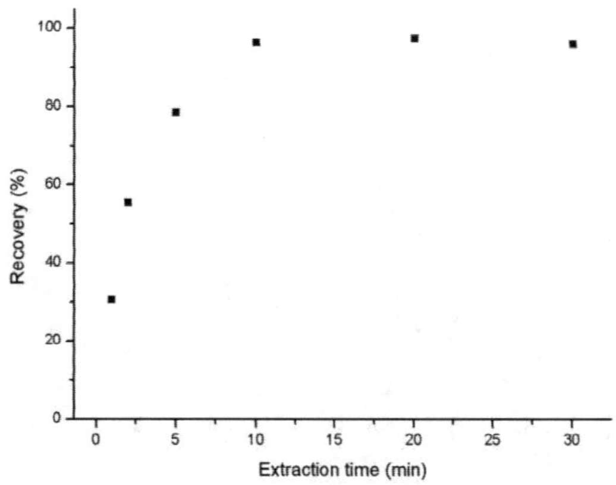

Figure 2. Effect of extraction time.

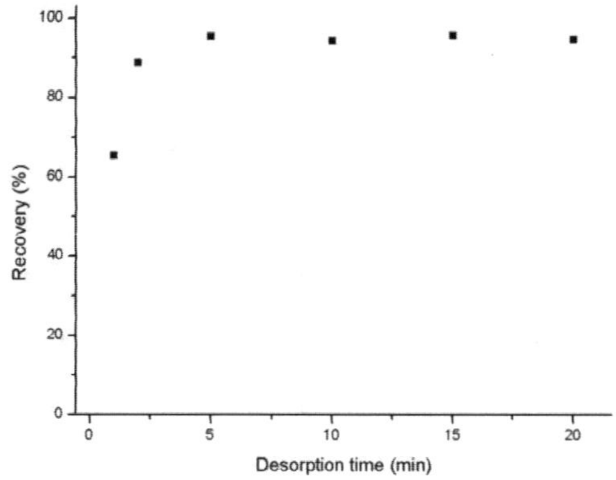

Figure 3. Effect of desorption time.

3.5 Analytical performance

Under the optimum experimental conditions, the calibration curve of melamine was linear in the concentration range from 0.3 to 30 (R^2 =0.9956).The limit of detection (LOD) were 0.15 μg/L The relative standard deviation (RSD) for 5 replicate measurements of 2.0 μg/L melamine was 3.6%.

3.6. Water sample analysis

To evaluate the capability of the proposed method for analysis of real samples with different matrices containing various amounts of melamine, three environmental water samples were analyzed, and the analytical results were given in Table 1.

Table 1. Analytical results of melamine in water samples.

Sample	Add/ (μg/L)	Found/ (μg/L)	Recovery (%)	RSD (%)
Water sample 1	0	---	---	4.2
	1.0	0.92	92.0	3.3
Water sample 2	0	8.65	---	3.5
	10.0	17.98	93.3	2.1
Water sample 3	0	65.42	---	1.2
	50.0	113.3	95.8	2.6

4. Conclusions

This study developed a new method for determination of melamine in aqueous solution, using MGO based SPE coupled with HPLC-DAD. The adsorption results obtained from this study suggest that the prepared MGO can act as a suitable sorbent for melamine. Recoveries of melamine were in the range of 92–95.8% with RSDs of 1.2–4.2%, showing the practicability of the proposed technique.

5. References

[1] Fashi A, Yaftian M R and Zamani A 2015 *Food Chem.* **188** 92.
[2] Zhang Y, Ma X and Fan Y 2014 *Food Anal. Methods* **7** 1763.
[3] Li R, Yang J, Han J, et al 2017 *Physica E* 88 164.
[4] Wackett L P, Sadowsky M J, Martinez B, et al 2002 *Appl. Microbiol. Biotechnol.* **58** 39.
[5] Nong Y, Ma X, Fan S and Yu Y 2014 *Anal. Methods* **6** 4124.
[6] Zhao Z, Chen L, Bai B, et al 2018 *Environ. Sci. Pollut. R.* **25** 2285.
[7] Chen J, Ma X and Wu Y 2014 *Food Anal. Methods* **7** 1083.
[8] Li C, Ma X, Zhang X., Wang R, ChenY and Li Z 2016 *Anal. Bioanal. Chem.* **408** 7857.
[9] Barbosa A, Segatelli M and Pereira A 2007 *Talanta* **71** 1512.
[10] Zou J, Ma X, Dang Y and Chen Y 2014 *J Anal At Spectrom.* **29** 1692.
[11] Fan Y, Ma X and Li Z 2016 *Food Anal. Methods* **9** 16.
[12] Heidari N,Ghiasvand A and Abdolhossaini S 2017 *Anal. Chim. Acta* **975** 11.
[13] Ma X and Li Y 2006 *Anal. Chim. Acta* **579** 47.
[14] Ma X, Huang P and Chen M 2007 *Rare Metals* **26** 541.
[15] Faghihi A, Vakili M and Hosseinzadeh G 2016 *Desalin. Water Treat.* **57** 22655.
[16] Ma X and Zhang Z 2004 *J. Anal. At. Spectrom.* **19** 738.
[17] Krawczyk M, Akbari S, Jeszka-Skowron M, et al 2016 *J. Anal. At. Spectrom.* **31** 1505.
[18] Ma X and Zhang Z 2003 *Anal. Chim. Acta* **485** 233.
[19] Zhang X, Ma X, Li X and Chen M 2018 *Water Air Soil Pollut.* **229** 270.
[20] Huang R, Ma X and Li X 2018 *J. Colloid Interf. Sci.* **514** 544.
[21] Faghihi A., Vakili M.H., Hosseinzadeh G., et al. 2016 *Desalin. Water Treat.* **57** 22655.
[22] Xie X, Ma X, Guo L, Fan Y, Zeng G, Zhang M and Li J 2019 *Chem. Eng. J.* **357** 56.

[23] Li M, Liu Y and Liu S 2018 J. *Colloid Interf. Sci.* **521** 150.
[24] Zhang M, Ma X, Li J et al. 2019 *Chemosphere* **234** 196.
[25] Zhao B, He M, Chen, B and Hu B 2015 *Spectrochim. Acta* B **107** 115.
[26] Li X, Ma X, Huang R, Xie X, Guo Li and Zhang M 2018 *J. Sep. Sci.* **41** 2837.
[27] Kazemi E, Shabani A, Dadfarnia S, et al 2016 *Anal. Chim. Acta* **905** 85.
[28] Li D, Ma X, Wang R and Yu Y 2017 *Anal. Bioanal. Chem.* **409** 1165.
[29] Li J , Ma X, Zhang M, et al. 2019 *J Electrochem. Soc.* **166** **B**84.
[30] Chen Y, Ma X, Huang M, Peng J and Li C 2016 *Anal. Methods* **8** 824.
[31] Wang R, Ma X, Zhang X, et al 2019 *Ecotox. Environ. Safe.* **170** 789.
[32] Kazemi E., Shabani A. M. H., Dadfarnia S., et al 2016 *Anal. Chim. Acta* 905 85.
[33] Hua X., You H., Luo P., et al 2017 *Anal. Bioanal. Chem.* **409** 6885.
[34] Li C, Ma X, Huang R, et al 2017 *J. Sep. Sci.* **41** 2837.
[35] Guo L, Ma X, Xie X, et al 2019 *Chem. Eng. J.* **361** 245.
[36] Mahlangu T, Das R, Abia L. K, 2019 *Chem. Eng. J.* **360** 423.

AUTHOR INDEX